PSA 1978

VOLUME ONE

PSA 1978

PROCEEDINGS OF THE 1978 BIENNIAL MEETING OF THE PHILOSOPHY OF SCIENCE ASSOCIATION

volume one

Contributed Papers

edited by

PETER D. ASQUITH

&

IAN HACKING

1978
Philosophy of Science Association
East Lansing, Michigan

Library of Congress Catalog Card Number 78-61248

Cloth Edition: ISBN 0-917586-06-9
Paperback Edition: ISBN 0-917586-05-0

Manufactured in the United States of America
by Edwards Brothers, Ann Arbor, Michigan

CONTENTS

PREFACE

Among the charges of the By-Laws of the Philosophy of Science Association is furthering "free discussion from diverse standpoints in the field of philosophy of science." To achieve this end PSA engages in a variety of activities including the holding of Biennial Meetings and arranging for the publication of the Proceedings of these meetings. In an effort to achieve timely and affordable publication of the Proceedings PSA itself published the Proceedings of the 1976 Biennial Meeting. PSA 1976 appeared in two volumes - Volume I containing the contributed papers and some special session papers and Volume II containing the symposia papers. Volume I was published in September 1976 prior to the Biennial Meeting and Volume II was published in September 1977 less than a year after the Biennial Meeting. The combined volumes have 930 pages and the cost to purchase a single cloth covered copy of both volumes from the first printing is US $18.25. Thus the Governing Board's goal of timely and affordable publication of the Proceedings was achieved. Because of the success of PSA 1976, PSA is continuing to publish its own Proceedings. PSA 1978 is also being published in two volumes - Volume I containing the contributed papers and Volume II containing papers from the symposia and other sessions. As was the case with the previous Proceedings PSA 1978 is utilizing "camera-ready typewritten copy supplied by the authors."

Printing "camera-ready copy supplied by the author" means that, after submission, there is no "proof stage" during which errors can be corrected. Thus responsibility for accuracy, proof-reading, and the appearance of the copy rests with those producing the camera-ready copy - in this case the authors themselves. The authors were informed that they had this final responsibility and that PSA could not guarantee that any mistakes would be corrected. Errors did slip by, and when time and technical feasibility allowed, the editors attempted to correct deficiencies in the author supplied copy. Known errors or infelicities do remain which could not be corrected without encountering substantial delays in the production of the volume. There are also undoubtedly undiscovered errors that remain, as well as some errors that might have been introduced in the process of correcting ones already there. Responsibility for errors in the individual contributions rests with the authors. Responsibility for the portions of the book produced by the editors, of course, rests with the editors.

Our debts are numerous. Thanks go to the 1978 PSA Program Committee - Ernest Adams, Morton Beckner, Ronald Giere, Clark Glymour, and Peggi Marchi and to the contributors to this volume. Lucy Williams had the responsibility for the mechanics of taking a collection of individual papers and organizing them in a format that would enable them to become a book. David Boersema proofread all of the copy and assisted in the necessary library work.

Peter D. Asquith
Department of Philosophy
and Lyman Briggs College
Michigan State University

Ian Hacking
Department of Philosophy
Stanford University

July 31, 1978

1978 PSA PROGRAM[1]

FRIDAY, OCTOBER 27

9:00-11:30 a.m. - Symposium: *The Unity of Science*

Moderator: C.G. Hempel, Philosophy, Pittsburgh

Speakers: Ernest Nagel, Columbia, Emeritus

9:30-11:30 a.m. - Special Session: *Game Theoretic Semantics for the Language of Science*

Moderator: Dana Scott, Oxford, England

Speakers: E.W. Stachow, Theoretische Physik, Köln, Germany

Esa Saarinen, Academy of Finland

9:30-11:00 a.m. - Contributed Papers Session #1: *History and Philosophy of Science*

Moderator: Peter Machamer, History and Philosophy of Science, Pittsburgh

9:30 a.m. Nancy Maull, Philosophy, Yale, "Perception and Primary Qualities"

10:00 a.m. Edward MacKinnon, Philosophy, California State-Hayward, "The Development of Kant's Conception of Scientific Explanation"

10:30 a.m. John V. Strong, Philosophy, Boston College, "John Stuart Mill, John Herschel, and the 'Probability of Causes'"

11:00-12:30 p.m. - Contributed Papers Session #2: *Philosophy of Physics*

Moderator: John Winnie, History and Philosophy of Science, Indiana

[1]Program arrangements as of July 31, 1978.

11:00 a.m.	Jeffrey Bub and Vandana Shiva, Philosophy, Western Ontario, Canada, "Non-local Hidden Variable Theories and Bell's Inequality"
11:30 a.m.	Nancy Cartwright, Philosophy, Stanford, "The Only Real Probabilities in Quantum Mechanics"
Noon	Carlo Giannoni, Philosophy, Rice, " A Universal Axiomatization of Kinematical Theories"

Noon-2:30 p.m. - Symposium: *Philosophy of Technology*

Moderator:	Alex C. Michalos, Philosophy, Guelph, Canada
Speakers:	Paul Durbin, Philosophy, Delaware
	S.C. Thakur, Philosophy, Surrey, England
	Henryk Skolimowski, Engineering, Michigan

12:30-2:30 p.m. - Special Session: *Physical Randomness*

Moderator:	Peter Kirschenmann, Vrije Universitet, Amsterdam
Speaker:	Paul Benioff, Argonne National Laboratory, Illinois
Commentators:	Geoffrey Hellman, Philosophy, Indiana
	Paul Humphries, Philosophy, Arizona

1:00-3:00 p.m. - Contributed Papers Session #3: *Observation and Theory*

Moderator:	Patricia Smith Churchland, Philosophy, Manitoba, Canada
1:00 p.m.	Kathleen Okruhlik, Philosophy, Pittsburgh, "The Interplay Between Theory and Observation in the Solar Model of Hipparchus and Ptolemy "
1:30 p.m.	William B. Jones, Philosophy, Old Dominion, "Theory-ladenness and Theory Comparison"
2:00 p.m.	P. William Bechtel, Philosophy, Northern Kentucky, and Eric Stiffler, Philosophy, Western Illinois, "Observationality: Quine and the Epistemological

Nihilists"

2:30 p.m. Mark G. Tamthai, Chulalongkorn, Thailand, "Analyticity Without an Observational-Theoretical Distinction"

2:45-5:30 p.m. - Symposium: *The Purposes of Biological Classification*

Moderator: Marjorie Grene, Philosophy, Tulane

Speakers: Robert Sokal, Biology, SUNY-Stony Brook

Norman Platnick, American Museum of Natural History, New York

Steven Farris, Biology, SUNY-Stony Brook

David Hull, Philosophy, Wisconsin-Milwaukee

2:45-5:30 p.m. - Symposium: *Statistical Mechanics*

Moderator: Linda Wessells, History and Philosophy of Science, Indiana

Speakers: John Earman, Philosophy, Minnesota

David Malament, Philosophy, Chicago

Commentator: Lawrence Sklar, Philosophy, Michigan

3:30-5:00 p.m. - Contributed Papers Session #4: *Case Studies with Applications to the Philosophy of Science*

Moderator: Richard Burian, Humanities-Communication, Drexel

3:30 p.m. Barbara Klein, Philosophy, Yale, "The Role of Psychology in Functional Localization Research"

4:00 p.m. Thomas Nickles, Philosophy, Nevada-Reno, "Scientific Problems and Constraints"

4:30 p.m. Lindley Darden, History and Philosophy of Science, Maryland, "Discoveries and the Emergence of New Fields in Science"

8:00-9:00 p.m. - Business Meeting

9:00-12:00 p.m. - Reception

SATURDAY, OCTOBER 28

9:00-Noon - Symposium: *Philosophical Consequences of the Recent Revolution in Geology*

Moderator: Henry Frankel, Philosophy, Missouri

Speakers: David Kitts, History of Science, Oklahoma

Rachel Laudan, History and Philosophy of Science, Pittsburgh

Michael Ruse, Philosophy, Guelph, Canada

9:00-11:30 - Invited Lecture: *New Foundations for Quantum Mechanics*

Moderator: Michael Friedman, Philosophy, Pennsylvania

Speaker: Simon Kochen, Mathematics, Princeton

Commentator: Richard Healey, History and Philosophy of Science, Cambridge, England

9:00-11:00 a.m. - Invited Lecture: *The Psychology of Resemblance*

Moderator: Ned Block, Philosophy, M.I.T.

Speaker: Amos Tversky, Psychology, Stanford

12:15-2:45 p.m. - Symposium: *Lakatos' Philosophy of Mathematics*

Moderator: Paul Benacerraf, Philosophy, Princeton

Speakers: William Berkson

Solomon Feferman, Mathematics and Philosophy, Stanford

Arthur Fine, Philosophy, Illinois-Chicago Circle

12:30-2:30 p.m. - Special Session: *Reduction*

Moderator: William Wimsatt, Philosophy, Chicago

Speakers: Paul A. Bogaard, Philosophy, Mount Allison, Canada

Arthur Caplan, The Hastings Center, New York

Commentator: Kenneth Shaffner, History and Philosophy of Science, Pittsburgh

1:30-3:00 p.m. - Contributed Papers Session #5: *Explanation*

Moderator: Alex Rosenberg, Philosophy, California-Santa Cruz

1:30 p.m. Robert Cummins, Philosophy, Wisconsin-Milwaukee, "Explanation and Subsumption"

2:00 p.m. J.K. Derden, Philosophy, Humboldt State, "Reasons, Causes,and Empathetic Understanding"

2:30 p.m. T.R. Girill, California State-Fresno, "Approximative Explanation"

1:30-3:00 p.m. - Special Session: Discussion of the PSA Research Problems Project

Speakers: Henry E. Kyburg, Jr., University or Rochester

Peter D. Asquith, Michigan State University

3:00-5:30 p.m. - Symposium: *Habermas*

Moderator: John Searle, Philosophy, California-Berkeley

Speakers: Mary Hesse, History and Philosophy of Science, Cambridge, England

Thomas McCarthy, Philosophy, Boston University

Gary Gutting, Philosophy, Notre Dame

3:30-5:30 p.m. - Symposium: *Statistical Evidence*

Moderator: Roger Rosenkrantz

Speakers: Glenn Shafer, Mathematics, Kansas

Isaac Levi, Philosophy, Columbia

Teddy Seidenfeld, Philosophy, Pittsburgh

3:30-5:30 p.m. - Contributed Papers Session #6: *Aspects of Rationality*

Moderator: William Harper, Philosophy, Western Ontario, Canada

3:30 p.m. Sandra Harding, Philosophy, Delaware, "Four Contributions Values Can Make to the Objectivity of Social Science"

4:00 p.m. Caroline Whitbeck, Medical Humanities, Texas-Galveston, "Four Basic Concepts of Medical Science"

4:30 p.m. Paul Thagard, Humanities, Michigan-Dearborn, "Why Astrology is a Pseudoscience"

5:00 p.m. Maurice Finocchiaro, Philosophy, Nevada-Las Vegas, "Rhetoric and Scientific Rationality"

SUNDAY, OCTOBER 29

9:00-12:30 p.m. - Symposium: *Laudan's Progress and its Problems*

Moderator: Frederick Suppe, History and Philosophy of Science, Maryland

Speakers: Robert Westman, History, UCLA

Noretta Koertge, History and Philosophy of Science, Indiana

Hugh Mellor, Philosophy, Cambridge, England

Response: Laurens Laudan, History and Philosophy of Science, Pittsburgh

9:30-12:30 a.m. - Symposium: *Conditionals*

Moderator: Richard Jeffrey, Philosophy, Princeton

Speakers: Brian Skyrms, Philosophy, Illinois-Chicago Circle

Alan Gibbard, Philosophy, Michigan

John M. Vickers, Philosophy, Claremont

9:00-10:30 a.m. - Contributed Papers Session #7: *Quantum Logic*

Moderator: Ed Levy, Philosophy, British Columbia, Canada

9:00 a.m. Peter Mittelstaedt, Theoretische Physik, Köln, Germany, "The Metalogic of Quantum Logic"

9:30 a.m. David Holdsworth, Philosophy, Western Ontario, Canada, "A Role for Categories in the Foundations of Quantum Theory"

10:00 a.m. James McGrath, Philosophy, Indiana-South Bend, "Only if Quanta had Logic"

11:00-12:30 p.m. - Contributed Papers Session #8: *Applications of Statistical Ideas*

Moderator: Hartry Field, Philosophy, Southern California

11:00 a.m. B. Loewer, R. Laddaga and R. Rosenkrantz, "On the Likelihood Principle and a Supposed Antinomy"

11:30 a.m. Andre Mirabelli, Philosophy, Boston University, "Belief and Incremental Confirmation of One Hypothesis Relative to Another"

Noon David Papineau, Philosophy, Macquarrie, Australia, "Salmon, Statistics, and Backwards Causation"

SYNOPSIS

The following brief summaries provide an introduction to each of the papers in this volume.

1. *Perception and Primary Qualities.* Nancy Maull. The doctrine of primary qualities is commonly explained as science's return to a former ideal of mathematical intelligibility and as a sacrifice of the notion that we can be certain about what we perceive. According to the standard chronicle modern scientific explanations appeal to geometrically intelligible, yet theoretically imperceptible, particles. This thesis gains plausibility only by suppressing the role of physiological optics in the development of modern science. Descartes presented an original and significant theory of scientific observation in his Dioptrics; according to which, correct perceptual judgments can be made about the spatial qualities of any body, even about bodies so small as to be imperceptible under ordinary circumstances. This theory insured the perceptual accessibility of the spatial qualities of minute bodies but failed, as Locke discovered upon adding solidity to the list of primary qualities, to explain access to non-spatial qualities. As a result of the failure of Descartes' theory of perception to give an adequate account of scientific observation, Locke took the position that solidity or impenetrability cannot ordinarily be perceived, but rather that it can be and is regularly detected by experiment. This Lockean shift from direct perceivability of spatial qualities to the experimental detection of "new" primary qualities marks an important step in the methodological development of early modern science.

2. *The Development of Kant's Conception of Scientific Explanation.* Edward MacKinnon. In the course of his long development, Kant's concept of matter changed somewhat, while his concept of scientific explanation changed considerably. Both developments achieved a coherent integration in Kant's Metaphysical Foundations of Natural Science. Using this developmental background, the present paper argues that the Foundations should be interpreted as an attempted rational reconstruction of the mechanics of Newton and Euler. Kant attempted to do this by constructing a concept of matter that would confer a Leibnizian intelligibility on Newtonian mechanics, and also accord with Kant's theories on the nature of concepts and their role in scientific explanation.

3. *John Stuart Mill, John Herschel, and the "Probability of Causes".* John V. Strong. While historians of scientific method have recently called attention to the views of many of John Stuart Mill's contemporaries on the relation between probability and inductive inference, little if any note has been taken of Mill's own vigorous attack on the received

"Laplacean" interpretation of probability in the first (1843) edition of the System of Logic. This paper examines the place of Mill's critique, both in the overall framework of his philosophy, and in the tradition of assessing the so-called "probability of causes". It also offers an account of why, in later editions of the work, Mill appears to adopt a much more sympathetic stance toward the received view.

4. *Non-local Hidden Variable Theories and Bell's Inequality.* Jeffrey Bub & Vandana Shiva. Bell's proof purports to show that any hidden variable theory satisfying a physically reasonable locality condition is characterized by an inequality which is inconsistent with the quantum statistics. It is shown that Bell's inequality actually characterizes a feature of hidden variable theories which is much weaker than locality in the sense considered physically motivated. We consider an example of non-local hidden variable theory which reproduces the quantum statistics (and hence violates Bell's inequality). A simple extension of the theory, which preserves the non-local character, alters the statistics in such a way that Bell's inequality is satisfied.

5. *The Only Real Probabilities in Quantum Mechanics.* Nancy Cartwright. Position probabilities play a privileged role in the interpretation of quantum mechanics. The standard interpretation has it that $|\Psi(r)|^2$ represents the probability that the system is at (or will be found at) the location $\underline{r}$. Use of these probabilities, however, creates tremendous conceptual difficulties. It forces us either to adopt a non-standard logic, or to be saddled with an intractable measurement problem. This paper proposes that we try to eliminate position probabilities, and instead to interpret quantum mechanics through the use of energy transition probabilities. Energy transitions, unlike particle positions, either occur, or they do not. Their probabilities are unproblematic, and they do not require either a deviant logic or a nonclassical probability structure. They are thus good candidates to serve as the fundamental interpreted quantity of the quantum theory.

6. *A Universal Axiomatization of Kinematical Theories.* Carlo Giannoni. After a consideration of Reichenbach's and Winnie's axiomatizations of Special Relativity, three synchrony-free axioms which can be used to simultaneously axiomatize Special Relativity, Classical Aether Theory, and a threefold continuum of theories in between are suggested. The specific theory obtained from the axioms depends on the value of a parameter in each axiom. It is shown that the values of the three parameters are empirically determined by the Michelson-Morley, Kennedy-Thorndike, and Ives-Stilwell experiments.

7. *The Interplay Between Theory and Observation in the Solar Model of Hipparchus and Ptolemy.* Kathleen Okruhlik. Attempts by twentieth-century historians to account for the successes and failures of the Hipparchian-Ptolemaic solar model provide valuable case studies for philosophers who are studying the relationship between observational data and theoretical constructs. A brief survey of recent literature on the solar model reveals that in some cases results which appear to

be the product of highly accurate observation are, in fact, based on rather crude observations aided by a large measure of theoretical presupposition. On the other hand, mistaken results, which have sometimes been explained as the outcome of theoretical bias or conscious deceit, can plausibly be accounted for as a result of honest mistakes at the observational level.

8. *Theory-ladenness and Theory Comparison*. William B. Jones. Conditions are set forth which, if satisfied, would provide a rational basis for believing, even in a context in which observation is theory-laden, that (1) the two theories are rivals and (2) one of them enjoys observational support that is superior to that enjoyed by the other.

9. *Observationality: Quine and the Epistemological Nihilists*. P. William Bechtel & Eric Stiffler. Quine has proposed an alternative criterion for identifying observation sentences which has not yet received serious evaluation. We investigate this new criterion, showing how it differs from more traditional criteria and measuring it against the major objections to traditional criteria. Our judgment is that it meets Suppe's and Achinstein's objections and one version of the theory-ladenness objection offered by Hanson, Feyerabend, and Kuhn. We suggest how it might also provide an answer to the more serious version of the theory-ladenness objection. To determine whether it meets this final objection, though, requires actual analysis of scientific cases, which has not yet been performed.

10. *Analyticity Without an Observational-Theoretical Distinction*. Mark G. Tamthai. Most attempts at separating out those consequences of an empirical theory which play a purely linguistic role (i.e., the "analytic" consequences) from the rest have assumed a distinction between the observation and theoretical terms in the language of the theory. Since there has been much debate concerning this distinction and since there do not seem to be any intrinsic reasons why "analyticity" should be tied to it, we propose in this paper an explication of "theoretical analyticity" which does not rely upon the distinction.

11. *The Role of Psychology in Functional Localization Research*. Barbara Klein. The "method of deficit analysis" in functional localization research is a method for inferring localization of brain function on the basis of evidence of abnormal behavior plus evidence concerning brain damage location. Such inferences involve two steps: first, a functional deficit analysis of the neurologically impaired patient, and second, localization of function in the normal brain. It is argued that psychological theory is required for adequate functional deficit analysis, both for the identification of the relevant functional components of the system as well as for the explanation of the behavioral pathology in terms of a deficit in one of these components.

12. *Scientific Problems and Constraints*. Thomas Nickles. In this paper the relation between scientific problems and the constraints on their solutions is explored. First the historical constraints on the solution to the blackbody radiation problem are set out. The blackbody

history is used as a guide in sketching a working taxonomy of constraints, which distinguishes various kinds of reductive and nonreductive constraints. Finally, this discussion is related to some work in erotetic logic. The hypothesis that scientific problems can be identified with structured sets of constraints is interesting; however, a full defense of the identification thesis requires the resolution of some unsolved problems.

13. *Discoveries and the Emergence of New Fields in Science*. Lindley Darden. This paper analyzes features of the emergence of new fields in science by examining the cases of cytology and biochemistry. The first step in the emergence of these new fields was the discovery of a new entity. A subsequent claim was made that entities of this kind are found more generally; making this generalization constituted the construction of a new theory. As a line of research to test the theory began, a new domain was formed and the new field emerged. In each case the theory proposed a new solution to an old problem in science. Some implications of this analysis for understanding the emergence of other fields are indicated.

14. *Explanation and Subsumption*. Robert Cummins. The thesis that subsumption is sufficient for explanation is dying out, but the thesis that it is necessary is alive and well. It is difficult to attack this thesis: non-subsumptive counter-examples are declared incomplete, or mere promissory notes. No theory, it is thought, can be explanatory unless it resorts to subsumption at some point. In this paper I attack this thesis by describing a theory that (1) would explain every event it could describe, (2) does not explain by subsumption, and (3) is fundamental in that it is understood to be irreducible (hence there are no unstated laws waiting in the wings).

15. *Reasons, Causes, and Empathetic Understanding*. J.K. Derden. In this paper an attempt is made to establish that the parties on both sides of the disputes concerning whether reasons are causes have mischaracterized or misdescribed what is involved in acting from a reason. A characterization of acting from a reason is provided, and, as a result, it is shown why contradictory positions have been and still are maintained by opposing parties. As a consequence of this account, an attempt is made to show why all parties have missed the nature and significance of empathetic understanding.

16. *Approximative Explanation*. T.R. Girill. This paper develops an explicit, pragmatic solution to the problem of deciding when an explanans which only approximately "covers" a desired event-explanandum E is acceptable as an adequate explanation. It shows in detail how comparisons made by the explanation's audience with the numerical value in E are what determine how closely E must be approximated for success. With the aid of several physics examples, it spells out the principles that govern these explanandum-comparisons, the conditions under which those principles hold, and the audience traits that influence which comparisons they take seriously.

17. *Four Contributions Values Can Make to the Objectivity of Social Science*. Sandra Harding. Carnap reports that while all of the members of the Vienna Circle "were strongly interested in social and political progress," except for Neurath, they all insisted that the "intrusion" of political points of view into the methodology of science would violate the purity of scientific method. In opposition to this still dominant view of the relationship between moral/political values and objective inquiry, this paper specifies four ways in which certain moral/political values are necessary for maximizing objective inquiry in social science. If this argument is correct, a better conception of objectivity is needed.

18. *Four Basic Concepts of Medical Science*. Caroline Whitbeck. It is claimed that medicine is concerned with the prevention and treatment of certain types of psychophysiological processes and states which frequently compromise health, namely with disease, injuries, and (occasionally) impairments, rather than with health. It is argued that the normative component in the concepts, disease, injury and impairment, consists in each being a type of process or state which people wish to be able to prevent or effectively treat, because it interferes with the capacity to do something that people commonly want to be able to do, without also being necessary for such a capacity. One advantage of viewing the normative component as consisting in the judgment that it is desirable to collectively possess the means to control certain types of processes and states, is that it renders non-paradoxical the fact that it may not be in anyone's interest to prevent or treat a given case of disease, injury, or impairment.

19. *Why Astrology is a Pseudoscience*. Paul Thagard. Using astrology as a case study, this paper attempts to establish a criterion for demarcating science from pseudoscience. Numerous reasons for considering astrology to be a pseudoscience are evaluated and rejected; verifiability and falsifiability are briefly discussed. A theory is said to be pseudoscientific if and only if (1) it has been less progressive than alternative theories over a long period of time, and faces many unsolved problems, but (2) the community of practitioners makes little attempt to develop the theory towards solutions of the problems, shows no concern for attempts to evaluate the theory in relation to others, and is selective in considering confirmations and disconfirmations. This criterion has the interesting consequence that a theory can be scientific at one time but pseudoscientific at another.

20. *Rhetoric and Scientific Rationality*. Maurice Finocchiaro. Feyerabend's views are construed as formulating the problem of determining the role of rhetoric in scientific rationality and posing the solution-theory that scientific rationality is essentially rhetorical. He is taken to give three arguments against reason, of which the one from the insufficiency of reason and the one from incommensurability are shown to presuppose his historical argument; his historical argument is based on his account of Galileo, which hinges essentially on Feyerabend's analysis of the tower argument. This analysis is insightful in certain important ways but misconceived in others. Feyerabend's main error is to see a conflict between reason and rhetoric, where none exists; it is

argued that rhetorical devices have their own standards of propriety and impropriety, different from those of rational arguments, and that at the same time sound rhetorical analysis presupposes sound logical analysis.

21. *The Metalogic of Quantum Logic.* **Peter Mittelstaedt.** The logic of quantum physical propositions can be established by means of dialogs which take account of the general incommensurability of these propositions. Investigated first are meta-propositions which state the formal truth of object-propositons. It turns out that the logic of these meta-propositions is equivalent to ordinary logic. A special class of meta-propositions which state the material truth of object-propositions may be considered as quantum logical modalities. It is found that the logic of these modalities contains all the quantum logical restrictions and thus differs essentially from the modal calculi of ordinary logic.

22. *A Role for Categories in the Foundations of Quantum Theory.* **David Holdsworth.** A framework is introduced in which categories play an essential role in the foundations of physics. The work of F.W. Lawvere is exploited to provide axiomatizations for the relevant mathematical objects in a way which does not depend on set-theoretical foundations. An axiomatization of the category of complete atomic Boolean algebras is sketched as an example. These results are then used to give an explication of "truth in a (classical) physical theory" in terms of a commuting diagramme. This explication is then generalized to quantum theory and shown to capture, not truth, but necessity in the sense of van Fraassen.

23. *Only if Quanta had Logic.* **James McGrath.** Putnam and others have argued that an examination of structures arising in quantum mechanics can lead to a non-classical propositional logic, quantum logic. In this paper it is argued that the procedure by which quantum logic is said to be discovered, a process called reading off, is fundamentally flawed. A parody fable shows that reading off leads to absurd consequences. The fable also leads one to cast doubt on the following claims central to the quantum logic program: (i) logic can be reformed locally; (ii) classical logic is read off classical physics; and (iii) reading off is a justifiable process because quantum logic resolves quantum paradoxes.

24. *The Likelihood Principle and A Supposed Antinomy.* **B. Loewer, R. Laddaga & R. Rosenkrantz.** Allan Birnbaum has alleged that use of a likelihood criterion can find strong evidence against a true hypothesis with probability one. It is shown that, correctly applied, use of the likelihood function does not lead to any such result. Specifically, Birnbaum's example involves composite hypotheses, and, from a Bayesian point of view, the support of a composite hypothesis can be adequately assessed only by averaging the likelihoods of its constituent simple hypotheses.

25. *Belief and the Incremental Confirmation of One Hypothesis Relative to Another.* **Andre Mirabelli.** A formalism for the incremental confirmation of one hypothesis relative to another hypothesis and the

requirement for adding a hypothesis to a belief system is presented as a sufficient explication of confirmation. The formalism is contrasted with the explications of confirmation offered by Ian Hacking and Karl Popper. It is shown to solve some of the problems which are most often put forth as foils of similar approaches to confirmation theory. Finally, the method is indicated by which the formalism avoids all other standard problems.

26. *Salmon, Statistics, and Backwards Causation*. David Papineau. In order to explain why falling barometers don't cause rain, a "no-eclipsing" requirement needs to be added to the regularity account of causation. This refinement of the regularity account allows us to see how conclusions about deterministic causes can be based on statistical premises, and thus indicates a criticism of Wesley Salmon's "statistical relevance" account of causation. The refinement also casts some light on the problem of backwards causation.

Part I

History of Philosophy of Science

Perception and Primary Qualities

Nancy L. Maull[1]

Yale University

Early modern science, according to a misleading and now widely accepted thesis, introduced a split or schism between a world of colorless, imperceptible particles, on the one hand, and the familiar world of perception, on the other.[2] One of the most important dilemmas of modern philosophy, of course, seems to follow directly from this alleged rupture: For how are the two seemingly incongruous worlds to be "reconciled"? This way of formulating the problem, however, seems to be based on a misunderstanding of early modern science, on an interpretation that underestimates the very real difficulties that scientists faced in putting forward hypotheses about unseen minute bodies. If we start to pay attention to the way in which knowledge claims about the properties of atoms or corpuscles were advanced in the sixteenth and seventeenth centuries, in fact, it becomes increasingly obvious that such fundamental properties were first attributed to minute bodies on grounds of strict analogy, that is, on the grounds that all _perceptible_ bodies exhibited precisely the properties of size, shape, motion, and so forth. Descartes went so far as to say that correct perceptual judgments could be made about spatial qualities, no matter what size the body. According to Descartes' influential optical theory, as I argue in Part 1.,all such minute bodies are in principle perceptible. In the light of attempts by early modern science to make its knowledge claims plausible by appeals to physiological optics, the purported rupture between the world revealed by perception and a world revealed by science seems to disappear.

Instead of an abrupt discontinuity between perceptual experience and imperceptible particles, we find (as I argue in Part 2.) that certain "new" primary qualities, like Locke's--and Newton's--solidity were attributed to minute bodies on the basis of experimental findings, and not, as the spatial qualities had been, on the grounds of strict perceptual analogy. As I go on to show in Part 3. a new principle, according to which (in Newton's words) the primary qualities are those characteristic of all bodies "within the reach of our experiments"

PSA 1978, Volume 1, pp. 3-17

([25], p. 3), does not eliminate, but rather presupposes the perceivability of real spatial qualities. This new understanding of the relationship between perceptual experience and science's primary qualities, I believe, calls strongly into question the "two worlds" view of the emergence of modern science (a view held by philosophers as dissimilar as Sellars [26] and Husserl [17]). The problem for early modern science, I am suggesting, was not to reconcile two incongruous worlds, but to make reasonable its attempts to attribute fundamental qualities to unseen bodies, and at the same time, to avoid reintroducing "occult powers".

1. Descartes' Position

Nothing less than the very certainty Descartes claimed for science is at issue in the question as to whether micro-entities are perceivable. My central thesis concerning Descartes will be that minute bodies are, according to his theory, quite perceivable. Furthermore, the perceivability of very small bodies is not just the interesting outcome of his optical theory, but is necessary for Cartesian science. In order to establish this, I shall briefly outline Descartes' theory of observation, that is, his theory of correct perceptual judgment.

Descartes' theory of perception begins as a "representational" theory. We read in the *Dioptrics* [8], and in his *World* [14] and *Principles* [12], that *the content of our sensation need not resemble what it represents*. Descartes invokes an analogy to explain the perception of color (to explain, for that matter, any of the sensations proper to a single sensory modality--textures, heat and cold, tastes, odors, and sounds): just as words "are capable of making conceivable for us things to which they have no resemblance, why may not nature also have established a certain sign which should make us feel the sensation of light, although this sign should have nothing in itself resembling sensation?" ([14] pp. 312-13).[3] According to the analogy, then, a color (or any such quality) resembles nothing in a physical object.

If our sensations of, say, color or heat taken by themselves--as mere sensations--afford us no information at all about bodies, how then are we to obtain such information? Descartes introduces a second analogy to explain our access to independently existing physical objects. On the basis of this second analogy, he shows how we perceive those qualities we would associate with "depth." Descartes thinks that perceptions of the shape, size (or "magnitude"), distance, and position of a body are all "distance" perceptions. According to the analogy he uses to explain them in the *Dioptrics*, images may resemble objects in a limited way, as engravings "represent to us forests, towns, men, and even battles and storms," by resembling them only in shape. And even with respect to shape, Descartes continues, the resemblance is "very imperfect," for engravings represent a three-dimensional array on a flat, two-dimensional surface so that, "following the rules of perspective, circles are often better represented by ovals rather than by circles; and squares by diamonds rather than by other squares... ."([8], pp. 89-90).

Thus flat shape or figure can represent "outness," "depth" or "distance"; this much is unmistakably suggested by Descartes' analogy with engravings. In the sensation of figure, so said Descartes, we may expect the same kind of resemblance between the object of our immediate awareness and the independently existing physical object as we find between an uncolored engraving and the reality it depicts. Thus, having experienced flat figure in sensation, according to Descartes' analogy, a percipient will be able to generate judgments about the three-dimensional size, shape, position, and distance of a body. Visual cues, such as the relative distinctness of images, may suggest relative distances. Furthermore, judgments of spatial magnitude purportedly utilize a "natural geometry" which Descartes claims to be innate in perceivers. The distance judgments that utilize this natural geometry are not, according to Descartes' view, *a priori*, even though they employ an *a priori* Euclidian geometry.[4]

To summarize: for Descartes perceptual access to bodies is *indirect* and limited to information concerning the spatial characteristics and relationships of bodies. Even though our immediate visual sensations are of colored figures, the sensation of figure is to be distinguished from the sensation of color. For figure there is an algorithm (natural geometry) for generating judgments about the location and size of bodies in space. Furthermore, any distance judgment (whether it utilizes natural geometry or visual cues) can provide indirect access to independently existing physical objects ("indirect" because of the intervention of geometry or some reasoning about "direct" sensation). Because bodies *are* purely spatial for Descartes, this account explains how correct mathematical reasoning about nature is possible, while not guaranteed.

Descartes had assumed throughout his account of how we obtain reliable perceptual judgments that extension and its "modes"--distance, shape, size, position and motion--are the real and explanatory characteristics of bodies.[5] They are the essential qualities of bodies, which is to say that for Descartes, being extended is necessary *and* sufficient for being a body.

At the core of Descartes' theory of perception is a radical distinction between sensation and judgment. Mere visual sensation is of colored figures. Although sensations are "clear" (as he says in the *Principles*, i, LXVIII) sensation alone tells us nothing about the bodies (VIII-i, p.33).[6] Sensation *qua* sensation is still irrelevant to science. Judgments, by contrast, are knowledge claims, and the perception of depth is always judgmental for Descartes, whether it employs natural geometry or relies on reasoning from the experience of visual cues. Lacking the immediacy of sensation, however, such judgments are open to error. But error can be corrected by new, improved judgments. (As Descartes puts it, error is a flaw of judgment, not of sensation [VII, 438-39].) Correct perceptual judgment about the spatial magnitudes of bodies is then possible.[7]

Although Descartes holds that we can make correct perceptual judgments (and therefore attain certainty) about the spatial qualities of the particular bodies that we perceive, he also claims that we can only formulate <u>hypotheses</u> about the determinate spatial qualities of unseen minute bodies. Cartesian hypotheses are plausible, as we shall see, only if the properties and interactions of micro-entities are conceived as analogous to the bodies that we perceive. To hypothesize a new kind of explanatory property (or law) in very small bodies--a property that we fail to encounter in ordinarily perceived bodies--would be, in Descartes' words, "to invent all sorts of novelties that have no relation to those that we perceive."(IX, 319).[8] (Indeed, it would come close to the Scholastic maneuver of conjuring up new "powers" for each and every explanation.) Thus for Descartes, it is only reasonable to hypothesize micro-properties and interactions that are also characteristic of the middle-sized objects that we perceive, in short, to hypothesize on the basis of what is "perceptually universal." For the purposes of explanation the only admissible perceptually universal qualities are spatial. Color, as Berkeley would point out, is also perceptually universal; but according to Descartes no correct perceptual judgment ever locates color <u>in</u> a body. The only correct perceptual judgments are about the spatial qualities of bodies. Such judgments can, at least in principle, as we shall now see, be made even for very small and previously unperceived bodies.

Indeed, it is my central point about Descartes that his idea of a complete and finished science of nature is built on the premise that all minute bodies could in principle be perceived. Remember that Descartes' own optical writings (which make the eye appear to be an optical instrument of limited range) were intended primarily to provide a theoretical justification for the reliability and accuracy of mechanically augmented telescopic and microscopic observation. Descartes even makes the very strong claim in the <u>Dioptrics</u> that there is no limit to the size or distance of bodies perceptible via the proper combination of lenses, and hence no limit to the bodies which can, in principle, be known (VI, 155-56). Without reliable and accurate information about the size, shape, position and motion of corpuscles, for Descartes, our explanations of micro-phenomena would remain forever "hypothetical." By affirming the perceivability of minute bodies, as a result, Descartes believes himself to be laying the foundation for attaining certainty in science beyond mere hypothesis.

However, the certainty that Descartes seeks for science seems to be undermined by his hypothetical appeal to the primary qualities of unperceived particles. Much has been made, of course, of the "breakdown" of Descartes' deductive model for certainty in the sciences.[9] Descartes himself even admits in the <u>Discourse</u> that while empirical laws must be consistent with the <u>a priori</u> laws of motion and rules of impact, the former are not deducible from the latter. Indeed, Descartes' remarks about the formulation of hypotheses concerning the workings of the microstructure (for example, at the conclusion of Part iv of the <u>Principles</u>) have often been interpreted as signaling his own final scepticism with regard to the possibility of our ever being certain about the inner workings of nature. According to this interpretation, the probabilism associated with Descartes' hypotheses (or as he sometimes calls them, suppositions) is permanent.[10]

This interpretation of Descartes' hypotheses, although it is a misunderstanding, is understandable enough. Traditionally, Descartes' methodology has been narrowly construed in line with his supposedly one-sided "rationalism," as a sign of his concern for the deductive structure of a completed science. Eventually, of course, it was recognized that Descartes had indeed made a place for observation and hypothesis in his method. In fact, because Descartes openly attributes the essential qualities to bodies by hypothesis, it has seemed quite natural to interpret him as claiming that those properties are "hypothetical" or impossible to observe. However, this interpretation of the status of hypotheses fails to explain Descartes' ideas about the possibility of correct perceptual judgment. It fails, in other words, to explain why Descartes wrote the *Dioptrics*.

The optical writings (especially Discourses 7-10 of the *Dioptrics*) make it clear that Descartes does *not* have in mind that micro-entities are "theoretically" or "logically" unobservable. He thinks rather that while the micro-entities cannot be observed by the naked eye, they are in principle observable by means of magnifying devices. He cautions that microscopes are difficult to construct but claims that they are more useful than telescopes, for, he says, "by means of them we will be able to see the diverse mixtures and arrangements of the small particles which compose animals and plants, and perhaps also the other bodies which surround us and thereby derive great advantage *in order to arrive at the knowledge of their nature*."([8], p. 172; my italics).

(An additional line of thought helps explain why Berkeley's "logical" objection that the particles must be uncolored [since they are invoked in the explanation of color] and are therefore unperceivable cannot be brought against Descartes. He is not an atomist and for him the color of any small body could be explained in terms of the primary qualities of even smaller bodies. For Locke, by contrast, the secondary qualities of atoms are a problem, though not one of which he seems to have been aware. In any case--for reasons that I shall discuss below--Locke's atoms would, as it turns out, at least be perceptible by touch.)

Thus, a remark Descartes makes in a letter of 1637 is quite typical of his assertions about the scientific understanding of particle interaction: "It is perfectly reasonable to judge of things which are too small for the senses to perceive on the model of those we see."([11], p. 38). And he does not hesitate (for example in the *Principles*, iv, CCIII) to think about the interactions between unperceived bodies on the model provided by visible machines. Indeed, as his remarks make clear, it is the "mere minuteness" of small bodies that makes them "insensible". Only the factual limitations of our sensory organs render the particles indiscernible, and the exact science of nature certainly cannot submit itself to arbitrary boundaries imposed by human anatomy. "Our nerves," he says in the Principles (iv, CCI), "are not [themselves] the smallest possible bodies." However, it is the purpose of optics, as he claims at the end of the Seventh Discourse of the Dioptrics, to remedy the deficiencies of sight "through the application of other organs that are artificial." Even very small bodies, as a result, are not *in principle* imperceptible. No doubt, Descartes intends this to be the lesson drawn from physiological optics, which treats as instruments of science the telescope, the microscope, and the human eye: there is no permanent or insurmountable barrier to our observation and understanding of nature (although we may always fall short of God's understanding). According to Descartes' methodology, in other words, our hypotheses about micro-entities are not irreducibly hypothetical, since even the unseen minute bodies (the causal sources of *visibilia*) can in principle be perceived.[11]

Descartes did not (as the "two worlds" view would have it) claim for science a world of theoretically imperceptible small bodies. Rather, he concentrated his efforts in optics and the theory of perception on justifying certain knowledge claims about what we perceive, for the purpose of making plausible hypotheses about what we do *not* perceive as a consequence of mere size and human anatomy.[12]

I shall now go on to show that Locke attempted to follow a similar procedure in making plausible his hypothesis about the "new" primary quality solidity.[13] However, for Locke, the experimental (rather than perceptual) universality of resistance turns out to be the underlying rationale for a plausible hypothesis about ultimately resistant or impenetrable atoms.

2. Locke's Position

Locke rejects the Cartesian "space theory of matter" and argues, following Boyle, that extension and matter are conceptually distinguishable.[14] Matter, unlike mere extension, is solid. And furthermore, according to Locke, solidity is a proper object of touch (2.4.1). Solidity is not, like a spatial quality, a "common sensible," accessible either to sight or to touch. Where Descartes' primary qualities were all common sensibles, Locke's solidity is not. For this reason, as might be expected, Locke has great difficulty justifying the addition of solidity to his list of primary qualities. After all, solidity, as a proper sensible, is much more like a secondary quality than it is a primary one. Like a secondary quality, solidity is *given* in sensation, and like a secondary quality it seems to resemble nothing *in* a physical object.

Indeed, Locke begins his special chapter on solidity in the *Essay* with the claim that solidity is an idea derived from sensation, connected "most intimately with and essential to body" and arising from or being nothing more than (Locke says both) the resistance of bodies. We get our idea of solidity from large bodies, or so says Locke, but solidity (like all other primary qualities) is also characteristic of the smallest bit of matter (2.4.1).

Here Locke has a problem. The fact is, as he recognizes, that we do not experience everything physical as solid (whereas we do experience everything physical as extended) and we certainly do not experience objects with similar spatial characteristics as equally solid. Air does not seem to be solid at all and water seems to be less solid than a diamond. On closer examination, Locke is forced to admit, sensation only acquaints us with relative hardness or softness. Yet experiment can convince us, he says, of the fact that all bodies, even relatively soft ones like air, resist other bodies:

> For, though the two flat sides of two pieces of marble will more easily approach each other, between which there is nothing but water or air, than if there be a diamond between them: yet it is not that the parts of the diamond are more solid than those of water or resist more, but because, the parts of water being more separable from each other, they will, by a side motion, be more easily removed and give way to the approach of the two pieces of marble. (2.4.4)

Locke offers up as evidence for his claim an experiment purportedly made in Florence involving the resistance of water contained in a hollow globe of gold. Under extreme pressure, the water was said to have been squeezed through the pores of the metal before the globe collapsed (2.4.4). In accounting for this and other similar experiments, Locke is forced to distinguish solidity from hardness: solidity, properly speaking "consists in repletion, and so an utter exclusion of other bodies out of the space it possesses: but hardness, in a firm cohesion of the parts of matter making up masses of a sensible bulk, so that the whole does not easily change in figure." Hardness is a secondary quality. Locke says explicitly that it is relative "to the constitutions of our own bodies" (2.4.4).

Now although Locke begins his discussion of solidity by saying that it is given in sensation and a proper sensible, he concludes (but without ever explicitly acknowledging the contradiction) with the claim that true solidity is only properly attributable to the unperceived smallest parts of perceived bodies.[15] Significant for natural science, however, is the use to which Locke puts the concept in explaining perceived "facts" and experimental results in terms of unperceived entities, their properties and interactions. It is surprising only when viewed against the background of Descartes' reasoning about spatial qualities, that hardness, a secondary quality, becomes a primary quality (solidity) when abstracted to the limit in atoms.

It will be helpful, I believe, to recapitulate here the important differences between the spatial or Cartesian primary qualities on the one hand, and Locke's solidity on the other.

A. The spatial qualities are found by means of sight and touch in all bodies within the range of "normal" perceptual judgment. Bodies are always perceived as extended, positioned, in motion or at rest, and so forth. The spatial qualities are in this regard "perceptually universal."

B. Determinate spatial qualities are attributed, by hypothesis, to bodies that are, due to their size, in fact unperceived. Micro-entities and their interactions are conceived as being strictly analogous to macro-entities and their interactions.

A'. By means of experiment, according to Locke's account of solidity, resistance is found to be characteristic of all bodies within the range of normal perceptual judgment. We discover, then, that resistance is a property of all bodies that we perceive, or that it is "experimentally universal."

B'. Locke explains the experimental finding of (A') by attributing solidity (in the sense of impenetrability) to the smallest bodies (which are, in fact, unperceived). According to this reasoning, the atoms alone are impenetrable and atomic aggregates only relatively hard. Micro-entities and macro-entities are not strictly analogous, because the hardness of atomic aggregates is abstracted to a limit in the atoms.

3. Perceptual Experience and Primary Qualities

In conclusion, I want to restate and defend my two central claims about the status of primary qualities in early modern science. First, the doctrine of primary qualities depends on the following commitment: it is possible to make correct perceptual judgments about the spatial qualities of <u>any body whatsoever</u>. Both Descartes and Locke thought that the microscope would indefinitely extend the range of perceptual judgment.[16] We would be able, so they thought, to make correct perceptual judgments about the spatial qualities of bodies causally responsible for what we in fact do perceive. In sum, nature is spatially "the same" throughout and there is no principled objection to its being perceivable throughout. For Locke and Descartes, in other words, correct perceptual judgment might be said to be the basis for certainty in our active investigation of nature's spatial organization.

Second, perceptual universality fails to explain why solidity is a primary quality. Solidity, in its "limit" sense of impenetrability, is not characteristic of all bodies. In fact Locke attributes solidity only to the smallest bodies, and thus qualifies the idea that micro-entities and processes are analogous to macro-entities and processes. Locke's difficulties with solidity, in fact, stem from his attempt to force it into the previous standard of perceptual universality. In truth, solidity is a "new" type of primary quality, characteristic of the further development of modern science. "New" properties of this sort cannot be "read off" normal perception, but must be discovered by experiment. However, Descartes' idea that correct perceptual judgments provide observational certainty is not abandoned, for experiment <u>presupposes</u> the possibility of correct perceptual judgment.

In order to reinforce my first point, I shall recall briefly the concept of 'observation' which applies most naturally to Descartes' theory of correct perceptual judgment. A point that bears repeated emphasis is this: for Descartes, sensation, although it is direct, is not observation, for sensation is not *of* anything. Sensation is just given to awareness. When Descartes says that sensation is "clear" he means only that we are sure about having our own sensations. The net worth of this solid but scientifically vacuous certainty is summed up in the *Cogito*, the one clear and distinct idea to emerge from all clear awareness. But neither the *Cogito* nor the sensations provide information about physical objects and neither would qualify as reliable observation of nature.

By contrast, perceptual judgments are knowledge claims *about* bodies. In Descartes' language, moreover, our judgments about bodies can be *made* clear and distinct. I shall call such *correct* perceptual judgments 'observations'. Cartesian observation, then, always involves reference to the spatial or geometrical qualities of the perceived body.

Because the small component parts of perceived bodies are themselves unperceived, judgments concerning the shape, size, position, and motion of the micro-parts are not observations, but hypotheses. In this sense, the spatial qualities of unperceived bodies are "hypothetical"; their observation is the predicted outcome of procedures that would render them visible (or possibly, according to Descartes' theory, tangible, although the procedure is difficult to imagine). Quite simply, what Locke and Descartes have in mind for such a procedure is the division of large bodies into component parts and the microscopic observation of those parts.

Neither Locke nor Descartes actually rests his doctrine of primary qualities on claims about what is intelligible in our concept of matter. Descartes does argue (for example, in *Meditation* V) that extendedness alone is intelligible in our thinking about matter, and therefore, that matter is nothing more than extended. Locke has the counterclaim that solidity too must be thought when we think of bodies (2.8.11). However, it would be a fundamental and pervasive error to think that these appeals to "what we understand by the concept of body" or (to couch the idea in current jargon) to "what we mean by the term 'body'" are appeals to either introspective criteria of intelligibility or to the language of everyday experience. Nevertheless, it is now popularly alleged that the distinction between primary and secondary qualities is a "philosophical" distinction, to be defended (as if this were enough) by reference to linguistic intuitions about everyday situations. Riding this fashionable wave, Jonathan Bennett writes (in support of W. C. Kneale's claim about Locke) that the distinction between primary and secondary qualities is "a philosophical thesis whose support involves no *recherché* scientific information, no appeals to microscopy or the like but only to...unexciting...empirical material." ([2], p. 105).

This claim, as our analysis has shown, is simply indefensible. Although Descartes establishes his doctrine of the observation of spatial qualities by appealing only to "normal" perception, neither he nor Locke would _have_ a doctrine of primary qualities without the idea that primary qualities _explain_ what we normally perceive. Thus the distinction between primary and secondary qualities cannot be supported merely by appeal to what we normally perceive (by appeal to Bennett's "unexciting empirical material") but must include reference to the causal _explanation_ of what we perceive. I have argued that Descartes' conception of scientific knowledge includes a pivotal commitment to the microscopic perceivability of minute bodies. In short, Descartes' doctrine of primary qualities (and the same argument could be made for Locke) requires that the intelligibility of ordinary observation be preserved at the micro-level. The doctrine of primary qualities, in fact, requires hypotheses about what would be observed under conditions other than those of everyday perceptual experience, but it retains observation in normal perceptual circumstances as its model of observational certainty. The primary qualities of the micro-parts are thus "theoretical" (in that they are "hypothetical") but they are also held to be in principle "observable."

My second claim concerns a shift of emphasis in early modern science from observation to experimentation, a shift reflected in the addition of new, non-spatial primary qualities. Remember that the reasoning pattern characteristic of Descartes' thinking about the spatial qualities was roughly this: since certain qualities can be observed (in the sense, again, that we can form correct perceptual judgments about them) in large bodies, so they can also be observed in very small and hitherto unobserved bodies. But this line of reasoning does not apply neatly in the case of solidity. For solidity, in Locke's final sense of impenetrability, is not observed in the bodies with which we are commonly acquainted. Solidity is a "hypothetical" characteristic of ultimate particles; it is hardness abstracted to a limit. It is attributed to atoms on the grounds that experiments yield the information that all bodies resist. These facts could be coherently explained by ultimately resistant or impenetrable atoms.

I suspect that one motivation for the erroneous claim (characteristic of the "two worlds" view that I have criticized) that the micro-particles of early modern science were theoretically unobservable is that these particles very quickly took on "idealized" properties like impenetrability (or inertia), properties that cannot be observed under normal perceptual circumstances. However, the erroneous idea that modern science introduced unobservable particles and thereby contested the primacy of perception does register an important change of emphasis. It simply mislocates or misinterprets that change. Within early modern science is the implicit discovery that not all primary qualities can be "read off" perception; "new" primary qualities are experimental. This shift meant that the issue of our perceptual access to the primary qualities of minute bodies (treated by Descartes in the _Dioptrics_) eventually became largely _irrelevant_ to their understanding and was thus no longer a live issue for the theory of science. The eventual outcome of the shift was this: in order to justify hypotheses about the microstructure, we defend our experimental procedures and reasoning but we are not necessarily called upon to look at the microstructure.[17]

By the same token, however, detection-by-experimentation does not replace correct perceptual judgment, but rather presupposes it.[18] To put the matter bluntly: experiment depends on counting and measuring--on the determinations of size, shape, position, and motion that Descartes thought intelligible.

Notes

[1]I would like to thank Lindley Darden, Charles Larmore, and Dudley Shapere for their suggestions and criticisms.

[2]I shall not give a detailed account of the "two worlds" thesis here. I intend only to show that the idea that early modern science set up its claims in opposition to the truth claims of perceptual judgments is a misunderstanding. I have in mind, of course, Wilfrid Sellars as a prominent proponent of the "two worlds" view. For Sellars, a very important difference between the "manifest and scientific images of man-in-the-world" is that the latter holds physical objects to be complexes of imperceptible particles. Sellars understands the "initial challenge of the scientific image" in early modern science to be its construal of physical things "as systems of imperceptible particles, lacking the perceptible qualities of manifest nature" ([26], p. 26). Among other proponents of this view are E. A. Burtt ([6], pp.89-90), P. K. Feyerabend [4], and Edmund Husserl ([17], pp. 48-49). For a detailed account and defense of the "two worlds" view, see Larmore [19].

[3]See also the Dioptrics, Fourth Discourse (VI, 112) and the Principles iv, CXCVIII (VIII-i, 320-321). The analogy with words is familiar from medieval nominalism. Descartes, however, puts it to new use: to oppose Kepler's and Witelo's "picture theory" of vision. (Except where English translation is quoted directly, references to Descartes are to the standard edition [10] and the volume and page references are given in Roman and Arabic numerals respectively.)

[4]For a detailed discussion of Descartes' theory of natural geometry, see [24]. According to Descartes, natural geometry is not always used in judging distances. Distance may also be judged by using information provided by lens accommodation or by the distinctness of the retinal image. See the Treatise of Man (XI, 159-160) and the Dioptrics, Sixth Discourse (VI, 137-138).

[5]Like distance, position, shape, and size, motion is for Descartes a "mode" of extension (even though extension cannot be the ultimate source of motion). For Descartes, geometrical properties include motion; a line, for example, is defined by the motion of a point. (Further, motion was traditionally included among the common sensibles: see note 7 below.)

[6]See also Meditation III (VII, 37).

[7]Although Descartes almost always writes about the visual apprehension of primary qualities, he also holds that these qualities are accessible to touch. In short: the primary qualities are Aristotelian common sensibles, since they can be apprehended either by sight or by touch (XI, 158). The special or proper sensibles are, for Descartes, secondary qualities. See Principles iv, CXVIII.

[8]See also Principles iv, CCI.

[9]See, for example, R. M. Blake [4], Gerd Buchdahl [5], and Olscamp's introduction in [8].

[10]See Laudan [20] and [21] for his similar view of Locke's hypotheses.

[11]Daniel Garber [15] has argued, and I think convincingly, that by the time he wrote Part iv of the Principles, Descartes was discouraged by the practical difficulties of doing away with hypotheses and surrendered some of his earlier confidence about the attainment of certainty in science. An acute awareness of practical difficulties is seen too in Locke's Essay. Locke does not think that real essences are unknowable, but he never minimizes the shortcomings of contemporary science in this regard. See M. R. Ayers [1], and M. Mandelbaum ([23], Chap. 1).

[12]Descartes' theory of perception, in other words, is his theory of scientific observation. Because it is only a theory of perception and not a theory of measurement or experimentation, it was to prove vulnerable to Berkeley's attacks and ultimately inadequate as a philosophical account of scientific knowledge. Descartes' theory of perception is primarily a theory of vision with touch regarded as analogous in its workings to vision. According to this theory, geometry is in some way applicable to the sensory manifolds of sight and touch. But Descartes does not say how geometry is to be applied in these cases. (Is the sensory manifold itself a two-dimensional Euclidean space to which the rules of perspective are to be applied in "reverse"? Or is it only "like" a Euclidean plane?) Unfortunately, Descartes does not have even the rudimentary theory of measurement that might help in understanding his claim that the visual extendedness and the tactile extendedness of an object are the same or equal. He is therefore defenseless against Berkeley's attack on the commonsensibility of primary qualities, for if visual and tactile extendedness are not demonstrably equal, there are no common sensibles, and following Berkeley's argument, no primary qualities can be detected in the object ([3], Section 61, p. 47).

[13]Committed as he is to the intelligibility of the geometrical in perception, Descartes overlooks the possibility that a primary quality might be measurable, but not spatial. Indeed, one of the chief obstacles which prevents Descartes from formulating the correct laws of motion was the absence, among the fundamental properties he attributed to matter,

of the concept of the resistance of a body.

[14]However, as is well known, Locke follows the general outlines of the Cartesian distinction between (what Locke now calls) primary and secondary qualities (2.8.7, and 2.8.15). Locke concurs with Descartes once again when he writes that, although we get our ideas of the primary qualities from large bodies, we must assume that similar qualities are characteristic of the bodies that we do not, in fact, see. The primary qualities, he writes, are in bodies, "whether we perceive them or no; and when they are of that size that we can discover them, we have by these an idea of the thing as it is in itself, as is plain in artificial things" (2.8.23; see also 2.23.10-14 and Ayers [1], pp. 12-13). (References are to Locke's Essay [22] with the book, chapter, and section numbers given consecutively in Arabic numerals.)

[15]Hume correctly points out the confusion in Locke's account of solidity: "'Tis easy to observe," says Hume, "that 'tho bodies are felt by means of their solidity, yet the feeling is quite a different thing from the solidity and that they have not the least resemblance to each other."([16], p. 230). Hume's point is not (merely) that no feeling can resemble a quality. He is saying that we feel bodies because they resist, although we do not feel their resistance or solidity. We have, in Hume's language, no simple impression of solidity, or in Locke's terms, no sensation of solidity ([16], p. 231). Solidity, according to this line of thought, is not a proper object of touch. Indeed it seems only properly attributable to unperceived bodies. Locke's final conception of solidity, in other words, is identical with Hume's understanding of solidity as a causal concept.

[16]For Locke, see 2.23.11-13.

[17]I should hasten to add that although instrumentalism has been a live issue for the theory of science, it concerns the purported reference by theories to unobservables and not (what is my concern here) the way in which those theories come to be supported.

[18]In [27] Dudley Shapere shows how the concept of direct observation has changed to include detection-by-experimentation.

References

[1] Ayers, M.R. "The Ideas of Power and Substance in Locke's Philosophy." The Philosophical Quarterly 25(1975): 1-27.

[2] Bennett, Jonathan. Locke, Berkeley, Hume. Oxford: Clarendon Press, 1971.

[3] Berkeley, George, An Essay Towards a New Theory of Vision. In Works on Vision. Edited by C.M. Turbayne. Indianapolis: Bobbs-Merrill, 1963. Pages 7-102. (Originally published in 1709.)

[4] Blake, R.M. "The Role of Experience in Descartes' Theory of Method." In Theories of Scientific Method. Edited by R.M. Blake, C.J. Ducasse, and E. Madden. Seattle: University of Washington Press, 1960. Pages 75-103.

[5] Buchdahl, Gerd. Metaphysics and the Philosophy of Science. Cambridge: M.I.T. Press, 1969.

[6] Burtt, E.A. The Metaphysical Foundations of Modern Science. Garden City, NY: Doubleday, 1954.

[7] Descartes, René. Descartes: Selections. Edited by R. Eaton. New York: Charles Scribners' Sons, 1927.

[8] ---------------. Discourse on Method, Optics, Geometry and Meteorology. Edited by P. Olscamp. Indianapolis: Bobbs-Merrill, 1965. (Originally published as Discours de la Méthode pour bien conduire sa raison, et chercher la vérité dans les sciences; plus la Dioptrique, les Météores et la Géométric, qui sont des Essais de cette Méthode, 1637. Reprinted in [10], Volume VI.)

[9] ---------------. Meditations on First Philosophy. (Originally published as Meditationes de prima Philosophia, in qua Dei existentia et animae immortalitas demonstrantur, 1641. Reprinted in [10], Volume VII. English selections printed in [7]. Pages 84-266.)

[10] ---------------. Oeuvres. Edited by C. Adam and P. Tannery. New Edition. Paris: J. Vrin, 1964-1974.

[11] ---------------. Philosophical Letters. Edited by A. Kenny. Oxford: Clarendon Press, 1970.

[12] ---------------. The Principles of Philosophy. (Originally published as Principia Philosophiae, 1644. Reprinted in [10], Volume VIII, Part 1. English selections in [7]. Pages 267-311.)

[13] ---------------. Treatise on Man. (Originally published as Traité de l'homme, 1664. Reprinted in [10], Volume XI. Pages 119-202. English selections printed in [7]. Pages 350-354.)

[14] ---------------. The World. (Originally published as Le Monde, traité de la lumière, 1664. Reprinted in [10], Volume XI. Pages 1-118. English selections printed in [7]. Pages 312-349.)

[15] Garber, Daniel. "Science and Certainty in Descartes." Unpublished manuscript.

[16] Hume, David. A Treatise of Human Nature. Edited by L.A. Selby-Bigge. Oxford: Oxford University Press, 1951. (Books I and II originally published 1739 and Book III 1740. Reprint of 1888 edition.)

[17] Husserl, Edmund. The Crisis of European Science. Translated by David Carr. Evanston, Illinois: Northwestern University Press, 1970. (Originally published as Die Krisis der europäischen Wissenschaften. The Hague: M. Nijhoff, 1954.)

[18] Feyerabend, P.K. Explanation, Reduction, and Empiricism." In Scientific Explanation, Space and Time. (Minnesota Studies in Philosophy of Science, Volume III). Edited by H. Feigl and Grover Maxwell. Minneapolis: University of Minnesota Press, 1962. Pages 28-97.

[19] Larmore, Charles. "Scientific Theory and Experience." Unpublished Ph.D. Dissertation, Yale University, 1978.

[20] Laudan, Laurens. "The Clock Metaphor and Probabilism." Annals of Science 22(1966): 73-104.

[21] ---------------. "The Nature and Sources of Locke's Views on Hypotheses." In Locke on Human Understanding. Edited by I.C. Tipton. Oxford: Oxford University Press, 1977. Pages 149-162.

[22] Locke, John. An Essay Concerning Human Understanding. New York: Dutton, 1961. (Originally published 1689).

[23] Mandelbaum, M. Philosophy, Science and Sense Perception. Baltimore: Johns Hopkins University Press, 1964.

[24] Maull, Nancy, "Cartesian Optics and the Geometrization of Nature." The Review of Metaphysics. Forthcoming.

[25] Newton, Isaac. "Rules of Reasoning in Philosophy." Philosophiae Naturalis Principia Mathematica, Book III, 1686. (As reprinted in Newton's Philosophy of Nature. Edited by H.S. Thayer. New York: Hafner, 1953. Pages 3-4.)

[26] Sellars, Wilfrid. Science, Perception and Reality. London: Routledge & Kegan Paul, 1963.

[27] Shapere, Dudley. "The Concept of Observation in Science and Philosophy." Unpublished manuscript.

The Development of Kant's Conception of Scientific Explanation

Edward MacKinnon[1]

California State University-Hayward

Kant's *Metaphysical Foundations of Natural Science* [19] is undoubtedly the most neglected and misunderstood of all his major works. It should, I believe, be interpreted as an attempted *rational reconstruction* of the mechanics of Newton and Euler. What Kant attempted to reconstruct, however, were the conceptual foundations of science rather than the logical structure. The reasons for this as well as the particular way in which it was done can best be understood in the light of Kant's own philosophical development. In summarizing this I will focus on two interrelated issues: Kant's concept of matter, which changed somewhat in the course of his long development; and his interpretation of scientific explanation, which changed considerably.

Kant's earliest concerns with scientific explanation involved three aspects of interest here: Christian Wolff's (1679-1754) systematization of philosophy, which served as his point of departure; the ideas on scientific methodology, which he derived from Isaac Newton; and Kant's own work in science. Since, in Kant's view, the last two fit together, we will consider them first. From Newton's work and writings Kant accepted the idea of analysis and synthesis as *the* method responsible for advancement in science. Analysis proceeds from effect to cause or from a whole to its constituent parts; synthesis from cause to effect or from part to whole. As Newton had explained it:

> As in mathematics, so in natural philosophy, the investigation of difficult things by the method of analysis ought ever to precede the method of composition. The method of analysis consists in making experiments and observations, and in drawing conclusions from them by induction, and admitting of no objections against the conclusions but such as are taken from experiments, or other certain truths... . By this way of analysis we may proceed from compounds to ingredients, and from motions to the forces producing them; and in general from effects to their causes, and from particular causes to more general ones, till the argument and in the

---

PSA 1978, Volume 1, pp. 18-30

> most general. This is the method of analysis; and the synthesis consists in assuming the causes discovered, and established as principles, and by them explaining the phenomena proceeding from them and proving the explanations ([24], p. 543).[2]

The young Kant attempted to implement this method in a number of his early publications. He argued from Newton's law of gravity to tidal friction as a cause producing a gradual decrease in the Earth's rotation [6]. He argued to the evolution of the solar system by handling the constituent parts of this system exclusively in terms of Newton's laws [7]. He argued from the phenomenon of fire to an atomic composition of solids and fluids and to the cohesive force of ether [8]. He gave a causal analysis of prevailing wind patterns [12] and speculated on wide-ranging subterranean cavities as a cause of earthquakes [10].

Kant's earliest philosophical work, _Gedanken von der wahren Schätzung der lebendigen Krafte_ [5], foreshadowed his abiding concern with developing a concept of matter in which inner forces, accepted as basic, ground extension; developing an explanation of the differences between the concepts proper to physics and mathematics, and explaining why mathematics should be applicable to physical quantities. In 1755, nine years later, he wrote a little treatise, _Principiorum Primorum Cognitionis Metaphysicae Nova Dilucidatio_ [9], revising the Wolffian system [26] to accomodate the reasoning actually employed in science. Though the revision proved inadequate (cf., [25]), it manifested two features which came to characterize Kant's ideas on scientific explanation. First, his attempt to implement the Leibnizian notion of intelligibility took the form of trying to determine the grounds of the possibility of the knowledge we already have. The second is Kant's contention that the fundamental object of our knowledge is, though only in a confused way, the universe as a whole. To advance in knowledge is to clarify this obscure background knowledge: "Without doubt the infinite perception of the whole universe, which is always present to mind, although extremely obscure, already contains within itself whatever reality must be in the cognitions afterwards to be sufficed with more light, and the mind merely by turning attention afterwards to certain of these, while withdrawing an equal degree from others, and illuminating the former with intenser light, acquires greater knowledge day by day." ([9], p. 241).

Kant's _Monadologia Physica_ [11], his most successful early attempt to endow Newtonian physics with a Leibnizian intelligibility, replaced Newton's doctrine of hard, massy, impenetrable atoms, by a doctrine of physical monads. A physical monad, in this account, is an inextended point surrounded by two force fields. The force by which it occupies space, or is impenetrable, is a repulsive force. There is also a long range attractive force obeying an inverse square law. If the repulsive force is an inverse cube force, then the radius at which the two forces balance defines the sphere of impenetrability and should be the same for monads with different masses.

Kant's first tentative step from a distinction between analytic and synthetic methods towards a distinction between analytic and synthetic judgments occurred in his little treatise on negative quantities [13]. His focus, however, was not on the classification of individual judgments, but on the difference between inferences based on conceptual entailment, and inferences based on real physical relations: "I understand quite well how a consequence will follow from a ground in accord with the rule of identity, because it will be found through the dissection of the concepts contained in it. Thus necessity is a ground for unchangeableness, composition a ground for divisibility... . But how one thing can alter another thing, which does not follow from it by the principle of identity, that is something which I would be very happy to have made clear to me." ([13], p. 202).

Kant's _Prize Essay_ of 1764 [14], though ostensibly a treatise on the principles of natural theology and ethics, is in fact his most developed pre-critical work on scientific method. By this time Kant had concluded that all previous attempts to develop metaphysics must be dismissed as failures. To understand the reasons he assigns for this it is important to distinguish two different senses of 'mathematical'. The first, the use of mathematics in the treatment of physical quantitites, Kant strongly endorses. The second, stemming from Descartes' _mathesis universalis_ and embodied in the rationalistic conception of philosophy as logical deduction from self-evident principles, Kant now rejects as most harmful for philosophy. In its place he proposed an adaption of the Newtonian method: "The genuine method of metaphysics is, in fundamentals, identical with that which was introduced into natural science by Newton and which had such useful consequences there. It says that, by means of certain experiments and always with the aid of geometry, a search should be conducted for the rules according to which particular appearances of nature occur." ([14], p. 271). Kant's _Inaugural Dissertation_ [17] aimed at implementing this method. Its stress on synthesis complements the stress on analysis given in the _Prize Essay_: "In a substantial composite, just as analysis does not come to an end until a part is reached which is not a whole, that is to say a _simple_, so likewise synthesis does not come to an end until we reach a whole which is not a part, that is to say _a world_." ([17], p. 47).

Then the rememberance of Hume aroused Kant from his dogmatic slumber. What I wish to present here is, not one more reconstruction of this disputed event, but the bearing of the particular criticism that Kant claimed influenced him on the issues we have been considering. This dogmatic slumber is widely interpreted as referring to a relatively uncritical acceptance of dogmatic metaphysics. I find this interpretation, even when safeguarded with suitable qualifications, quite untenable. Kant had already rejected all previous developments of metaphysics as total failures.[3] Yet, he thought that a method existed for the development of a valid metaphysics, his own adaption of the Newtonian method of analysis and synthesis. This method, as Kant interpreted it, differed from the method of logical deduction in its dependence on _real_, rather than merely logical connections. Though aware that he had not explained the grounds by which such real inferences are justified, Kant did not yet consider this a serious obstacle to progress. The reasons for this confidence were indicated in the _Inaugural Dissertaion_ ([17], p. 79). In science practice precedes method. To discover the true method one must determine which proposed methods actually worked. In metaphysics, however, method must precede practice. The method of analysis and synthesis that Kant was proposing as basic for metaphysics seemed to be justified by its success.

Hume never denied the reasonableness and inevitability of practical reliance on cause-effect relations in science as in ordinary practice. Nor did he deny that our notion of 'cause' includes the idea of 'necessary connection'. What he did deny, and criticize quite effectively, is that there is any adequate non-circular theoretical justification for this doctrine of necessary connections between real objects. If this criticism is correct, then Kant's proposed method for developing a new metaphysics also lacks theoretical justification. Accordingly, to determine the possibility of metaphysics as a science it was now necessary to make a prior determination of the intelligibility, i.e., the conditions of the possibility, of the methods that worked in physics. It is only this aspect of the first Critique [18], the submerged reinterpretation of physics, that will concern us here. I will consider three points related to Kant's interpretation of physics, and treat each only to the degree necessary to understand what Kant is up to in the Foundations. These points are: the role of physics in the overall argument; the status of synthetic a priori judgments; and the presuppositions that ground physicalistic reasoning.

The first point will simply be sketched, on the presumption that the basic argument of the Critique is, or at least should be, familiar to philosophers. Newtonian physics still serves as a prototype of a successful empirical science and as a testing ground for any method proposed for advancing knowledge. Kant's Copernican Revolution, asking how objects conform to our way of knowing, leads to the conclusion that any science whose laws have necessity and universality must be based on axioms which are synthetic a priori judgments. An examination of the conditions of the possibility of such judgments leads to the conclusion that metaphysics, as a science concerned with objects which are not objects of possible experience, is impossible.

The key assumption, anchoring Kant's transcendental argument in the stream of developing science, is the novel idea that the axioms of any empirical science must be synthetic a priori judgments. Kant's position on this is best understood as a subordinate part of a general doctrine interrelating analytic and synthetic methods, concepts, and judgments. The brief summary that follows is based chiefly on the logic lectures Kant gave in 1782 [21], the year after the first edition of the Critique, and on the discussion of transcendental method in the first Critique.[4] The general method of analysis and synthesis has a restricted significance within logic. Analysis makes concepts distinct; synthesis makes distinct concepts. To make the concept of a thing distinct one begins on a descriptive level by collecting representative judgments in which other concepts, functioning as predicates, are attributed to the thing as subject. No given thing-concept can be strictly defined, for one cannot exhaust the concepts that are possible predicates attributable to the thing. However, one can approximate a definition by picking out the characteristics (*nota*, *Merkmal*) that are most basic in the sense that other characteristics presuppose them, but not *vice versa*. Thus, 'impenetrability' is a characteristic of 'material body' from which other characteristics can be derived.

In addition to analytic concepts, or concepts which are given either a priori, as 'substance', or a posteriori, as 'metal', there are also synthetic, or constructed, concepts. In mathematics, as Kant interpreted it, one relies on intuitive constructions. In physics one often replaces approximative definitions, developed by analysis, by strict definitions. Science, in fact, is characterized by its reliance on such constructed concepts. Any judgment explicating a defined or synthetic concept is necessarily analytic. The net result is that for a *given* concept both analytic and synthetic judgments can serve in the process of concept clarification. On an expositional level, a judgment attributing a concept to a thing may be considered analytic, in the loose sense proper to this level, if the characteristic represented by the predicate is part of the actual concept represented by the subject. Scientific usage goes beyond ordinary language usage in its reliance on the use of selected synthetic concepts to replace analytic concepts. If Kant's doctrine is interpreted in this way, then what the first *Critique* is using as a norm is an *ideal* of scientific explanation. Functioning science is related to this ideal in much the same way that an approximative analytic definition of a given concept is related to the synthetic concept that may replace it.

After his Copernican Revolution Kant still accepted the method of analysis and synthesis as the basic method of advancement within science, but he no longer accorded this method an unrestricted applicability.[5] Within the discipline of physics, the physicist should act as an empirical realist. He is concerned with studying real bodies having an objective existence independent of the observer. The time and space in which these bodies exist are also treated as having an objective reality independent of the observer. This empirical realism also extends to inferred or postulated entities, provided that their postulation has observable consequences. Thus, to illustrate the postulate of empirical thought that the actual must be connected with sense perception, Kant instances the inference from the attraction of iron filings to a magnetic material pervading all matter. He accepts such a magnetic field as real on the ground that the grossness of our senses is not the criterion of reality. ([18], p. 243).

The external questions arise only when one steps out of the framework of physics as a functioning science and inquires, in a critical way, into what it means for *anything* to be an object of knowledge. This, the main thrust of the *Critique*, will be omitted here. Instead, I will simply list in a simplified form some presuppositions about objects which Kant develops in the *Critique* and extends in the *Foundations*:[6]

1. Both the properties and activities of physical bodies can be expressed through mathematical laws which are necessary and universal;

2. Whatever is basic in physical reality is conserved;

3. Any alteration in the state of a physical system can be explained in terms of antecedent causes; and

4. The universe is an ordered totality of mutually interacting objects.

The presuppositions, if accepted, warrant inferences from the existence, state, or activity of one body to the existence, state, or activities of other bodies, without any modification in logic.

Kant's distinction between inner and outer sense led to a distinction between two complementary systems of metaphysics. Thus, after distinguishing between 'nature' in the formal sense--the nature of some thing; and 'nature' in the material sense--the sum total of all things insofar as they can be objects of our senses, Kant says of the latter: "Nature taken in this signification of the word has two main parts according to the two main distinctions of our senses: the one contains the objects of the external sense, the other the objects of the internal sense. Therefore a twofold doctrine of nature is possible; a doctrine of body and a doctrine of soul. The first considers extended nature, and the second, thinking nature." ([19], p. 3). A second type of complementarity is also implicit in the *Foundations*, a complementarity between functioning science and science as reconstructed in accord with philosophical ideals. As Kant explained for chemistry in the *Foundations* ([19], p. 4), and in a more general way in his *Critique of Judgment* ([20], pp. 18-26), the laws scientists discover are contingent with respect to our way of knowing. Such laws lack the necessity and certainty which true scientific laws should have. In the *Foundations* and later in the *Opus Postumum* [22], Kant attempted to fill the gap between the philosophical ideal and functioning science through rational reconstruction of the mechanics developed by Newton and Euler.[7] For Kant, however, such a reconstruction focuses, not on the logic of a deductive system, but on the *concept* of matter which plays a foundational role in mechanics. He intended to confer an a priori necessity on the laws of mechanics, and the inferences these laws sanction, by replacing the *analytic* concept of matter by a synthetic concept.

The replacement of such a foundational concept is, in Kant's account, subject to three different constraints. First, the synthetic concept must be able to replace the analytic concept in its basic scientific usages, *salve veritate*. Secondly, the synthetic concept must supply an intelligible basis for mechanics as a mathematical science of nature. Thirdly, the most complex requirement of all, the synthetic concept of matter must accord with Kant's general account of how man comes to know a universe in which he is a part, a knower, and a moral agent.

Kant's lectures on logic explained what is involved in meeting the first requirement. The nominal essence of a thing is represented, for functioning science, by an approximative analytic definition involving the essential notes. Thus 'material body' is characterized by 'extension', 'impenetrability', and 'figure', with 'impenetrability' basic.[8] Though this analytic concept of matter worked adequately *within* mechanics, it did not suffice to give the science a truly intelligible basis. The protracted debates over forces, inertia, action at a distance, and the method of motion's transmission, as well as Kant's own Second Antinomy, brought this out.

The second requirement is more easily met. Philosophy is concerned with knowledge through concepts; mathematics with knowledge through the construction of concepts. The to-be-constructed concept of matter is based on an a priori intuition of spatial extension. This grounds the application of mathematics to physical quantities. The more elaborate justification given for the application of mathematics to the quantities of qualities will not be treated here. Kant's construction of the concept of matter has, not surprisingly, four steps corresponding to the fourfold division of the categories into quantity, quality, relation, and modality. Phoronomy, corresponding to quantity, gives the first stage: "Matter is the moveable in space." ([19], p. 18). Since internal structure is not considered, the moving space can be treated as a moving point, as in Euler's *Mechanica*, which was undoubtedly Kant's source for the science of mechanics.

Dynamics, the next stage and the one corresponding to quality, adds one further characteristic to the concept of matter: "Matter is the moveable insofar as it fills a space." ([19], p. 40). Filling a space, or resisting anything else attempting to occupy the same space, requires a special moving force. As in the *Monadologia*, Kant again postulates an inverse cube repulsive force. However, he does not now have a doctrine of physical monads as individual units. The forces themselves are accepted as basic. What he is postulating here is, in modern terminology, a unified field theory of matter, but one which includes only central force fields, the only type then known.

The concept of matter constructed on this postulated foundation can be treated in three steps: supplying a ground for mathematics; synthetic reconstruction of the characteristics basic to the analytic concept of matter; and preparing a domain for empirical investigation and determination. Here, I will only consider one aspect of the third step. In the synthetic process he is following, Kant can use his postulated foundational concepts to construct strict definitions of such terms as 'body', 'cohesion', 'solid', 'fluid', etc. Both Newton and Euler had begun their treatises on mechanics with a series of similar definitions. Neither, however, had a developed theory on the nature of definitions. When Newton, for example, defined 'quantity of matter' and 'quantity of motion' he did not discuss the relation between stipulating meanings and using terms to explain reality. Kant's definitions, on the other hand, must be understood in the light of his distinction between analytic and synthetic concepts. The defined concepts should replace the *given* concepts, i.e., concepts whose meanings are determined by their use, for the purposes of *re*constructing a science of matter. The defined meanings are not more basic *simpliciter*.

The third stage, the *metaphysical* foundations of mechanics, adds the next characterisitc to the concept of matter: "Matter is the moveable insofar as it is something having a moving force." ([19], p. 95) Mechanics, in Kant's new technical sense, differs from dynamics inasmuch as it is concerned with the force one body exerts on other bodies. Kant is arguing that his new foundations cannot only reproduce established mechanics; it can also confer on it an intelligibility it previously lacked. This is done in three ways. The first is through a further justification of the applicability of mathematics to physical concepts. Kant's synthetic concept of matter relies exclusively on extensive magnitudes and, accordingly, affords an intelligible basis for mathematical integration. Secondly, the Newtonian concept of hard, massy, impenetrable atoms supplied no basis, Kant concluded, for understanding how motion is communicated from one body to another. If, as in Kant's reconstruction, force is basic, so that a body's boundaries are simply equilibrium points of opposing forces, then such a moving force is quite intelligible.

Thirdly, Kant can derive three laws of mechanics which give Newton's laws an intelligible foundation. In Kant's familiar four-fold distinction, mechanics comes under the general heading of <u>Relation</u>. In the first <u>Critique</u> the temporal schematization of this category led to the three Analogies of Experience. Now Kant is concerned with schematizing this category for <u>spatial</u> as well as temporal intuitions. This leads to mechanical counterparts of the Three Analogies. The first law for Kant's Mechanics is: "With regard to all changes of corporeal nature, the quantity of matter taken as a whole remains the same, unincreased and undiminished." ([19], p. 102). When 'substance', in the vague sense proper to the B edition of the First Analogy, is replaced by a purely extensive concept constructed in accord with a priori spatial intuitions, the result is a law of mass conservation. It is, in fact, the first clear statement of such a law. Kant's Second and Third Laws of Motion concern changes of motion as caused by forces and mutual interactions.

The final section, the "Metaphysical Foundations of Phenomenology," relates to the category of modality and adds the final characteristic to the synthetic concept of matter: "Matter is the moveable insofar as it can be an object of experience." ([19], p. 119). Something becomes an object of experience only through the unification of: a given, forms of sensibility, schematization in imagination, concepts of understanding, and the final synthesis in the transcendental unity of apperception. Kant has constructed a concept of matter as the moveable in space and explained how this a priori construction can structure intuition. To determine an object of experiential knowledge, this constructed concept must be predicated of a subject which involves a representation of a given. The crucial issue here is the question of how motion inheres in a subject. Since Kant has constructed matter out of space, this becomes a question of how one moving space is understood relative to other spaces and leads to Kant's doctrine of the relativity of motion.

The development we have been considering had two decisive turning points. The first was Kant's early realization of the significance for science of the distinction between inferences based on conceptual entailment--or the purely logical deduction basic to the Wolffian theory of science--and inferences based on real connections between physical objects. The second was Kant's much later realization that a justification of these real inferences required an account of what it means for anything to be an object of knowledge. To give Kant himself the final word on this issue we will conclude with his final public statement on philosophy, contained in his open letter to Fichte: "...I hereby declare that I regard Fichte's Theory of Science [Wissenschaftslehre] as a totally indefensible system. For the pure theory of science is nothing more or less than mere logic, and the principles of logic cannot lead to any material knowledge. Since logic, that is to say, pure logic, abstracts from the content of knowledge, the attempt to cull a real object out of logic is a vain effort and therefore a thing that no one has ever done." ([23], p. 253).

Notes

[1]This is a digested version of material that will be presented more fully as chapter two in my forthcoming Atomic Physics and Scientific Explanation.

[2]The historical background to the method of analysis and synthesis is treated in [3]. Kant does not seem to have been familiar with the way this method was used in Greek geometry.

[3]On this point see especially Kant's letter to Moses Mendelssohn, April 8, 1766 (in [23], pp. 54-77) and [15], Part I, chapter 3.

[4]The summary account given here is partially based on the excellent account of this matter given in the long "Introduction" to the English version of [21].

[5]Kant's most explicit presentation of this internal-external distinction as basic to his doctrine of scientific realism is in the first edition development of the Fourth Paralogism of Pure Reason in [18], p. 347.

[6]The basis for these presuppositions is chiefly Kant's treatment of the schematization of the categories and the Analogies of Experience.

[7]Since Kant has so few references to sources it is difficult to determine influence by counting citations. The works of Euler's that Kant cites explicitly are: "Recherches Physiques sur la Nature des Moindre Parties de la Matière," (in [1], Series III, Volume 1, pp. 1-15); and Euler's immensly popular, "Lettres à Une Princess d'Allemagne," (in [1], Series III, Volumes 11, 12). He does not explicitly cite Euler's Mechanica (in [1], Series II, Volume 1), though he must have been familiar with it. Few philosophers cite introductory physics texts as their sources. Often Kant simply cites the 'well-known Euler'. More significant than the number of citations is Kant's willingness to accept Euler's position on disputed issues, even when Euler disagrees with Newton on Newtonian science. Thus in [19], p. 72, note, and in [20], p. 66, Kant endorses Euler's wave theory of light and color, though

Newton's particle theory was still the accepted position. He also used Euler's ideas on point masses, the ether, and even his attack on physical monads, though Kant himself had made the most significant contribution to this doctrine. I know of only one place where Kant cites Euler to disagree with him, and then it is on a philosophical rather than a scientific issue. In [16], p. 57 Kant refuses to accept Euler's a posteriori proof of absolute space.

[8]Kant gives an analytic explication of 'material body' in [18], p. 49, and in [15], pp. 32-43, where he developed the notion of 'body' as a counter to a possible notion of 'spirit'. The key distinction is that spirits, if such there be, could not be impenetrable. If they were impenetrable, they would be bodies for all practical purposes.

References

[1] Euler, L. Opera Omnia. Lausanne: Swiss Society of Natural Science, 1911-.

[2] Heidegger, M. Kant and the Problem of Metaphysics. Translated by J.S. Churchill. Bloomington: Indiana University Press, 1962. (Translated from Kant und das problem der metaphysik. Frankfurt: V. Klastermann, 1951.)

[3] Hintikka, J. and Remes, U. The Method of Analysis. Dordrecht, Holland: D. Reidel, 1974.

[4] Kant, I. Kant's gesammelte Schriften. Berlin: Könglich Preussischen Akademie der Wissenschaften, 1902-1972.

[5] -------. Gedanken von der wahren Schätzung der lebendigen Kräfte und Beurteilung der Beweise (1746), In [4] Volume I. Pages 1-181. (English selections: Thoughts on the True Estimation of Living Forces. In Kant's Inaugural Dissertation. (trans.) J. Handyside. Chicago: Open Court, 1929. Pages 1-15.)

[6] -------. Die Erde in ihrer Umdrehung um die Achse (1754), In [4] Volume I. Pages 183-191. (English translation: Examination of the Question Whether the Earth has Undergone an Alteration of its Axial Rotation. In Kant's Cosmogony. (trans.) W. Hastie. Glasgow: James MacLehose and Sons, 1900. Pages 1-11.)

[7] -------. Allegemeine Naturgeschichte und Theorie des Himmels (1755). In [4] Volume I. Pages 215-368. (English translation: Universal Natural History and Theory of the Heavens. In Kant's Cosmogony. (trans.) W. Hastie. Glasgow: James MacLehose and Sons, 1900. Pages 13-167.)

[8] -------. De Igne (1755). In [4] Volume 1. Pages 369-384.

[9] -------. Principiorum Primorum Cognitionis Metaphysicae Nova Dilucidatio (1755). In [4], Volume I. Pages 385-416. (English translation: A New Exposition of the First Principles of Metaphysical Knowledge. In Kant's Conception of God. F.E. England. New York: The Dial Press, 1929. Pages 213-252.)

[10] -------. Ursachen der Erderschutterungen (1756). In [4], Volume I. Pages 417-427.

[11] -------. Monadologia Physica (1756). In [4], Volume I. Pages 473-487.

[12] -------. Theorie der Winde (1756). In [4], Volume I. Pages 489-503.

[13] -------. Versuch, den Begriff der negativen Grössen in die

Weltweisheit einzuführen (1763). In [4], Volume II. Pages 165-204.

[14] -------. Untersuchung über die Deutlichkeit der Grundsätze der natürlichen Theologie und der Moral (1764). In [4], Volume II. Pages 273-301. (English translation: The Distinction of the Principles of Natural Theology and Morals. In Kant's Critique of Practical Reason and Other Writings in Moral Philosophy. (trans.) L.W. Beck. Chicago: University of Chicago Press, 1949. Pages 261-285.)

[15] -------. Träume eines Geistersehers, erläutert durch Träume der Metaphysik (1766). In [4], Volume II. Pages 315-373. (English translation: Dreams of a Spirit Seer. (trans.) J. Manolesco. New York: Vantage Press, 1969. Pages 29-98.)

[16] -------. Von den ersten Grunde des Unterschiedes der Gegenden im Raume. (1768). In [4], Volume II. Pages 375-383. (English translation: Concerning the Ultimate Foundation of the Differentiation of Regions in Space. In Kant: Selected Precritical Writings and Correspondence with Beck. (trans. and intro.) G.B. Kerferd and D.E. Walford. Manchester: Manchester University Press, 1968. Pages 36-43.)

[17] -------. De Mundi Sensibilis atque Intelligiblis Forma et Principiis. (1770). In [4], Volume II. Pages 385-419. (English translation: Dissertation on the Form and Principles of the Sensible and the Intelligible World. In Kant's Inaugural Dissertation. (trans.) J. Handyside. Chicago: Open Court, 1929. Pages 33-85.)

[18] -------. Kritik der reinen Vernunft. 2nd. ed. 1787. In [4], Volume III. (English translation: Critique of Pure Reason. (trans.) N.K. Smith. London: MacMillan, 1963.)

[19] -------. Die metaphysischen Anfangsgründe der Naturwissenschaft (1786). In [4], Volume IV. Pages 465-565. (English translation: Metaphysical Foundations of Natural Science. (trans.). J. Ellington. Indianapolis: Bobbs-Merrill, 1970. Pages 1-134.)

[20] -------. Kritik der Urteilskraft (1790). In [4], Volume V. Pages 165-485. (English translation: The Critique of Judgment. (trans.) J.C. Meredith. Oxford: Clarendon Press, 1952.)

[21] -------. Vorlesungen Kants Über Logik. (ed.) G.B. Jäsche (1800). In [4], Volume IX. Pages 1-150. (English translation: Logic. (trans.) R. Hartmann and W. Schwarz. Indianpolis: Bobbs-Merrill, 1974.)

[22] -------. Opus Postumum. In [4], Volume XXI and XXII.

[23] -------. Philosophical Correspondence, 1759-99. (ed. and

trans.) A. Zweig. Chicago: University of Chicago Press, 1967.

[24] Newton, I. _Opticks_. 1st ed. (1704) 2nd ed. (1706). (As reprinted in _Great Books of the Western World_, Volume 34. Edited by Robert M. Hutchins. Chicago: Encyclopedia Britannica, 1952. Pages 375-544.)

[25] Reuscher, J.A. "A Clarification and Critique of Kant's _Principiorum Primorum Nova Dilucidatio_." _Kantstudien_ 68(1977): 18-32.

[26] Wolff, Christian. _Gesammelte Werke_. (ed.) Jean Ecole. Hildesheim: Georg Olms Verlagsbuchhandlung, 1962.

John Stuart Mill, John Herschel, and the 'Probability of Causes'

John V. Strong[1]

Boston College

1. Introduction

Any discussion of probability theory in early nineteenth-century Britain requires a specially vigorous exercise of the historical imagination. A century of subsequent thought about the foundations of the subject has made both historians and philosophers of science only too aware of how elusive and polyvalent a term 'probability' is; indeed, since Carnap, it has scarcely been possible to use the word without a qualifying adjective or subscript. Before the nineteenth century, on the other hand, during the 'emergence' of formalized probabilistic thinking, writers from Pascal and Fermat onwards were keenly conscious that the concept they used was often ill-defined and equivocally employed.

There is a sensation, therefore, of sailing unexpectedly into a patch of flat calm when one turns to such early Victorian writers (I use the term in its common, extended sense to refer to the period from around 1830 till the middle of the century) as Lubbock, Drinkwater-Bethune, Galloway, and, above all, De Morgan.[2] For them, probability _is_ Laplace's _Essai philosophique_ [16] and _Théorie analytique_ [17] with such semi-popularized presentations of Laplace's ideas as Lacroix's _Traité élémentaire_ [15] and the writings of Quetelet ([23]; [24], English translation [25]). Even a mathematician of the stature of De Morgan devotes virtually all his efforts in this field to making Laplace's results accessible to English-speaking readers.

Given the state of British mathematics during these years, this is not too surprising. Only a short while had passed since the subject had been wakened from its slumbers by the labors of the Young Turks of the Cambridge 'Analytical Society': John Herschel, George Peacock, Charles Babbage and their contemporaries. Much of the best British mathematics was still an import from the Continent (and indeed the early efforts of the Cambridge group had gone largely into translating standard French treatises). In

PSA 1978, Volume 1, pp. 31-41

probability theory (as in celestial mechanics), Laplace towered over his peers, and hence the obvious task for the British innovators was the exposition and dissemination of his ideas and, as their confidence grew, the further development and application of them. Both precedent (in the form of Laplace's own works, as well as those of earlier writers like Condorcet[3]), and acquiescence in Laplace's serene assurance that the theory as a whole was no more than "good sense reduced to calculation," encouraged such steps; after all, who would be so perverse as to insist that *his* field of interest required no common sense?

At this point events took a turn of a kind familiar in the history of ideas. Just as in the seventeenth century a much-simplified version of the philosophy of Bacon's *Novum organum* was adopted as 'Baconianism', and just as doctrines about nature and scientific method far less nuanced than Newton's own views became the 'Newtonianism' of the Enlightenment; so Laplace's richly ambivalent view of the subject matter of the 'doctrine of chances' became, in the nineteenth century, a much more univocal 'Laplacean' interpretation of probability. "Chance," says De Morgan in an early essay, "is merely an expression of our ignorance of the chain of events which have led to any particular occurrence." ([7], p. 103). In modern parlance, probabilities are rational degrees of belief, and any question of probability is simply "an inquiry into the number of ways in which, under given conditions, an event may happen or fail"; when these are enumerated "the proportion of these two is that of the probabilities for and against." (*ibid*.). The ground is thus laid for a critic dissatisfied with the received interpretation of Laplace's ideas, to question whether Laplace himself was a 'Laplacean' in this sense; and as we shall see, this is in fact what happens.

When these criticisms did develop they were (as might be anticipated) directed against both the presuppositions of the Laplacean doctrine and its applications. That some objections were based simply upon misconceptions or bias is clear from the fact that writers like De Morgan took pains on repeated occasions (see [7], [8], [9]) to insist that the theory was neither irreligious nor an inducement to gambling! Other criticisms were aimed chiefly at the applications of the theory: for instance, whether in particular cases like the calculations of Condorcet on judicial verdicts, the mathematics had not run away from its users, giving impressively precise but totally unrealistic conclusions.[4] One line of criticism, though--that which challenged the legitimacy of the so-called 'inverse' use of the probability calculus to estimate which of several causes, each capable of producing an observed effect, had in fact done so--bore much more directly on the fundamental assumptions of the Laplacean interpretation. As I have argued elsewhere [29], the attacks of George Boole (most notably in his *The Laws of Thought* [4]) and of John Venn (in *The Logic of Chance* [31]) drastically undermined the whole project of merging degree-of-belief accounts of probability with the method of hypothesis, to produce a probabilistic logic of scientific inference. Indeed, after the appearance in 1874 of that curiously anachronistic work, Jevons' *The Principles of Science* [13], such efforts were almost entirely abandoned for nearly half a century, until revived (in a variety of very different guises) by Keynes [14] Reichenbach [27], Ramsey [26], and Carnap [6].[5]

Though a really adequate account of the developments noted in the last paragraph has yet to be written, the major events, and the identities of the principal contributors, seem well enough established.[6] What I should like to argue here is that, even as a crude sketch, the picture I have drawn is flawed, omitting as it does (and as does every discussion of the topic of which I am aware), one of the most interesting treatments of probability, both 'pure' and applied, to be found in the nineteenth century. I refer to the following root-and-branch attack on Laplace, which begins with this gloss on the initial definition of probability found in the Essai philosophique:

> If his [i.e., Laplace's] unrivaled command over the means which mathematics supply for calculating the results of given data, necessarily implied an equally sure judgment of what the data ought to be, I should hardly dare give utterance to my conviction, that in this opinion he is entirely wrong; that his foundation is altogether insufficient for the superstructure erected upon it; and that there is implied, in all rational calculation of the probabilities of events, an essential condition, which is either overlooked in Laplace's statement, or so vaguely indicated as neither to be suggested to the reader, nor kept in view by the writer himself. ([21], pp. 1140-1141).

The author of this passage is John Stuart Mill; the book is the System of Logic. But these views were likely not familiar to many of Mill's contemporaries, since they appeared only in the first (1843) edition of the Logic, being replaced in all editions from 1846 onwards with what was (or could at least easily be read as) a much more sympathetic appreciation of the standard 'Laplacean' account of probability. What I would like to do in the remainder of this paper, after describing briefly Mill's position in the first edition of the Logic (hereafter SL_1), is suggest some answers to the following questions: (1) Why is Mill talking about probability at all in the Logic ? (2) What is the source of his initial, very negative views on the received interpretation? (3) Why did these views alter so markedly in the interval between 1843 and 1846? and (4) What was his final position on the matter? Finally, I shall indicate why the whole subject seems to me not only interesting in itself, but also significant for understanding how British philosophy in the nineteenth century tried to come to grips with the alluring but recalcitrant 'doctrine of chances'.

2. Probability and the Structure of the System of Logic

For those accustomed to think of the relationship between probability and scientific inference in twentieth-century terms, the fact that Mill devotes three chapters of the System of Logic (in all editions) to the subject of probability may come as a surprise. Unlike, say, Jevons, Mill allows only a minor and subsidiary role to hypotheses in science. For Mill, scientific explanation is, broadly speaking, subsumption under causal laws, with the latter interpreted as invariable sequences of phenomena ([21], pp. 323-369); the logic of hypotheses is a set of procedures for eliminating all but one candidate among a set of laws, all of which purport to cover the same phenomena.

Probability does find a natural place in Mill's discussion, however. Central to Mill's treatment of causality (and hence to the argument of the *Logic* as a whole) is the doctrine of the plurality of causes ([21], pp. 434-453). Events are commonly the effect of an intersection of causal chains, of a 'collocation' of causes. In order to find if there is in any given case a deeper order underlying such collocations, we require means for deciding whether observed regularities among phenomena are in fact mere coincidences, or whether they betray the presence of hitherto unsuspected causes at work.

Now this is an issue which had already been raised by several generations of writers. Originally discussed in connection with a kind of argument *to* design in the early eighteenth century by John Arbuthnot (see [11], pp. 166ff.), it emerged in a full-blown, secularized form in a 1735 essay by Daniel Bernoulli on the origin of the solar system ([3]; *cf.* [30], pp. 222-224). The question, used as an illustration by virtually every writer on probability for the next hundred and fifty years, was this: given the fact that all the known planets revolve in the same sense about the sun, and that their orbits all lie more or less in a plane, it seems sensible to ask whether the observed order could have arisen simply by coincidence, rather than from some single cause. In other words, prior to theorizing about the mechanism which might account for the regularity, one ought first to ask whether there need *be* any mechanism.

This question about the 'probability of causes'--in Mill's language, the problem of "how to distinguish coincidences which are casual from those which are the result of law" ([21], p. 1148)--is clearly crucial to any account of scientific explanation of the sort Mill proposes. And in fact Mill accepts the validity of the calculation done by Daniel Bernoulli (and, following him in essentials, by Laplace). What Mill objects to strongly in SL_1 is the implication that somehow such an assessment of probabilities can go beyond what is given in experience; "that there can be a measurement of probability where there is no experience."([21], p. 1143). For Mill, Laplacean probability calculations are valid so far, but only so far, as their foundations are empirical: "Conclusions respecting the probability of a fact rest not upon a different, but upon the very same basis, as conclusions respecting its certainty; namely, not our ignorance, but our knowledge: knowledge obtained by experience, of *the proportion between the cases in which the fact occurs, and those in which it does not occur*. Every calculation of chances is grounded upon an induction..." ([21], p. 1142; emphasis added). It seems quite clear from this passage (and the impression is further strengthened when it is taken in context) that Mill is arguing for a frequency interpretation of probability, almost a quarter century before Venn.[7] I think the likeliest reason for this, and for Mill's stand against the near-unanimous acceptance of Laplaceanism by his contemporaries, is implicit in what has already been said about the depth of his empiricist commitments (though 'empiricism' was not for him an attractive label). The received version of probability smacked too much of apriorism, of a device whereby truths about the world could be spun out of the human mind without observation or experiment, and "by a system of operations upon numbers, our ignorance...coined into a science."(*ibid.*).[8]

Nonetheless, I would like to suggest one source of influence on Mill's ideas that may have been of significance. The writer to whom Mill seems to have turned most regularly for methodological inspiration is the Scottish Common Sense philosopher Dugald Stewart.[9] Stewart has relatively little to say on probability, but he does devote one long note to the subject in his Philosophy of the Active and Moral Powers of Man (1828). A large part of this note is given over to quotations from a paper of Prevost and L'Huillier (1796)--characterized by Stewart as "very able"-- in which the authors assert that our judgments about 'Laplacean' probabilities, and those about observed causal regularities among natural events, "dépendent de deux ordres de facultés différents": in particular, that the latter rely wholly on "la liason des idées" found in our experience ([28], VII, pp. 116-118). Given the paucity of criticism of the foundations of probability prior to Mill's time, and the accessibility of this passage to Mill while he was writing the Logic, the arguments of Prevost and L'Huillier could have supported, even if they did not inspire, Mill's misgivings about the Laplacean view.[10]

3. Herschel, Mill, and the Second Edition of the System of Logic (1846)

If the above account is even roughly correct, especially in its emphasis on the ease with which a non-Laplacean, frequency interpretation of probability finds a place in Mill's philosophy of science, it becomes all the more surprising to find him apparently turning away sharply from this interpretation only three years later, in the 1846 edition of the Logic (SL_2). "The theory of chances," he writes there, "as conceived by Laplace and by mathematicians generally, has not the fundamental fallacy which I had ascribed to it."([21], p. 535). And on another point on which he had taken issue with Laplace in SL_1 (the rules for estimating the credibility of witnesses testifying to the occurrence of an improbable coincidence), Mill now writes, "This argument of Laplace's, though I formerly thought it fallacious, is irrefragable in the case which he supposes..." ([21], p. 636).

Fortunately for the historian, it is possible to trace in great detail the stages in Mill's change of heart.[11] There is abundant evidence that, for a writer preparing a treatise of unprecedented scope on the 'methods of scientific investigation' (as the subtitle of the Logic ran), Mill knew painfully little about the natural sciences of his day.[12] Even after the publication of SL_1, he was conscious that the book might well contain errors or omissions that he had been unable to detect; and accordingly, when Sir John Herschel wrote a cordial letter of thanks for a presentation copy of the Logic which Mill had sent him, the latter made bold to ask if Herschel would suggest any corrections that seemed necessary.[13]

The choice of critic was a shrewd one. If Mill knew surprisingly little science, Herschel had mastered an astonishingly large amount. (His only rival would have been William Whewell, with whose Philosophy of the Inductive Sciences Mill had dealt quite sharply in the Logic.) Even more impressive than Herschel's accomplishments was his reputation; as one recent commentator has put it, for the early Victorians, to be scientific was to be as much like Herschel as possible ([5], p. 219). Though five years returned from his epochal survey of the southern skies at the Cape of Good Hope, Herschel was still deep in the monumental task of preparing his observations for publication; but he readily turned his attention to the Logic when Mill began serious work on a new edition in 1845. In December of that year, Herschel sent a long letter to Mill, detailing his objections to Mill's strictures against Laplace's statement of the theory of probabilities, as presented in SL_1. "With these objections I can no ways agree," he wrote, "and I will not conceal from you that I read them with great concern and an earnest wish that you would give them a full reconsideration."[14] Granted, says Herschel, that Laplace shares the tendency of many mathematicians sometimes "to prefer mathematical calculation by rule and formula to the plain exercise of common sense." But this is no more than a lapse on Laplace's part. Mill's criticism misses the point that Laplace was aware that only through experience could the formal calculus of probabilities--which is, as Herschel notes, "simply the theory of combinations"--be linked with the real world. Probability is, to be sure, epistemic, "a matter of opinion & judgment" that certain events are equally possible "as regards our limited knowledge or conceptions." But the basis of all such assessments is experience, whether in the form of the direct evaluation of relative frequencies, or of more generalized and tacit use of observation and experiment. "How do we know," asks Herschel, "that in a forest of trees half of which are oaks it is equally possible to stumble on an oak and another tree? Has anybody ever tried it?" And yet such a judgment is in no sense a priori: "A total absence of all knowledge of the connexion of events--or so to speak, of the mechanism of the events is incompatible with that state of mind which leads us to assert an equal probability....We must see enough of the case to get or make for ourselves an impression (perhaps an erroneous one) that there does exist a similarity of circumstances. In short an opinion or a judgment to be worth the name must be grounded on something not merely negative."

The 1846 revision of the Logic incorporates all these criticisms. After summarizing his original view (with an amplified quotation from Laplace's Essai to open the chapter), Mill devotes several pages to a detailed recantation of his position in SL_1. Probability is indeed "probability to us," that is, the warranted degree of expectation we have that an event will occur ([21], p. 535). This degree of expectation can be calculated by the "common theory" of probability even in those cases in which we are unable to appeal to any "special experience"; under such conditions, the "logical ground of the process [of calculation] is our knowledge, such knowledge as we then have, of the laws governing the frequency of occurrence of the different cases; but in this case our knowledge is limited to that which, being universal and axiomatic, does not require reference to specific experience, or to any considerations arising out of the special nature of the problem under discussion."([21], p. 537).

4. Conclusion

It is clear that one wants to ask at least two questions about Mill's revised estimate of the Laplacean (or perhaps Laplacean-Herschelian) account of probability: is it fairer to Laplace's own views, and is it a valid interpretation of the doctrine of chances in its own right? The answer to the first question is surely yes, since only $\underline{SL_2}$ brings out the centrality of the epistemic interpretation of probability, and of the concept of equipossibility, in the *Essai*. But precisely insofar as it does capture Laplace's thought, the revised version of Mill's presentation likewise reflects the deep double-mindedness of the *Essai* about these concepts. Are possibilities attributes of things (and hence the objects of empirical investigation like other physical properties)? Or do they rather refer, like probabilities, "in part to our ignorance, in part to our knowledge"? And if probabilities are to be defined in terms of (equi-)possibilities, can either interpretation leave us any the wiser about the true nature of probability?

As can often happen, Mill was brought round by good advice to a position that was soberer, better-balanced, and considerably less interesting than his initial stance. But his original rejection of Laplace was not without its sequel. Twenty years after the second edition of Mill's *System of Logic* appeared, John Venn published the first thorough-going frequentist treatise on probability, *The Logic of Chance* [31]. "You seem to go farther in rejecting the doctrines of the mathematicians than even I do," Mill wrote in a thank-you letter to Venn when the latter sent him a copy of the work as a gift ([20], pp. 1360-1361). Mill's particular problem, and indeed the whole tradition of the probability of causes, was by that time well on its way to being taken up, transformed, and in a very short time dissolved altogether, to become an intriguing but minor episode in the history of statistics. The underlying issues about the nature of probability which Mill and Herschel had raised and then skirted round, on the other hand, had as we know a very different fate; left pretty much as they were found, they were passed along as a perplexing legacy to the philosophy of science of our own time.

Notes

[1] I wish to thank Larry Laudan for his continuing advice and encouragement in the larger research project of which this essay forms a part. I am also grateful to the library staff at The Royal Society, London, for their kind assistance in examining the Herschel Papers deposited there.

[2] For these writers, see John Herschel's review of Quetelet [23] and [24], reprinted in [12].

[3] Todhunter gives an unsympathetic account of such applications, especially those proposed by Condorcet, in [30]. A very full treatment of the background to Condorcet's ideas may be found in [1], pp. 225-244.

[4] This tradition was still very much alive in Laplace's time; see the chapter on the 'probability of testimonies' in the 1820 edition of the *Essai* ([16], pp. 109-125).

[5] For an introduction to Jevons' thought, see the chapter on him by E.H. Madden in Blake *et al.* [2].

[6]In America the project was going forward, in a quite different form, in the work of C.S. Peirce.

[7]Mill's criticisms in $\underline{SL_1}$ may thus antedate those of Robert Leslie Ellis, whose brief, difficult papers in the _Transactions_ of the Cambridge Philosophical Society for 1843 and 1856 also argued for the frequency interpretation.

[8]Hacking ([10],[11]) has shown that Laplace's views uneasily embrace _both_ degree-of-belief and relative-frequency accounts of probability, and that this duality can be traced back to the writings of James Bernoulli and Leibniz on the subject.

[9]See Olson ([22], esp. pp. 94-124) for an extensive discussion of Stewart's views and of their influence on Victorian philosophy of science.

[10]The importance of this Continental tradition for the overall development of methodology in Britain has been stressed in a recent paper by Laudan [18] .

[11]This process has been noted by J.M. Robson in his "Textual Introductio to [21]; see esp. pp. lxxx-lxxxi.

[12]Mill was engagingly aware of his limitations in this regard, as his remarks on the matter in his _Autobiography_ ([19], pp. 133-134) testify.

[13]Although I have concentrated here on what seem to me to be the chief points of Herschel's criticisms regarding Mill's views on probability, Herschel also provided a number of corrections on other topics; see Mill's letter to Herschel of 28 February 1846 ([20], p. 695).

[14]Herschel to Mill, 22 December 1845; manuscript (apparently a fair copy by an amanuensis) at the Royal Society, London. I am grateful for permission to quote from this letter, to which the Society holds copyright.

References

[1] Baker, K.M. Condorcet: From Natural Philosophy to Social Mathematics. Chicago and London: University of Chicago Press, 1975.

[2] Blake, R.M., Ducasse, Curt J. and Madden, E.H. Theories of Scientific Method: The Renaissance Through the Nineteenth Century. Seattle: University of Washington Press, 1960.

[3] Bernoulli, D. "Recherches physiques et astronomiques... ." In Pièces qui ont remporté le prix double de l'Adadémie royale des sciences en MDCCXXXIV. Paris: Imprimérie royale, 1735. Pages 93-122.

[4] Boole, G. An Investigation of the Laws of Thought. London: Macmillan, 1854.

[5] Cannon, W.F. "John Herschel and the Idea of Science." Journal of the History of Ideas 22(1961): 215-239.

[6] Carnap, R. Logical Foundations of Probability. 2nd ed. Chicago: University of Chicago Press, 1962.

[7] De Morgan, A. "Quetelet on Probabilities." Quarterly Journal of Education IV(1832): 101-110.

[8] ------------. "Theory of Probabilities." In Encyclopaedia Metropolitana. London: B. Fellowes, 1817-1845.

[9] ------------. An Essay on Probabilities. London: Longman and Taylor, 1838.

[10] Hacking, I. "Equipossibility Theories of Probability." British Journal for the Philosophy of Science 22(1971): 339-355.

[11] ----------. The Emergence of Probability. Cambridge: Cambridge University Press, 1975.

[12] Herschel, J.F.W. Essays from the Edinburgh and Quarterly Reviews. London: Longman, Brown, Green, Longmans, and Roberts, 1857.

[13] Jevons, W.S. The Principles of Science. 2 vols. London: Macmillan, 1874.

[14] Keynes, J.M. A Treatise on Probability. London: Macmillan, 1921.

[15] Lacroix, S.F. Traité élémentaire du calcul des probabilités. 2ème éd. Paris: Bachelier, 1822.

[16] Laplace, P.S. Essai Philosophique sur les Probabilités. (Included as the "Introduction" to [17]. Pages V-CLXIX. English translation: A Philosophical Essay on Probabilities. (trans.)

F.W. Truscott and F.L. Emory. New York: J. Wiley & Sons, 1902. Reprint by Dover, 1951.)

[17] --------------. Théorie Analytique des Probabilités. 3rd edition. (1820). In Oeuvres de Laplace, Volume VII. Paris: Gauthier-Villars, 1878-1912.

[18] Laudan, L. "The Sources of Modern Methodology." In Historical and Philosophical Dimensions of Logic, Methodology and Philosophy of Science. (University of Western Ontario Series in Philosophy of Science, Volume 12.) Edited by R.E. Butts and J. Hintikka. Dordrecht: D. Reidel, 1977. Pages 3-19.

[19] Mill, J.S. Autobiography. Indianapolis: Bobbs-Merrill, 1957. (Originally published 1873.)

[20] ---------. The Earlier Letters of John Stuart Mill 1812-1848. 2 volumes. Edited by F.E. Mineka. Toronto: University of Toronto Press. London: Routledge and Kegan Paul, 1963.

[21] ---------. A System of Logic Ratiocinative and Inductive. Volumes 7 and 8 of the Collected Works of John Stuart Mill. Edited by J.M. Robson. Toronto: University of Toronto Press, 1974.

[22] Olson, R. Scottish Philosophy and British Physics 1750-1880. Princeton: Princeton University Press, 1975.

[23] Quetelet, A. Instructions populaires sur le calcul des probabilités. Bruxelles: Tarlier et Hayez, 1828.

[24] -----------. Lettres...sur la théorie des probabilités. Bruxelles: Hayez, 1846.

[25] -----------. Letters...on the Theory of Probabilities. (trans.) O.G. Downes. London: Layton, 1849. (A translation of [24].)

[26] Ramsey, F.P. "Truth and Probability." In The Foundations of Mathematics and Other Logical Essays. (ed.) R.B. Braithwaite. Paterson, NJ: Littlefield, Adams, & Co., 1960. Pages 156-211.

[27] Reichenbach, H. Wahrscheinlichkeitslehre. Leiden: Sijthoff, 1935. (Translated and revised as The Theory of Probability. Berkeley: University of California Press, 1949.)

[28] Stewart, D. The Collected Works of Dugald Stewart. Edited by Sir William Hamilton. Farnborough: Gregg International, 1971. (A reprint of the Edinburgh edition of 1854-1860.)

[29] Strong, J.V. "The Infinite Ballot Box of Nature: De Morgan, Boole, and Jevons on Probability and the Logic of Induction." In PSA 1976, Volume 1. Edited by F. Suppe and P.D. Asquith.

East Lansing, Michigan: Philosophy of Science Association, 1976. Pages 197-211.

[30] Todhunter, I. *A History of the Mathematical Theory of Probability*. London and Cambridge: Cambridge University Press, 1865.

[31] Venn, J. *The Logic of Chance*. London and Cambridge: Macmillan, 1866.

Part II

Philosophy of Physics

Non-local Hidden Variable Theories and Bell's Inequality

Jeffrey Bub and Vandana Shiva

University of Western Ontario

1. Introduction

The conceptually puzzling features of quantum mechanics as a statistical theory all have their source in the impossibility of relating the probability assignments defined by the quantum state to distributions over determinate values of the physical magnitudes. While some interpretations of quantum mechanics,[1] as well as "hidden variable" modifications of the theory, have proposed several constructions for assigning values to magnitudes (beyond the value assignments defined by the quantum state alone), a variety of theorems exist which impose severe restrictions on such theoretical constructions.[2]

This note will deal with Bell's proof (in [2]), that any hidden variable theory satisfying a physically reasonable locality condition is characterized by an inequality which is inconsistent with the quantum statistics. Bell's result has initiated a series of experiments designed to refute the entire class of local hidden variable theories.[3] We show that Bell's inequality actually characterizes a feature of hidden variable theories which is much weaker than locality in the sense considered physically motivated. We consider an example of a non-local hidden variable theory which reproduces the quantum statistics (and hence violates Bell's inequality). A simple extension of the theory, which preserves the non-local character, alters the statistics in such a way that Bell's inequality is satisfied. This observation does not, of course, vitiate Bell's conclusion, that no local hidden variable theory can reproduce the quantum statistics. It does, however, considerably clarify the significance of Bell's result and its relation to other proofs concerning hidden variable reconstructions of the quantum statistics.

2. Phase Space Theories

Hidden variable theorems of von Neumann [14], Jauch and Piron [12], Gleason [10], and Kochen and Specker [13],[4] concern a class of hidden

PSA 1978, Volume 1, pp. 45-53

variable theories to be called <u>phase space theories</u>. The characteristic feature here is that the values of <u>all</u> quantum mechanical magnitudes are <u>determinate</u>--the hidden variables in fact merely parametrize these determinate values (as the phase point does in classical mechanics). The various results show that, while there exist assignments of values to the <u>maximal</u> (i.e., non-degenerate) magnitudes, no assignment of values to the maximal magnitudes can be extended to the non-maximal magnitudes without violating certain algebraic relations between the magnitudes which hold quantum mechanically. Bell [3] has countered this argument by questioning the necessity for preserving a relation point-wise on the phase space between the values of certain magnitudes on the grounds that the relation holds in quantum mechanics--preserving the relation in a statistical sense for distributions of hidden variables corresponding to quantum states might be sufficient.

Bub ([5], [6], [7]) has argued that, irrespective of how values are assigned to the non-maximal magnitudes, assignments of values to the maximal magnitudes will violate the quantum statistics for sequential measurements, unless the hidden variables are randomized between successive measurements (i.e., immediately after conditionalization on a first measurement). Exploiting Wigner's derivation (in [19]) of Bell's inequality Bub ([5], [6]) shows that in a phase space theory Bell's locality condition excludes randomization after conditionalization. It follows that the class of local hidden variable theories is included in the class of "strict (i.e., "no randomization") hidden variable theories.

This conclusion, while correct, misplaces the significance of Bell's result. Firstly, it should be emphasized that the Wigner and Bub derivations of Bell's inequality concern phase space theories only, and not hidden variable theories in general. Secondly, if we have two separated spin-1/2 systems, S and S', with spin magnitudes A,B,C,...and A', B', C', ...respectively, the locality condition in a phase space theory is that the phase space functions $f_{A\otimes A'}$, $f_{A\otimes B'}$, $f_{A\otimes C'}$,...corresponding to the the maximal magnitudes $A\otimes A'$, $A\otimes B'$, $A\otimes C'$,...all have the same A-value at the same phase point, so that a phase space function f_A is defined for the non-maximal ("locally maximal") magnitude A. Similarly, for the non-maximal magnitudes B,C...; A',B',C',....But Bub's derivation depends only on the <u>existence</u> of well-defined phase space functions f_A, f_B, f_C,...; $f_{A'}$, $f_{B'}$, $f_{C'}$,...for the locally maximal magnitudes, and the assumption that probabilities for the eigenvalue pairs (a_i, b'_j) of the maximal magnitudes of the form $A\otimes B'$ are to be computed as joint probabilities of the corresponding eigenvalues of the locally maximal magnitudes A and B', i.e.

$$p_\psi \quad (a_i, b'_j) \quad = \quad \mu_\psi \ (f_A^{-1}(a_i) \cap f_{B'}^{-1}(b'_j)),$$

where μ_ψ denotes the phase space measure which generates the quantum statistics for the quantum state ψ. The point-wise equality of f_A with the A-value of $f_{A\otimes A'}$, $f_{A\otimes B'}$, $f_{A\otimes C'}$,...(and similarly for f_B, f_C,...;

$f_{A'}$, $f_{B'}$, $f_{C'}$,...and the corresponding maximal magnitudes) is not required. It is sufficient to assume a condition of statistical consistency, viz., that for phase space measures corresponding to quantum states we have

$$\mu_\psi (f_A^{-1}(a_i) \cap f_{B'}^{-1}(b'_j)) = \mu_\psi (f_{A\otimes B'}^{-1}(a_i, b'_j))$$

Evidently, the statistical condition can be satisfied without the stronger assumption of locality, i.e., for some points x in the phase space X we may have, say

$$f_A(x) = a_1$$
$$f_{B'}(x) = b'_2$$
$$f_{A\otimes B'}(x) = (a_2, b'_1).$$

3. Hidden Variable Theories

From the standpoint of such hidden variable theorems as those of von Neumann, Jauch and Piron, Gleason, and Kochen and Specker, and Bub's analysis of Bell's proof, it would appear that the situation with regard to hidden variable theories is as follows: A hidden variable theory is a phase space representation of the quantum statistics, i.e., the phase points parametrize the determinate values of the quantum mechanical magnitudes, and appropriate measures over these parameters are introduced to recover the quantum statistics. Certain theorems show that no value assignment to the maximal magnitudes can be extended to the non-maximal magnitudes without violating some algebraic relations which hold between the operator representatives of the magnitudes in quantum mechanics. Bell's result shows that no extension of values assigned to the maximal magnitudes of a pair of spin-1/2 systems, S+S', is possible to the locally maximal magnitudes (i.e., spin magnitudes of S alone and spin magnitudes of S' alone), irrespective of how values are assigned to other non-maximal magnitudes. This is because the assignment of values to the locally maximal magnitudes as well as the maximal magnitudes excludes randomization of the hidden variables after conditionalization, and no such theory can reproduce the quantum statistics for sequences of measurements. We have argued that this representation of Bell's result is misleading because the extension of values assigned to the maximal magnitudes does not have to be consistent point-wise with the values assigned to the locally maximal magnitudes for the proof to go through. Only a statistical consistency is required, and this is a very weak notion of locality, allowing the possibility for some phase points of a value assignment such as

$$f_A(x) = a_1$$
$$f_{A\otimes A'}(x) = (a_1, a'_2)$$
$$f_{A\otimes B'}(x) = (a_2, b'_1).$$

This perspective is, however, inadequate on another count. There exists a wide class of hidden variable theories which are not phase space theories, for example, the hidden variable theories of Bohm and his collaborators.[5] For definiteness in the following, we shall discuss the Bohm-Bub theory [4] and its extension by Tutsch ([15],[16],[17])[6].

In the Bohm-Bub theory, hidden variables do not parametrize the determinate values of quantum mechanical magnitudes. Rather ψ, the quantum state, and λ, the hidden variable, jointly *determine* the values of all *maximal* magnitudes in the context of appropriate measurement interactions. This does not mean that the values of the maximal magnitudes are determinate in the hidden variable theory. An equation of motion involving ψ and λ, applicable during a measurement interaction, describes the projection of ψ onto an eigenvector of the magnitude measured (which eigenvector depends on ψ and λ). For our purposes here, we shall simply regard the equation of motion as a rule for assigning a value to each maximal magnitude *on measurement*, given ψ and λ . The transition of ψ and λ during the measurement interaction is a separate matter.[7] We shall use the term "hidden variable theory" to refer to those modifications of quantum mechanics in which the hidden variables determine the values of quantum mechanical magnitudes on measurement, as opposed to phase space theories.

In the Bohm-Bub theory, the values of non-maximal magnitudes are only determined via ψ and λ *as functions of maximal magnitudes*. Thus there is no prescription in the theory for the result of an "autonomous" measurement of a "locally maximal" magnitude A of S in a composite system S+S', even given ψ and λ . The *probability* of a particular result for such measurement, given any ψ state, is computed by averaging the results for the magnitudes $A\otimes A'$, $A\otimes B'$, $A\otimes C'$,.., for all λ , and the theory is so constructed that the probability is equal to the quantum mechanical probability. The extension of the theory proposed by Tutsch allows the pair ψ, λ to determine the value of *any* magnitude, not merely the maximal magnitudes. The idea is simply this: Each maximal magnitude defines a complete orthonormal set of eigenvectors. In the original theory, the equation of motion describes the projection of ψ onto one of the eigenvectors of the magnitude measured, depending on the relation between ψ and λ , which is represented as a vector in a dual Hilbert space. A non-maximal magnitude defines a set of hyperplanes, corresponding to the distinct eigenvalues of the magnitude. Tutsch modifies the equation of motion to describe the projection of ψ onto one of the hyperplanes, depending on the relation of ψ and λ . The correct quantum mechanical probabilities are obtained from a uniform measure over the unit hypersphere in λ-space, as in the original theory. Again, we shall simply regard the Bohm-Bub theory extended by Tutsch's rule as a prescription for assigning values to magnitudes--maximal or non-maximal--on measurement, given ψ and λ .

The Bohm-Bub theory yields the correct probabilities in the case of two separated systems, S+S', in the singlet spin state. The theory is non-local in the sense that a different A-value may be obtained if we measure $A\otimes B'$ rather than $A\otimes A'$. Tutsch's extension of the theory determines the result of an "autonomous" measurement of a locally maximal magnitude such as A on S, given ψ and λ. <u>But the result of A-measurement on S need not in general yield the same value for A as the result of a measurement of $A \otimes A'$ on S+S'</u>, and in general the A-result of a measurement of $A \otimes A'$ will differ from the A-result of a measurement of $A \otimes B'$. Thus, the modified theory is as non-local as the original theory.

Nevertheless, it is easy to see that Bell's derivation of the inequality purportedly characterizing local hidden variable theories goes through for Tutsch's extension of the Bohm-Bub theory. (Of course, the inequality is <u>not</u> satisfied in the original Bohm-Bub theory, which yields the quantum statistics, ignoring for the moment problems with sequences of measurements.) Let $f_A(\psi, \lambda)$ denote the value assigned by Tutsch's rule for an autonomous A-measurement on S, given ψ and λ. Here ψ is the singlet spin state of a pair of spin-1/2 particles. Let $\rho_\psi(\lambda)$ be the distribution of hidden variables yielding the quantum statistics defined by ψ. Let P(A,B') be the expectation value of the product of an A-measurement on S and a B'-measurement on S'. We assume, again, that probabilities for the eigenvalue pairs (a_i, b_j') of the maximal magnitudes of the form $A \otimes B'$ are to be computed as joint probabilities of the corresponding eigenvalues of the locally maximal magnitudes A and B'. Then

$$\begin{aligned} P(A,B') &= \int f_A(\psi, \lambda) f_{B'}(\psi, \lambda)\rho_\psi(\lambda)d\lambda \\ &= -\int f_A(\psi, \lambda) f_B(\psi, \lambda)\rho_\psi(\lambda)d\lambda \end{aligned}$$

since $f_A(\psi, \lambda) = -f_{A'}(\psi, \lambda)$, etc., except at a set of points of measure zero. Hence

$$\begin{aligned} P(A,B') - P(A,C') &= -\int [f_A(\psi, \lambda)f_B(\psi, \lambda) - f_A(\psi, \lambda)f_C(\psi, \lambda)]\,\rho_\psi(\lambda)d\lambda \\ &= \int f_A(\psi, \lambda)f_B(\psi, \lambda)\,[f_B(\psi, \lambda)f_C(\psi, \lambda)-1]\,\rho_\psi(\lambda)d\lambda \end{aligned}$$

since $f_B(\psi, \lambda) = \pm 1$ and hence $f_B(\psi, \lambda)f_B(\psi, \lambda) = 1$. It follows that

$$\begin{aligned} |P(A,B') - P(A,C')| &\leq \int [1 - f_B(\psi, \lambda)f_C(\psi, \lambda)]\,\rho_\psi(\lambda)d\lambda \\ &\leq 1 + P(B,C') \end{aligned}$$

since $\rho_\psi(\lambda)d\lambda = 1$. This is Bell's inequality, as derived by Bell in [2], which is not satisfied by the quantum mechanical expressions for P(A,B'), P(A,C'), P(B,C').

4. Conclusion

It appears from the above analysis that Bell's inequality characterizes any hidden variable theory of a composite system S+S' which determines values for the locally maximal magnitudes, so that the quantum mechanical probabilities for the values of these magnitudes are derived from distributions over the hidden variables, and in which probabilities for the eigenvalue pairs (a_i, b'_j) of the maximal magnitudes of the form $A \otimes B'$ are computed as joint probabilities of the corresponding eigenvalues of the locally maximal magnitudes A and B'. The locally maximal magnitudes are those non-maximal magnitudes which are maximal in the Hilbert space of S alone, or maximal in the Hilbert space of S' alone. From this point of view, Bell's result is a special case of a class of theorems concerning the impossibility of recovering the quantum statistics from a hidden variable theory which (a) determines values for a set of non-maximal magnitudes "autonomously", and (b) derives the probabilities for the eigenvalues of those maximal magnitudes defined by n-tuples of non-maximal magnitudes in the set as joint probabilities of the corresponding eigenvalues of the non-maximal magnitudes.

We may put the point this way: Bell's result appears to be about locality because the analysis purports to exclude the possibility of extending the values assigned to maximal magnitudes to locally maximal magnitudes, in such a way that the value assigned to a locally maximal magnitude A is equal to the A-value of a maximal magnitude $A \otimes B'$, for any locally maximal magnitude B'. Thus, the argument apparently shows the impossibility of <u>extending</u> the values determined for <u>maximal magnitudes</u> to a certain class of non-maximal magnitudes, the locally maximal magnitudes. In fact, the proof concerns the impossibility of determining values for a certain class of non-maximal magnitudes, the locally maximal magnitudes in such a way that the probabilities of eigenvalues of the maximal magnitudes are derived as joint probabilities of values of non-maximal magnitudes, <u>irrespective of how values are determined for the maximal magnitudes</u> (i.e., whether or not a locality condition is satisfied). All we require is a condition of statistical consistency,[8] viz., that for all distributions ρ_ψ over the hidden variables corresponding to quantum states

$$\int f_A(\psi, \lambda) f_{B'}(\psi, \lambda) \rho_\psi(\lambda) d\lambda \quad = \quad \int f_{A \otimes B'}(\psi, \lambda) \rho_\psi(\lambda) d\lambda$$

Bell's result is taken as enjoying two advantages over other proofs of the impossibility of a hidden variable reconstruction of the quantum statistics. Firstly, the locality condition is regarded as physically plausible, while the various conditions concerning the preservation of algebraic structure are regarded as questionable. The above analysis shows that Bell's proof actually involves a condition concerning joint probabilities which is quite artificial. If we consider Tutsch's extension of the Bohm-Bub theory as involving an equation of motion describing the transition of the state vector onto its normalized projection in the hyperplane corresponding to the result of a measurement (either of a maximal or a non-maximal magnitude) then the expression for P(A,B') in § 3 would not be correct. Thus, the objection to Bell's result as a proof of the impossibility of hidden variables underlying the quantum statistics is quite analogous to Bell's rejection of the proofs of von Neumann, Jauch and Piron, Gleason, and Kochen and Specker. Secondly, the locality condition is regarded as providing an experimental test of the hidden variable question, in the sense that a crucial experiment is possible between quantum mechanics and the only physically interesting class of hidden variable theories, the local hidden variable theories. We have shown that an experimental confirmation of the inequality would be consistent with a broad class of hidden variable theories, including such obviously non-local theories as that discussed in Section 3.

Notes

[1]The interpretation of Arthur Fine is the obvious example here. See Fine [9].

[2]For example, the theorems of von Neumann, Jauch and Piron, Gleason, Kochen and Specker, cited in Bell [3].

[3]An experiment was first proposed by J.F. Clauser, M.A. Horne, A. Shimony, and R.A. Holt [8]. A variety of experiments have since been performed, with somewhat conflicting results.

[4]The proofs of von Neumann, Jauch and Piron, and Gleason are all discussed by Bell [3]. Bell regards the Kochen and Specker theorem [13] as a corollary to Gleason's theorem.

[5]The various papers are all discussed by Max Jammer in [11], Chapter 7.

[6]For a discussion of Tutsch's work on the Bohm-Bub theory, see Belinfante [1].

[7]Actually, in order to obtain the correct quantum mechanical probabilities for sequences of measurements, we must assume the transition of ψ to an eigenvector corresponding to the eigenvalue measured, as well as the randomization of λ. Our use of the Bohm-Bub theory here is analogous to the earlier Wiener-Siegel theory [18].

[8]Our conclusion is perhaps an alternative way of presenting Fine's view (see [9]), that Bell's result shows the impossibility of maintaining this statistical condition in a hidden variable theory.

References

[1] Belinfante, F.J. *A Survey of Hidden Variable Theories*. New York: Pergamon Press, 1973.

[2] Bell, J.S. "On the Einstein-Podolsky-Rosen Paradox." *Physics* 1(1964): 195-200.

[3] ---------. "On the Problem of Hidden Variables in Quantum Mechanics." *Reviews of Modern Physics* 38(1966): 447-452.

[4] Bohm, D. and Bub, J. "A Proposed Solution of the Measurement Problem in Quantum Mechanics by a Hidden Variable Theory." *Reviews of Modern Physics* 38(1966): 453-469.

[5] Bub, J. "Randomness and Locality in Quantum Mechanics." In *Logic and Probability in Quantum Mechanics*. (Synthese Library Volume 78). Edited by P. Suppes. Dordrecht: D. Reidel, 1976. Pages 397-420.

[6] ------. "Hidden Variables and Locality." *Foundations of Physics* 6(1976): 511-525.

[7] ------. "The Measurement Problem of Quantum Mechanics." In *Problems in the Foundations of Physics*. (*Proceedings of the International School of Physics "Enrico Fermi", Varenna* (1977), Course LXVII). Forthcoming.

[8] Clauser, J.F., Horne, M.A., Shimony, A., and Holt, R.A. "Proposed Experiment to Test Local Hidden-Variable Theories." *Physical Review Letters* 23(1969): 880-884.

[9] Fine, A. "On the Completeness of Quantum Mechanics." In *Logic and Probability in Quantum Mechanics*. (Synthese Library Volume 78). Edited by P. Suppes. Dordrecht: D. Reidel, 1976. Pages 179-193.

[10] Gleason, A.M. "Measures on the Closed Subspaces of Hilbert Space." *Journal of Mathematics and Mechanics* 6(1957): 885-893.

[11] Jammer, M. *The Philosophy of Quantum Mechanics*. New York: Wiley, 1974.

[12] Jauch, J.M. and Piron, C. "Can Hidden Variables be Excluded in Quantum Mechanics?" *Helvetica Physica Acta* 36(1963): 827-837.

[13] Kochen, S. and Specker, E.P. "The Problem of Hidden Variables in Quantum Mechanics." *Journal of Mathematics and Mechanics* 17(1967): 59-87.

[14] Neumann, J. von. Mathematical Foundations of Quantum Mechanics. (trans.) R.T. Beyer. Princeton: Princeton University Press, 1955. (Originally published as Mathematische Grundlagen der Quantenmechanik. Berlin: Julius Springer-Verlag, 1932.).

[15] Tutsch, J.H. "Collapse Time for the Bohm-Bub Hidden Variable Theory." Reviews of Modern Physics 40(1968): 232-234.

[16] -----------. "Simultaneous Measurement in the Bohm-Bub Hidden Variable Theory." Physical Review 183(1969): 1116-1131.

[17] -----------. "Mathematics of the Measurement Problem in Quantum Mechanics." Journal of Mathematical Physics 12(1971): 1711-1718.

[18] Wiener, N. and Siegel, A. "The Differential-Space Theory of Quantum Systems." Nuovo Cimento (Supplementary Volume II, Series X). 1955: 982-1003.

[19] Wigner, E.P. "On Hidden Variables and Quantum Mechanical Probabilities." American Journal of Physics. 38(1970): 1005-1009.

The Only Real Probabilities in Quantum Mechanics

Nancy Cartwright

Stanford University

Position probabilities play a privileged role in the interpretation of quantum mechanics. The quantum wave function (usually designated '$\Psi(r)$') cannot itself represent a physical quantity, because it is complex valued. Its square -- $|\Psi(r)|^2$ -- however, is real; and it is $|\Psi(r)|^2$ which is the fundamental interpreted quantity of the theory. Schrödinger, who originated wave mechanics, interpreted $|\Psi(r)|^2$ as the density of a charged particle ([14],[15]); but this suggestion was inadequate. A principal reason is that, for most physical systems, $|\Psi(r)|^2$ spreads out in time, whereas particles themselves do not. Schrödinger's interpretation was replaced by Born ([1],[2],[3],[4],[5]), who introduced probabilities into quantum mechanics. Born urged that $|\Psi(r)|^2$ represents not the density of the system in space, but rather a probability density. This is almost universally how quantum mechanics is taught today: $|\Psi(r)|^2$ gives the probability that a system in state $\Psi(r)$ is located at $\underline{r}$.

Despite its fundamental role, however, there is no coherent way of expressing the Born interpretation. The position probabilities that it postulates are unintelligible, for there are no events of which they could be the probability. There are two candidates, and both run into serious problems. The first and obvious suggestion is that the events are what I have just said they are. $|\Psi(r)|^2$ represents the probability that the system is located at r. (Really, $|\Psi(r)|^2 dr$ represents the probability density that the system is located in a small region around $\underline{r}$.) The two slit experiment teaches, however, that in general quantum systems are not located anywhere. If we assume that the particle was positioned at either the top slit or at the bottom, we generate statistical predictions which strongly contradict experimental evidence. Thus Ψ, evaluated at the top slit, cannot represent the probability that the particle is at the top slit; nor Ψ, evaluated at the bottom slit, the probability that the particle is at the bottom slit.

One way of dealing with this situation is to try to block some step in the chain of inferences which starts with the assumption that the particle

PSA 1978, Volume 1, pp. 54-59

is located somewhere, and concludes with the erroneous statistical prediction. The ways of doing so are quite dramatic, however. To block the conclusion, it seems we must either adopt a deviant logic ([12],[13]) or admit a non-standard probability structure ([7],[8],[9] or [11]). To many this is too high a price to pay.

The alternative is to look for a different event space to let the probabilities range over. Although it leads to statistical contradictions to assume that particles in general have definite positions, whenever the position of a particle is measured, the particle is found in some well-defined region. This suggests that $|\Psi(r)|^2$ represents not the probability that the particle is <u>at r</u>, but rather the probability that the particle would be <u>found at r</u> if a position measurement were made.

This alternative, which is the choice of most physicists, is not any more satisfactory. Interpreting $\Psi(r)$ in this way is the source of the notorious and unresolved measurement problem: it leads to the anomalous 'reduction of the wave packet,' and in many treatments ([17],[18],[19], [20] or [16]) to the introduction of consciousness as a special non-physical force in nature.

I would like to urge a more radical alternative. $|\Psi(r)|^2$ does not represent a probability at all. Hence there is no problem of finding an event for it to be the probability of. Moreover, there is no probability distribution over the conjugate variable, momentum. The only probabilities there are in quantum mechanics are probabilities for energy interchanges (transition probabilities to energy eigenstates). These probabilities are special because there is no interference among different energy transitions; such events either occur, or they do not. Hence both the logic and the probability structure over energy transitions is completely standard.

On this interpretation, it is not the unintelligible position probabilities that are fundamental, but the completely straightforward energy transition probabilities. It makes good sense to take energy transitions as basic for the interpretation of quantum mechanics. For it is only through interchanging energy that quantum systems interact and can register their interactions in a macroscopically observable way. In a very well-known argument against reduction of the wave packet, Hans Margenau has urged that all measurements are ultimately measurements of position. But this should be pushed one step further. All position measurements are ultimately measurements of energy transitions. No matter that a particle passes by a detector -- the detector will not register unless it exchanges some energy with the particle. The exchange of energy is the basic event that happens in quantum mechanics; and the basic event whose effects are theoretically described and predicted. For example, although knowledge of the state of an electron has predicative power, knowing the actual position of a bound electron tells us nothing about the future behavior either of it, or of anything with which it interacts. But if the electron "jumps orbits" -- that is, if the atom makes an energy transition -- this event has effects which show up in spectroscopic lines, in the dissolution of chemical bonds, in the formation of ions, etc.

I say that energy transitions either occur, or do not, and that the probabilities for such transitions form the link between theory and observation. I am here taking the well-known case of radioactive decay as the model for interpreting all of quantum mechanics. In radioactive decay, the radioactive substance persists for a while in an excited state; then suddenly, indeterministically, in a way not predicted by the normal evolutionary equations of quantum mechanics, the atom decays. It makes a transition to a new state. The wave packet is reduced in radioactive decay; but not at the will of an observer. The process is completely indeterministic and not capable of being influenced.

So long as a radioactive atom is undecayed, all the possible decay states are represented in its excited state, with various weights. According to the Born interpretation, these weights represent probabilities in the excited state. But we have the same articulation problem here as when the Born interpretation is applied to positions: probabilities for what? Not probabilities that the atom is _in_ one of the decay states, for it is certainly not in any of them; nor probabilities that it would be found in one of these states on measurement, for this implies that by measuring, an observer could force the atom into its decay states. Rather, what is important in radioactive decay are the probabilities that the excited atom will _actually make_ a transition to one of its decay states. These probabilities can be calculated from the "weights" used in the Born interpretation; and I suggest that these are the only true probabilities that appear in the theory.

So far, what I have said has the force of a suggestion; it is a program for an interpretation, not an interpretation. It is no good just to claim there are no deviant probabilities in quantum mechanics; one must show how to do without them. What I will do in the concluding section of this discussion is to describe part of a first step in this program -- the elimination of position probabilities. Even for those who are not inclined to buy the interpretation I have been proposing, I think the result I will describe is interesting in its own right. For it raises an important and often ignored point about measurement: circumstances must be extremely fortuitous in quantum mechanics to guarantee that position measurements can measure position.

An immediate first and obvious response to the interpretation proposed here is to insist that not only are there position probabilities, contrary to my claim, but in fact we can see them. All we have to do is to erect a photographic plate. This is just what happens in a two slit experiment. A beam of similarly prepared particles passes through the two slits and falls on the photographic plate. A small point appears wherever a particle is detected. If enough particles are used, the pattern becomes dense enough that it is possible to see the probability distribution across the plate.

This is too simple a story, however, if one considers the details of what happens at the photographic plate -- or in any other position measuring device. Hans Margenau has long urged that all quantum measurements are measurements of position. It is equally true to say that position measurements are themselves ultimately records of scattering interactions, and hence of energy interchanges. Position measurements occur when the particle whose position is to be measured scatters from a detector. The scattering is inelastic: energy is not conserved in the particle and the detector is activated by the energy which the particle gives up as it scatters. In the common position measuring devices -- cloud chambers, scintillation counters, and photographic plates -- the relevant scattering interaction is the same. The particle scatters from a target in the detector, and the energy which is given up by the particle causes the detector to ionize. The devices differ in how they collect or register the ions, but in each case, the presence of the particle is registered only if the appropriate ionization interaction occurs. So at best, the probability of a count in a detecting device located at a particular place is equal to the probability of the designated ionization interaction at that location.

In photographic plate experiments, however, we know that the pattern on the plate resembles that predicted by $|\Psi(r)|^2$. Yet I claim that the fundamental events which are occurring are energy interchanges; and that analysis of the measurement interaction shows that the pattern on the plate reveals the distribution of these ionization events. The two claims are compatible, however. For it is possible to show that the two patterns are the same, under the conditions which standardly obtain in photographic plate experiments. That is, it is possible to show the probability of a particle "being at" $\underline{r}$ -- $|\Psi(r)|^2$ -- is proportional to the probability of inelastic scattering from a target located at $\underline{r}$. (For a proof see [6]. It follows as a fairly trivial consequence of the so-called "fundamental result of scattering theory.")

The lesson to be learned from this is two-fold. 1) There is no reason to think that photographic plate experiments measure position probabilities rather than energy-transition probabilities. Given the problems raised by the use of position probabilities, and in particular the difficulty of describing what these probabilities are probabilities of, it seems sensible to see how far we can go in eliminating them in favor of transition probabilities. 2) Even if this program runs aground, it is important to see that the patterns predicted by $|\Psi(r)|^2$ and the pattern predicted by a scattering analysis are the same, for this guarantees that the theory is consistent - different methods provided for calculating the same quantity give the same result.

References

[1] Born, M. "Zur Quantenmechanik der Stossvorgänge." Zeitschrift für Physik. 37(1926): 863-867.

[2] -------. "Quantemechanik der Stossvorgänge." Zeitschrift für Physik. 38(1926): 803-827.

[3] -------. "Zur Wellenmechanik der Stossvorgänge." Gottinger Nachrichten (1926): 146-160.

[4] -------. "Das Adiabatenprinzip in der Quantenmechanik." Zeitschrift für Physik 40(1926): 167-192.

[5] ------- and Fock, V. "Beweis des Adiabatensatzes." Zeitschrift für Physik 51(1928): 165-180.

[6] Cartwright, N. "Measuring Position Probabilities." Unpublished manuscript.

[7] Fine, Arthur. "On the Completeness of Quantum Theory." In Logic and Probability in Quantum Mechanics. (Synthese Library 78). Edited by P. Suppes. Dordrecht: D. Reidel, 1976. Pages 249-281.

[8] ------------. "Probability and the Interpretation of Quantum Mechanics." British Journal for the Philosophy of Science 24(1973): 1-37.

[9] ------------. "Some Conceptual Problems of Quantum Theory." In Paradigms and Paradoxes. (University of Pittsburgh Series in Philosophy of Science, Volume 5). Edited by R. Colodny. Pittsburgh: University of Pittsburgh Press, 1972. Pages 3-31.

[10] Margenau, H. The Nature of Physical Reality. New York: McGraw-Hill, 1950.

[11] ---------- and Cohen, L. "Probabilities in Quantum Mechanics." Edited by M. Bunge. Berlin: Springer-Verlag, 1967. Pages 67-89.

[12] Putnam, H. "How to Think Quantum Logically." In Logic and Probability in Quantum Mechanics. Edited by P. Suppes. Dordrecht: D. Reidel, 1976. Pages 47-53.

[13] ---------. "Is Logic Empirical?" In Boston Studies in the Philosophy of Science, Volume 5. Dordrecht: D. Reidel, 1969. Pages 216-241.

[14] Schrodinger, E. "Quantisierung als Eigenwertproblem." Annalen der Physik 79(1926): 361-376, 489-527; 80(1926): 437-490; 81 (1926): 109-139. (English translation in Collected Papers on Wave Mechanics. London: Blackie & Sons, 1928. Pages 1-40, 62-123.)

[15] --------------. "Der stetige übergang von der Mikro-zur Makromechanik." Die Naturwissenschaften 14(1926): 664-666. (English translation in Schrödinger, E. Collected Papers on Wave Mechanics. London: Blackie & Sons, 1928. Pages 41-44.

[16] von Neumann, J. Mathematical Foundations of Quantum Mechanics. (trans.) Robert T. Beyer. Princeton: Princeton University Press, 1955. (Originally published as Mathematische Grundlagen der Quantenmechanik. Berlin: Julius Springer-Verlag, 1932.)

[17] Wigner, Eugene, P. "Epistemological Perspective on Quantum Theory." In Contemporary Research in the Foundations and Philosophy of Quantum Theory. (The University of Western Ontario Series in Philosophy of Science, Volume 2.) Edited by C.A. Hooker. Dordrecht: D. Reidel, 1973. Pages 369-385.

[18] ----------------. "Physics and the Explanation of Life." Foundations of Physics 1(1970): 33-45.

[19] ----------------. "Remarks on the Mind-Body Question." In The Scientist Speculates. Edited by I.J. Good. London: William Heinemann, 1962. Pages 284-302.

[20] ----------------. "Two Kinds of Reality." The Monist 48(1964): 248-264.

A Universal Axiomatization of Kinematical Theories

Carlo Giannoni

Rice University

By a kinematical theory we mean a theory of space, time, and light signals as measured by rigid rods and clocks. In this paper we shall construct an axiomatization of possible kinematical theories which has two especially interesting properties: 1) it is universal in the sense that it presents a threefold continuum of kinematical theories ranging from Special Relativity (STR) on one extreme to Classical Aether Theory on the other; and 2) it is synchronization independent, that is, it does not assume any particular synchronization for clocks. Reichenbach's axiomatization [3] has the threefold continuum implicitly in it, but he does not explicitly present it as a universal axiomatization and he does not give it a synchrony free formulation. On the other hand, the synchrony free axiomatization of STR by Winnie [5] is not universal and it masks an important distinction within the light principle. However, both of these axiomatizations were central to developing the present axiomatization.

1. Reichenbach's Axiomatization

Reichenbach's axiomatization of STR consists of 13 Light Axioms, 6 Matter Axioms, 17 Light Definitions, and 4 Matter Definitions. Having defined a kinematical theory as a theory of space and time as measured by rigid rods and clocks, it turns out that none of the Light Axioms are relevant for our purposes. To put it another way, since Reichenbach has shown that the Light Axioms are common to both Galilean transformations and Lorentz transformations ([3], p. 76), they will all be common to our threefold continuum of theories as well.

We will retain all of the Light Definitions except Definition 8, which is Einstein's definition of simultaneity. For our purposes, Reichenbach's Definition 2 is sufficient.

Let a first signal, sent from O at time t_1 and reflected at P,

---

PSA 1978, Volume 1, pp. 60-70

> return to 0 at a time t_3. Then the instant of its arrival at P is to receive the time value
>
> $$t_2 = t_1 + \varepsilon(t_3 - t_1), \quad 0 < \varepsilon < 1,$$
>
> where ε is an arbitrary factor which must have the same value for all points P.

This definition is slightly defective since all points P of a space cannot be synchronized with the same ε unless $\varepsilon = 1/2$. I have considered arbitrary synchronization throughout the space in [2], but for the purposes of this paper we will consider only clocks which have a $\Delta y = \Delta z = 0$ coordinate difference between them as synchronized with arbitrary ε. We will continue to synchronize clocks whose $\Delta x = 0$ with $\varepsilon = 1/2$.

Therefore, we can focus our attention on the 6 matter Axioms. Of these 6 axioms, Axioms VI.1, VI.2, and VII will also be carried over into our axiomatization, for they are already universal and synchrony free. The relevant axioms for our purposes then are Axioms VIII, IX, and X, which we shall rewrite in a synchrony free manner and with appropriate parameters to obtain our threefold continuum. Let us now look at these three axioms individually.

> (Axiom VIII): Two intervals AB and AC, which are equal when measured by rigid rods, are also light-geometrically equal, when measured according to Definition 10.
>
> (Definition 10): If two signals depart simultaneously from A along the paths ABA and ACA and return simultaneously to A, then the interval AB = AC.

As Reichenbach points out, this axiom is essentially a formulation of the Michelson-Morley experiment as generalized to any inertial system. This axiom can be considered to be an intraframe isotropy light axiom, for it states that the travel time of light is independent of direction within any given inertial system.

> (Axiom IX): The unit transported by rigid rods into a uniformly moving system is equal to the unit transferred according to Definition 16.
>
> (Definition 16): The length of the unit at rest in K', measured in K, has to be equal to the length of the unit at rest in K, measured in K'. $1^{K}_{K'} = 1^{K'}_{K}$.

Two points about this axiom can be made immediately: first of all, it is not synchrony free for the measurement of the length of a moving rod is the measurement of the distance between simultaneous events at the endpoints of the rod. The equality of $1^{K}_{K'}$ and $1^{K'}_{K}$ presupposes Einstein synchronization of clocks in K and K'. Secondly, the axiom is an interframe relativity principle. Therefore, we shall be looking for a synchrony-free interframe relativity principle to replace Axiom IX.

> (Axiom X): The time unit of natural clocks is always of such a kind that the time ABA of a light signal measured by these clocks at A is the same everywhere if AB is the same when measured by rigid rods.

This axiom is very strong for as given it asserts isotropy, homogeneity, and invariance between frames. In fact, Axiom VIII appears to be a special case in that it also asserts isotropy. Therefore, for precision and clarity we should eliminate the isotropy claim from Axiom X and separate out the homogeneity and invariance claims into separate axioms.

> (Axiom X.1) Intraframe homogeneity light axiom: The time unit of natural clocks is always of such a kind that the time ABA of a light signal measured by these clocks at A is the same everywhere if AB is always in the same inertial system, AB always has the same orientation, and AB is always the same length when measured by rigid rods.

> (Axiom X.2) Interframe relativistic light axiom: The time unit of natural clocks is always of such a kind that the time ABA of a light signal measured by such clocks at A is the same everywhere if AB always has the same orientation, A has the same location, and AB is always the same length when measured by rigid rods.

2. Winnie's Axiomatization

Winnie uses three axioms in giving us a synchrony-free formulation of STR:

> (RTLP) Round-Trip Light Principle: The average round-trip speed of any light signal propagated (*in vacuo*) in a closed path is equal to a constant c in all inertial frames of reference. ([5], p. 229).

> (EPT) Principle of Equal Passage Times: Let K and K' be two inertial frames in relative motion, and let A and A' be arbitrary points on the x-axes of K and K', respectively. Let Δt be the time-interval in K of the passage of a rod at rest in K' of rest length s past the point A in K, and let $\Delta t'$ be the time interval in K' of the passage of a rod at rest in K of rest length s past the point A' in K'. Then $\Delta t = \Delta t'$. ([5], p. 230).

> (LIN) Principle of Linearity: For any two inertial frames K and K', regardless of the choices of ε in K and K', any point P in constant straight-line motion with respect to K is also in

> constant straight-line motion with respect to K', and conversely. ([5], p. 231).

The RTLP embodies Axioms VIII, X.1, and X.2. Once again we want to separate out independent components of our axioms. The homogeneity component of the RTLP can be dropped since the force of it is carried by LIN; if the velocity of light depended on location, the equations would not be linear. We are left, therefore, with two separate independent components in the RTLP: the isotropic light principle and the interframe relativistic light principle.

Winnie's axiomatization can, therefore, be seen to differ from Reichenbach's in that Winnie has EPT in place of Reichenbach's Axiom IX. Since the two axiomatizations are equivalent, EPT must be the synchrony-free counterpart of Axiom IX. Therefore, as our synchrony-free formulation of STR, we shall work with the following four axioms: LIN, Axiom VIII, Axiom X.2, and EPT. In order to obtain our universal axiomatization, we shall want to allow each of these axioms except LIN to vary. (LIN embodies the homogeneity of space and time which we want to assume in all of our theories.)

3. The universal synchrony-free axiomatization

We shall assume throughout that we are dealing with inertial systems having the usual Cartesian coordinates and in standard configuration, that is, the direction of motion of any system is always in the x direction, the x axes are all coincident, and the y and z axes, respectively, are all parallel to each other. Furthermore, in formulating the universal counterparts of EPT and Axiom X.2, we can make use of the fact that if either axiom fails, then the system is non-relativistic, that is, we can distinguish between inertial systems and pick one out as the Absolute Rest Frame. We shall signify by w the speed of an inertial system relative to the Absolute Rest Frame, and we shall assume that w is in the x-direction. Now, according to Axiom X.2, the round trip speed of light along the x axis of an inertial system is c, while, according to the Galilean transformation, it is $c(1 - w^2/c^2)$. The universal counterpart of Axiom X.2 should, therefore, have the velocities c and $c(1 - w^2/c^2)$ as special cases. We can achieve this with the following axiom.

> (Axiom X.2*): The round trip velocity of light along an interval AB oriented in the x direction as measured by rigid rods and clocks in an inertial system K is
>
> $$c = c_A\,(1 - w^2/c_A{}^2)^m, \quad 0 \leq m \leq 1,$$
>
> where w is the absolute speed of the inertial system, m is an empirical parameter whose value is to be empirically determined, and c_A is the round trip velocity of light as measured in the absolute rest frame.

An alternative to Axiom X.2* would be to drop the axiom completely, and simply represent the round trip velocity of light in each inertial system with a distinct symbol. In fact, the equations we derive below do not assume Axiom X.2*, but in order to get the Galilean transformations as a special case one must assume Axiom X.2* with m = 1.

The other potentially non-relativistic axiom is EPT. I have discussed the universal non-relativistic counterpart of this axiom at length elsewhere [1] and will here only state the outcome of these considerations. The statement of the axiom is the same as EPT except for the final clause $\Delta t = \Delta t'$ which now becomes:

$$\text{(EPT*)}\quad \Delta t' = \Delta t\ \ \Gamma^{*2n}\ c'/c, \quad 0 \leq n \leq 1,$$

where $\Gamma^* = \sqrt{((1 - w^2/c_A^{\ 2}) / (1 - w'^2/c_A^{\ 2}))}$, w and w' are the absolute velocities of K and K', and c_A, c, and c' are the round trip velocities of light as measured in the absolute rest frame, K, and K', respectively. While it is immediately clear that when m = n = 0, EPT* reduces to EPT, it will be seen below that when m = n = 1, the axiom holds in the Galilean system. (It is not immediately evident, but EPT* with n = 1 is in fact equivalent to zero clock retardation, regardless of the values of c and c'.)

To obtain the universal counterpart of Axiom VIII, we note that in Classical Aether Theory, the round trip velocity of light in a direction other than the x direction, $c_\emptyset$, is related to the round trip velocity of light in the x direction, c, as follows:

$$c_\emptyset = c \ / \ \sqrt{(1 - w^2 \sin^2\emptyset \ / \ c_A^{\ 2})},$$

where $\emptyset$ is the angle between the direction of the light signal and the x axis, whereas in STR $c_\emptyset = c$. Our universal counterpart is therefore:

$$\text{(Axiom VIII*)}\quad c_\emptyset = c \ / \ \sqrt{(1 - w^2 \sin^2\emptyset \ / \ c_A^{\ 2})^q}, \ 0 \leq q \leq 1.$$

Once again, as in the case of Axiom X.2*, we could simply drop Axiom VIII and represent the round trip velocity of light in the y direction with a distinct symbol from its velocity in the x direction. Our derivation actually uses Axiom VIII* only to the extent that it makes the latter assumption, but then in order to get the Galilean transformation as a special case we must assume the entire Axiom VIII* with q = 1.

Now that we have our three universal axioms, we want to derive the universal transformation equations. The equations should have the property that when m, n, and q = 0, the equations will be the ε-generalized Lorentz transformation equations; when m, n, and q = 1 the equations will be the ε-generalized Galilean transformations; and when in addition ε has the Galilean values, they will be the standard Galilean transformations. Then one can consider the experimental significance of any one of the parameters being other than 0, for STR can fail in any one of three independent ways.

In deriving the universal transformation equations for x and t, I very literally follow Winnie's derivation ([5], pp. 231-234) through step by step altering it by the addition of parameters and factors as necessary. There is a one-one correspondence between the equations of the two derivations, and the reader can easily duplicate the derivation for himself. The derivations for y and z which are not in Winnie are given in the Appendix. The resulting set of universal synchrony-free transformation equations are:

$$x' = (c'/c)\, \Gamma^{*n}\, \Gamma\, (x - \vec{v}_\varepsilon t),$$

$$y' = (c'_y / c_y)\, \Gamma^{*n}, \quad z' = (c'_z / c_z)\, \Gamma^{*n} z, \tag{1}$$

$$t' = \Gamma^{*n}\, \Gamma\, ((1 + 2\vec{v}_\varepsilon (1 - \varepsilon - \varepsilon')/c)t - x(2(\varepsilon - \varepsilon')/c + 4\vec{v}_\varepsilon (\varepsilon)(1 - \varepsilon)/c^2)),$$

where from Axiom X.2*

$$c = c_A (1 - w^2/c_A{}^2)^m, \quad 0 \leq m \leq 1,$$

from Axiom VIII*

$$c_y = c_z = c \,/\, \surd(1 - w^2/c_A{}^2)^q, \quad 0 \leq q \leq 1,$$

and where

$$\Gamma^* = \surd((1 - w^2/c_A{}^2) \,/\, (1 - w'^2/c_A{}^2)),$$

$$\Gamma = 1/\surd((1 - (2\varepsilon - 1)\,\vec{v}_\varepsilon/c)^2 - \vec{v}_\varepsilon{}^2/c^2).$$

The inverse transformations are obtained by replacing primed terms by unprimed terms and vice-versa. The resulting two sets of equations can be seen to be internally consistent by substitution from the second set into the first. We note that when m = n = q = 0 the equations reduce to the ε-generalized Lorentz equations found in Winnie, but it is more interesting to note that when m = n = q = 1, these equations reduce to:

$$x' = (\Gamma / \Gamma^*)\, (x - \vec{v}_\varepsilon t),$$

$$y' = y, \quad z' = z, \tag{2}$$

$$t' = \Gamma^* \Gamma\, ((1 + 2\vec{v}_\varepsilon (1 - \varepsilon - \varepsilon')/c)t - x((2(\varepsilon - \varepsilon')/c + 4\vec{v}_\varepsilon (\varepsilon)(1 - \varepsilon)/c^2))).$$

These are the ε-generalized Galilean transformations as can be seen by substituting in the Galilean values of ε and ε', viz., $(c_A + w)/2c_A$ and $(c_A + w')/2c_A$. With these values of ε and ε', Γ* = Γ, and we obtain:

$$x' = x - \vec{v}t, \qquad y' = y, \quad z' = z, \quad t' = t. \tag{3}$$

If in the universal transformations we insert the values $\varepsilon = \varepsilon' = 1/2$, and furthermore one assumes that $q = 0$, i.e., the STR version of Axiom VIII*, then one obtains:

$$\begin{aligned} x' &= (c'/c)\ \Gamma^{*n}\ (x - \vec{v}t) / \sqrt{(1 - v^2/c^2)}, \\ y' &= (c'/c)\ \Gamma^{*n}\ y, \qquad z' = (c'/c)\ \Gamma^{*n}\ z, \\ t' &= \Gamma^{*n}\ (t - \vec{v}x/c^2) / \sqrt{(1 - v^2/c^2)}. \end{aligned} \tag{4}$$

These last equations are the same as Reichenbach's equations (74) ([3], p. 110) with the exception that where Reichenbach has a coefficient β we have the coefficient $\Gamma^{*n} \sqrt{(1 - v^2/c^2)}$ with the explicit parameter n. These equations represent a system in which Axiom VIII holds but n and m are left open to be empirically determined.

Reichenbach then considers the special case of these last equations when $\beta = \sqrt{(1 - v^2/c^2)}$. This is equivalent to our letting $n = 0$. If we let $n = 0$ in (4), we obtain:

$$\begin{aligned} x' &= (c'/c)\ (x - \vec{v}t) / \sqrt{(1 - v^2/c^2)}, \\ y' &= (c'/c)\ y, \qquad z' = (c'/c)\ z, \\ t' &= (t - \vec{v}x/c^2) / \sqrt{(1 - v^2/c^2)}. \end{aligned} \tag{5}$$

These are identical to Reichenbach's equations (75) ([3], p. 111). If we then, furthermore, assume that $m = 0$, i.e., that $c' = c$, we would obtain the standard Lorentz transformation equations. (Note, incidentally, that in (5) the t-transformation is already the standard Lorentz transformation.)

A possibility which Reichenbach does not explicitly consider is where $m = q = 0$, but n is left open. In this case, the equations are:

$$\begin{aligned} x' &= \Gamma^{*n}\ (x - \vec{v}t) / \sqrt{(1 - v^2/c^2)}, \\ y' &= \Gamma^{*n}\ y, \qquad z' = \Gamma^{*n}\ z, \\ t' &= \Gamma^{*n}\ (t - \vec{v}x/c^2) / \sqrt{(1 - v^2/c^2)}. \end{aligned} \tag{6}$$

I have extensively considered this set of equations in [1]; however, I should like to point out a very interesting feature in comparing (5) and (6). If we look at the y-transformation, we see:

$$y' = (c'/c)\ y, \quad \text{and} \quad y' = \Gamma^{*n}\ y.$$

If we have either that $c' \neq c$ or $\Gamma^{*n} \neq 1$, we get a very clearly non-relativistic result. This indicates that a theory can fail to be relativistic in at least two independent ways, which shows that Einstein's Principle of Relativity contains within it at least two empirically distinct claims.

The concluding point which I would like to make relates our considerations above to the argument by Robertson [4] that the Michelson-Morley (MM), Kennedy-Thorndike (KT), and Ives-Stilwell (IS) experiments "do in fact enable us to replace the greater part of Einstein's postulates with findings drawn inductively from observation." ([4], p. 378). In our universal axiomatization we have three axioms containing empirical parameters which must be determined experimentally: Axiom VIII* with q, Axiom X.2* with m, and EPT* with n. The other axioms are common to all the theories. We have already pointed out that MM establishes that q = 0. We shall now show that KT "establishes" that m = 0, and that IS "establishes" that n = 0.

First, we consider KT, and we assume that q = 0 from MM. Therefore, $c_y = c$ and $c'_y = c'$. We take two rods A and B of unequal length a and b at right angles to each other. The round trip times along A and B in an inertial system are 2a/c and 2b/c and the difference is 2(a - b)/c. If they are placed in a different inertial system, the difference in round trip times will now be 2(a - b)/c'. An interferometer would show a shift in the interference fringes only if $c' \neq c$. No shift showed up in the experiment. Of course, according to our Axiom X.2*, $c' = c_A (1 - w'^2/c_A{}^2)^m$ and $c = c_A (1 - w^2/c_A{}^2)^m$. Therefore, if m is very small but not equal to 0, the difference between c and c' may be below the margin of error in the experiment.

Now we consider IS which we can interpret as stating that the round trip clock retardation factor is $\sqrt{(1 - v^2/c^2)}$. Substituting $\vec{v}_\varepsilon\, t$ for x in (1), we get:

$$t' = (\Gamma^{*n}/\Gamma)\ t.$$

Now if we assume that the moving clock traverses a closed path in each direction with constant v in each direction, we obtain:

$$t' - t \simeq -(1 - n)vx/c^2.$$

For a round trip clock retardation factor of exactly $\sqrt{(1 - v^2/c^2)}$, we

$$t' - t = (2x/v) \quad (\sqrt{(1 - v^2/c^2)} - 1) \simeq - vx/c^2.$$

Therefore, a round trip clock retardation of exactly $\sqrt{(1 - v^2/c^2)}$ corresponds to n being exactly equal to 0. Once again any slight deviation from $\sqrt{(1 - v^2/c^2)}$ yields $n \neq 0$, and consequently, a nonrelativistic theory.

In summary, STR is the limit of a threefold continuum of absolute space theories when the three parameters m, n, and q all take on precisely the value 0. Any deviation from 0 in any parameter is evidence for absolute space.

APPENDIX

In the derivation of the y and z-transformations, we must make use of the assumption that there is one isotropic reference frame K_A in which the velocity of light c_A is the same in all directions. We shall get the transformations between any two frames K and K' by taking the product of the two transformations from K and K' to K_A. We shall assume that ε_A of K_A is 1/2 since this does not enter the relationship between K and K' and makes the derivation simpler.

Let a light signal leave the origin of K and K_A when they cross and travel directly up the y axis of K. Let event E_1 be its arrival at y. The coordinates of E_1 in K and K_A are:

$$E_1\text{:}\ \ (x_A = w\ t_A;\ y_A;\ t_A),\ (x = 0;\ y = c_y t;\ t). \tag{7}$$

Now, $t_A = \sqrt{}(w^2\ t_A{}^2 + y_A{}^2)\ /\ c_A$, and therefore:

$$t_A = y_A/c_A \sqrt{}(1 - w^2/c_A{}^2). \tag{8}$$

The relationship between t_A and t is gotten from the t-transformation of (1) by substituting t for t', t_A for t, w for $\vec{v}_\varepsilon$, wt_A for x, 1/2 for ε, etc. The result is:

$$t = t_A \sqrt{}(1 - w^2/c_A{}^2)\ /\ \sqrt{}(1 - w^2/c_A{}^2)^n. \tag{9}$$

Combining (9), (8), and (7), we obtain:

$$y = c_y\ y_A\ /\ c_A\ \ \sqrt{}(1 - w^2/c_A{}^2)^n. \tag{10}$$

The same derivation applies between K' and K_A yielding:

$$y' = c'_y\ y_A\ /\ c_A\ \ \sqrt{}(1 - w'^2/c_A{}^2)^n. \tag{11}$$

Combining (10) and (11) into a transformation between y' and y, we obtain:

$$\begin{aligned} y' &= (c'_y/c_y)\ \ \sqrt{}((1 - w^2/c_A{}^2)\ /\ (1 - w'^2/c_A{}^2))^n\ y \\ &= (c'_y/c_y)\ \Gamma^{*n} y, \end{aligned} \tag{12}$$

where c_y and c'_y are the round trip velocities of light in the y direction in K and K'. Similarly, we obtain:

$$z' = (c'_z/c_z)\ \Gamma^{*n}\ z. \tag{13}$$

References

[1] Giannoni, Carlo. "Clock Retardation, Absolute Space, and Special Relativity." Foundations of Physics. Forthcoming.

[2] ---------------. "Relativistic Mechanics and Electrodynamics Without One-Way Velocity Assumptions." Philosophy of Science 45(1978): 17-46.

[3] Reichenbach, Hans. Axiomatization of the Theory of Relativity. Berkeley and Los Angeles: University of California Press, 1969. (Translation of Axiomatik der relativistischen Raum-Zeit-Lehre. Braunschweig: Friedrich Vieweg, 1924.)

[4] Robertson, H.P. "Postulate versus Observation in the Special Theory of Relativity." Reviews of Modern Physics 21(1949): 378-382.

[5] Winnie, John. "Special Relativity Without One-Way Velocity Assumptions." Philosophy of Science 37(1970): 81-99 and 223-238.

Part III

Observation and Theory

The Interplay between Theory and Observation In the Solar Model of Hipparchus and Ptolemy

Kathleen Okruhlik[1]

University of Pittsburgh

Accounts of ancient mathematical astronomy tend to err in either of two directions as regards the relationship between theory and observation. On the one hand, there are overly positivistic accounts according to which it is assumed that all theoretical progress made by the ancients was rendered possible by the increasing bulk and accuracy of observational data. On the other hand, there are those authors who, whenever they are confronted with a discrepancy between the observational report of an ancient and a modern result, are only too quick to conclude that the alleged observations were fudged (or even manufactured out of thin air) in order to support certain pre-existing theoretical commitments. It would appear from the case studies cited below that the truth about the relationship between observational data and theoretical presuppositions is somewhat more complicated than either of these two extreme accounts would allow.

My aim in this paper is to examine that relationship as exemplified in certain aspects of the solar theory developed by Hipparchus and accepted almost in entirety by Ptolemy [7]. The first part of the paper will comprise a very brief account of the solar theory, showing which are the relevant parameters and how these are to be determined. The remainder of the paper will deal with difficulties which arise in their ascertainment. I will treat in succession the determination of solstices and equinoxes, the eccentricity, the longitude of the apogee, the obliquity of the ecliptic, and the latitude of Alexandria. These last two parameters are closely related to one another, and though they do not figure as directly as the others in the solar model itself, they are essential to the compilation of Ptolemy's tables. For this reason, then, I include them in my account.

1. The Solar Model

The following is a brief summary of the solar theory as presented in

---

PSA 1978, Volume 1, pp. 73-82

Almagest III, 4.[2]

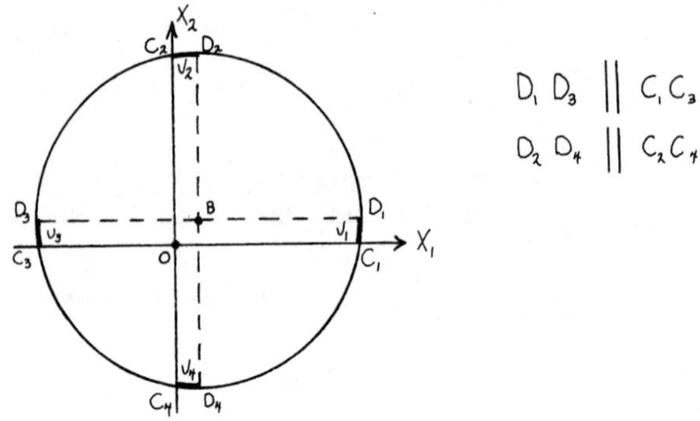

Ptolemy assumes that the sun's motion around the earth can be adequately described on an eccentric model according to which the sun moves with constant angular velocity $\underline{k}$ in a circular orbit which it completes in exactly one year. We on the earth observe the earth's motion, not from the center B of the orbit, but from another point O. The eccentric model is completely determined by three parameters: (1) the constant angular velocity $\underline{k}$ of the sun on its eccentric deferent; (2) the eccentricity $\underline{e}$ = OB; and (3) the direction of the apsidal line, given by the longitude of the apogee - i.e., $\underline{a} = \measuredangle C_1OA$.

The three parameters $\underline{e}$, $\underline{a}$, and $\underline{k}$ are determined from the observed quantities J, J_1, J_2, J_3, and J_4. J is the length of the tropical year measured in days; and J_1, J_2, J_3, and J_4 are the lengths of the four seasons, also measured in days. J_1 is the time interval from vernal equinox to summer solstice; J_2, the interval from summer solstice to autumnal equinox; J_3, the interval from autumnal equinox to winter solstice; and J_4, the interval from winter solstice to vernal equinox.

It is clear then that the observations most fundamental to this discussion will be the observations by Hipparchus and Ptolemy of solstices and equinoxes. These will comprise the subject matter of the next section. Before examining them, however, it would be well to see just how $\underline{e}$, $\underline{a}$, and $\underline{k}$ are computed.

Because $J_1 > J_2$ and because $J_1 + J_2 > J_3 + J_4$, it follows that B must be located in the first quadrant and that the longitude of the apogee is between 0° and 90°. It is clear that the constant angular velocity $\underline{k} = 360/J$. Now let arc $C_1D_1 = v_1$, and let arc $C_2D_2 = v_2$. Then, because the sun moves through arcs C_1C_2 and C_2C_3 in times J_1 and J_2, we know that

$$\underline{k}\, J_1 = 90 + v_1 + v_2, \text{ and } \underline{k}\, J_2 = 90 + v_1 - v_2.$$

So

$$v_2 = (1/2)(\underline{k})(J_1 + J_2) - 90, \text{ and } v_2 = (1/2)(\underline{k})(J_1 - J_2).$$

Using v_1 and v_2, we obtain

$$\underline{e} = \sqrt{\sin^2 v_1 + \sin^2 v_2}, \text{ and } \sin(90 - \underline{a}) = (\sin v_2)/\underline{e}.$$

Thus, all three parameters _k_, _e_, and _a_ are straightforwardly determined. Ptolemy calculates J_3 and J_4 from this model, but (interestingly enough for our purposes) never suggests checking these calculated values against observed values. This omission is not characteristic of Ptolemy's mode of procedure.

2. Observations of Solstices and Equinoxes

In _Almagest_ III, Ptolemy mentions 28 observations of solstices and equinoxes. Four of these are his own, and the remainder are taken from the works of Hipparchus. By comparison with modern values, it is evident that Ptolemy's observations are significantly and systematically in error. The times reported by Hipparchus exhibit relatively small errors, and they fall on either side of the computed times. Another difference is this: The observations of Hipparchus (except for the equinoxes of -161 and -142) form a consistent series exhibiting a systematic error in declination of about -0;7°; they are otherwise accurate to the nearest fraction Hipparchus could have observed. Ptolemy's observations, on the other hand, exhibit no systematic error in declination.

John Britton [2] suggests an interesting and very plausible explanation of the errors made by Hipparchus. This explanation involves an important use of theoretical presuppositions, and it ties in nicely with work by Asger Aaboe and Derek Price on "Qualitative Measurement in Antiquity" [1]. The errors made by Ptolemy cannot be explained in such a satisfying way. The account which one is forced to give of his observational errors involves theoretical commitments, too, but of a less interesting sort. I shall deal first with Britton's account of Hipparchus's error in declination.

There is a fair bit of evidence to indicate that Hipparchus's equinoxes were determined from measurements of the sun's declination and, hence, that he used a meridional armillary to make his measurements. If he did employ such an instrument, then he determined the sun's declination from its meridian altitude and the latitude of the place of observation or, if the equator were marked on the instrument, then he could read the sun's declination directly from the scale. An instrument of reasonable size would permit the times of the equinoxes to be estimated to the nearest quarter of a day.

Britton points out ([2], p. 26) that though it is possible that Hipparchus's error -0;7° was purely accidental, it is remarkable that this is precisely the error which would lead to an interval between spring and fall equinoxes of 187 days. This raises the possibility that Hipparchus might have adjusted the equator of his meridional armillary to yield this result _if he had prior reason to believe that the length of this interval was 187 days_.

If, in fact, Hipparchus had such a reason and made such an adjustment in his instrument, he would thereby have reduced the uncertainty which one always encounters in attempts to determine fundamental alignments by direct measurement. This procedure would also have decreased by half the average error in the times of individual equinoxes as compared to the length of the interval between spring and fall equinoxes. This is a case, then, in which prior theoretical commitment would stand to significantly improve the accuracy of observed data. The question which remains is whether Hipparchus _did_ have any reason to adjust the equator of his instrument to yield a 187-day interval.

Britton notes ([2], p. 28) that he may well have had such a reason, since the value 187 days for the interval between spring and fall equinoxes is the value given in an ancient solar scheme described by Geminus in his _Isagogue_. Aaboe and Price have pointed out that this value might have been easily determined by means of _skaphe_ sundials which were in use from about -300. So it is clear that the value 187 for the interval was around long before Hipparchus and that he might very well have _assumed_ this parameter rather than determine it from scratch. This would explain the relatively small errors in his determinations of individual equinoxes, and it would explain the symmetry of his error at spring and fall equinox. It would protect him somewhat from large systematic error and would have provided a good method for determining whether the sun's motion had a second inequality. (This was a problem of great interest.)

Having mentioned the work of Aaboe and Price, it would be well here to briefly summarize their thesis concerning "the derivation of accurate parameters from crude but crucial observations." [1] The argument is to the effect that the high apparent accuracy of ancient astronomy is due not so much to refined instrumental technique as to a "delicate choice" of which things ought to be observed and how these observations should be fed into the mathematical apparatus. Very often, argue Aaboe and Price, instruments in antiquity served convenience rather than precision. Measurement depended not on instrumental perfection, but on correct choice of crucial phenomena. The idea was to gather a number of these crucial phenomena which were consistent with one another and then to weld them into a mathematical matrix yielding exact agreement between theory and observation. The upshot of this argument is that very crude observations can be manipulated in such a way as to produce parameters which look like the result of very accurate measurement.

The chief example advanced by Aaboe and Price in support of their thesis is that which forms the subject matter of this paper, the solar model developed by Hipparchus.

The values which Hipparchus gives for the four seasons of the year are:

$92\frac{1}{2}$	$94\frac{1}{2}$
$88\frac{1}{8}$	$90\frac{1}{8}$

At first glance, one might think that ancient astronomers were able to determine equinoxes and solstices accurately to within three hours. This is not the case, however, because at the solstices, the sun's declination is going through a maximum or minimum; and it is impossible, therefore, to directly locate a solstitial point to an accuracy better than three or four days. Equinoxes can be fixed to within an accuracy of one-fourth day. But the impossibility of usefully locating a solstice implies, according to Aaboe and Price, that none of the four Hipparchian numbers was directly determined. So we must look elsewhere for their origin.

Because of the circular symmetry of the eccentric model, it is possible, given any two quarters of the year, to immediately deduce the other two. So our task is to determine how Hipparchus might have come up with two of the parameters.

Because equinoxes can be determined with reasonable accuracy, it is logical to suppose that one of the desired parameters is the time interval from spring equinox to autumnal equinox, 187 days. The complementary interval, $178\frac{1}{4}$ days is not as nice and was no doubt obtained by subtracting 187 from $365\frac{1}{4}$, rather than by direct observation. Even very crude observation would allow one to determine the interval from equinox to equinox to the nearest day. So we assume that the interval from spring to autumn was measured and the interval from autumn to spring deduced.

Now, how might Hipparchus have come up with another parameter which would allow deduction of the whole quartet of seasonal intervals? Aaboe and Price suggest that Hipparchus might well have had reason to assume that there was a two-day difference between the lengths of the two quarters whose sum is 187 days. If this is the case, of course, it follows that the two quarters have $94\frac{1}{2}$ and $92\frac{1}{2}$ days respectively. If we know this, we can easily deduce the values of the remaining intervals.

What reason might Hipparchus have had for assuming such a two-day difference? Again, the answer is to be found in the solar model described by Geminus. In this model, each month on the spring side is one day longer than the corresponding month on the autumn side. (Aries is a day longer than Virgo; Taurus is a day longer than Leo; and Gemini is a day longer than Cancer.)

Aaboe and Price contend that Hipparchus might simply have modified this convention of Geminus from three pairs of signs to two. The summer solstice lies between Gemini and Cancer, and it cannot be usefully measured. So, according to the authors, "we are free to assume with Geminus that these show the same one-day difference, or with Hipparchus that they add no further discrepancy to the two quarters of the year." ([1]. p. 12).

The moral, then, is that two extremely crude observational parameters (187 days and two days), correctly manipulated, yielded highly accurate results.

A great deal of space has been given over to discussing Hipparchus's determination of equinoxes and solstices; but the space seems justified because the speculations of Britton and of Aaboe and Price concerning the role of theoretical presuppositions seem to be highly plausible and certainly are very interesting. Accounts of Ptolemy's determinations of equinoxes and solstices, alas, are not so interesting.

R.R. Newton [4] has argued that Ptolemy's errors in his values for solstices and equinoxes are implausibly large given the techniques available to him and that only one hypothesis can account for these errors: Ptolemy did not observe these phenomena at all; rather he calculated them using the Hipparchian model and then presented them as confirming evidence for that same model. John Britton's discussion [2] of the same problem is more thoughtful; but in the final analysis he is unable to provide a satisfactory alternative explanation for Ptolemy's errors.

Ptolemy, unlike Hipparchus, seems to have used an equatorial ring in his (alleged) determinations of equinoxes and solstices. His errors, however, are not such as might have arisen from a systematic error in the declination of his ring. Equatorial rings permit one to observe equinoxes to within an apparent precision of one hour. Britton points out, however, that this apparent precision is more than offset by two serious difficulties which Ptolemy could not have avoided if he used such a ring. These difficulties concern (1) alignment of the ring and (2) variable effects of refraction. Britton's discussion of these two sets of difficulties is very long, and I shall not go into it here. The upshot, however, is that Ptolemy could not have made independent observations of the sort he claims to have made employing an equatorial ring. Furthermore, his data cannot even be interpreted as consistent observations designed to verify Hipparchus' solar parameters. (It would have been impossible to align a ring so that all three equinoxes might be observed.)

Britton, therefore, is finally forced to the unsatisfying conclusion that Ptolemy's equinox observations are nothing more than computations. This is the only way one can plausibly account for their agreement with Hipparchus's observations and his value for the length of the year. He must have simply adopted Hipparchus' solar parameters in order to proceed with his other observations. He could not have made these other observations in a theoretical vacuum and chose as his starting point the solar theory of Hipparchus.

3. Errors in the Eccentricity and in the Longitude of the Apogee

Ptolemy and Hipparchus both arrived at the same value for the eccentricity (1/24) and for the longitude of the apogee ($65\frac{1}{2}°$). The value for e is pretty poor for both astronomers; but Hipparchus's determination of a is remarkably good. The value of a increased by about 5° during the time between Hipparchus's observations and those made by Ptolemy. Consequently, Ptolemy's value is very poor; and his observational abilities have been unfavorably compared with those of his predecessor on this account. Viggo M. Petersen and Olaf Schmidt [6] have recently come to Ptolemy's defense in an article called "The Determination of the Longitude of the Apogee of the Orbit of the Sun according to Hipparchus and Ptolemy."

Their conclusions may be summarized as follows: It is not correct to infer from Ptolemy's error in the determination of a that he was unable to measure angles to an accuracy greater than 5°. This is because a is determined from J, J_1, and J_2. It is the accuracy of these values, therefore, that ought to be examined.

Petersen and Schmidt carefully compare the values obtained by Hipparchus and Ptolemy for J, J_1, J_2, e, and a. They conclude that the errors in Hipparchus's values for J_1 and J_2 are of the same order of magnitude as the errors in Ptolemy's measurements of J_1 and J_2. Therefore, Hipparchus's accuracy was just a matter of co-incidence. Furthermore, the similarity in the magnitude of the error explains why both Hipparchus and Ptolemy arrived at a poor value for e.

The moral of this little story seems to be that crude observations can sometimes lead to highly accurate values not because of a "delicate choice of crucial phenomena," but because of sheer stupid luck. In any event, it should stand as a warning to anyone who is tempted to conclude that behind every accurate parameter there lurks a body of highly refined data.

4. Determination of the Obliquity of the Ecliptic and the Latitude of Alexandria[3]

The obliquity of the ecliptic is the angle between the ecliptic and the celestial equator; it is designated by ε.

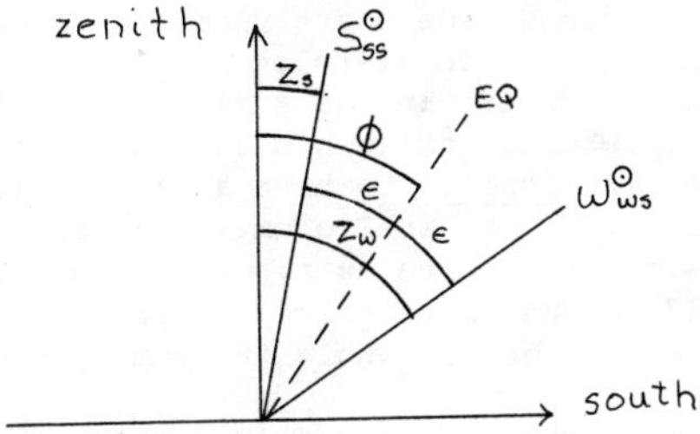

Let φ be the latitude of the place of observation. Let Z_w be the noon zenith distance at winter solstice; and let Z_s be the noon zenith distance at summer solstice. Then

$Z_w + Z_s = 2\phi$, and

$Z_w - Z_s = 2\varepsilon$.

Ptolemy says (Almagest I, 12) that his own observations indicate that the value of 2ε always falls between the limits 47;40° and 47;45°. He concludes that his observations confirm the value obtained by Eratosthenes: ε = 23;51,20°. (Adopting the mean between Ptolemy's limits would yield ε = 23;51,15°. The 0;0,5° difference is too small to warrant changing an apparently satisfactory value employed in the interpretation of previous observations.)

The modern value for the obliquity at Ptolemy's time (+130) is $\varepsilon = 23;40,50°$. This means that the error in the value adopted by Ptolemy is -0;10°. The error in the angle actually measured (2ε) is -0;21°. This latter error is four times the precision claimed by Ptolemy. It is, therefore, tempting to ask whether Ptolemy just fudged his results in order to achieve agreement with Eratosthenes or whether it is plausible that he found his alleged limits without adapting his observational procedures to yield the predetermined result. Britton argues [3] that there is indeed a likely source for Ptolemy's error if we assume that he honestly attempted to make independent observations.

There is good reason for believing that Ptolemy derived his value for the latitude of Alexandria from observations of noon zenith distances. If we employ his value for the latitude of Alexandria (30;58°) and his value for the obliquity, we obtain the following noon zenith distances:

Summer Solstice: $Z_s = 7;6,40°$, and

Winter Solstice: $Z_w = 54;49,20°$.

These are the only values consistent with Ptolemy's latitude of Alexandria. So there would seem to be good reason to believe that Z_w and Z_s are the distances Ptolemy actually measured. Comparison with modern data tells us that Z_w is very nearly correct, thus indicating that Ptolemy's errors in value for the obliquity arose from an error in his measurement of Z_s.

Ptolemy determined the zenith distances by observing the movement of the sun's shadow on a plinth. Since the shadow on the plinth loses definition as the sun crosses the meridian, it is easiest to extrapolate the noon value from readings made shortly before noon. Britton's very interesting contribution here consists in pointing out that the behavior of the shadow on the plinth in the summer is just the opposite of what we might expect. Whereas the sun's actual zenith distance always has a <u>minimum</u> at noon, it <u>appears</u> to have a <u>maximum</u> at noon when the sun is north of the equator. Thus, if Ptolemy made his observations some time before noon, he may well have extrapolated in the wrong direction. This hypothesis would account for Ptolemy's error. There are problems with this explanation, but on the whole it seems pretty plausible.

Certainly it is more interesting than Newton's quick conclusion that Ptolemy fudged his data in order to achieve agreement with Eratosthenes. It would seem on the whole wise to adopt a principle of charity in trying to account for the errors of ancient astronomers. Just as theory may sometimes improve the quality of the observational data, so also it may cause error in the interpretation of the data.

Since the first draft of this paper was written, its message has been dramatically underscored by the appearance of a much-touted book by Newton entitled <u>The Crime of Claudius Ptolemy</u> [5]. Working from a very naive picture of the relationship between theory and observation, Newton concludes that Ptolemy was a thoroughgoing fraud who yearned to be known as a great astronomer. "He may have found, early in his career, that he did not have the qualifications, so he turned to the only remaining way of satisfying his ambition, which was to replace ability by fraud." ([5], p. 376). Despite its philosophical naiveté, the book has received a great deal of attention in respectable circles.

5. Conclusion

What has been exemplified in this paper is the intricacy of the interplay between theoretical and observational elements in science. It is not the case that every accurate parameter is backed by a large body of highly refined observations. Nor is it the case that every error in the data indicates either a failure of personal integrity or gross incompetence at making observations. Rather the error may have its source in reasonable but mistaken theoretical presuppositions. We must, therefore, be very careful in ancient astronomy (as in other areas) to examine each case individually and on its own terms in order to assess the relationship between theory and observations. Blanket explanations cannot do justice to the variety and complexity of forms which that relationship manifests in concrete cases.

Notes

[1]I am grateful to Bernie Goldstein and to Jim Brown for their assistance.

[2]The following discussion and diagram are adapted from [6].

[3]Diagram and discussion are adapted from [3].

References

[1] Aaboe, Asger and Price, Derek J. de Solla. "Qualitative Measurement in Antiquity: The Derivation of Accurate Parameters from Crude but Crucial Observations." In *Mélanges Alexandre Koyré*. Volume 1. Paris: Hermann et Cie., 1965. Pages 1-20.

[2] Britton, John Phillips. *On the Quality of Solar and Lunar Observations and Parameters in Ptolemy's Almagest*. Unpublished Ph.D. Dissertation, Yale University, 1966. Xerox University Microfilms Publication Number 67-06997.

[3] ---------------------. "Ptolemy's Determination of the Obliquity of the Ecliptic." *Centaurus* 14 (1969): 29-41.

[4] Newton, Robert R. "The Authenticity of Ptolemy's Parallax Data-Part I." *Quarterly Journal of the Royal Astronomical Society* 14 (1973): 367-388.

[5] ----------------. *The Crime of Claudius Ptolemy*. Baltimore: Johns Hopkins University Press, 1977.

[6] Petersen, Viggo M. and Schmidt, Olaf. "The Determination of the Longitude of the Apogee of the Orbit of the Sun according to Hipparchus and Ptolemy." *Centaurus* 12 (1968): 73-96.

[7] Ptolemy (Claudius Ptolemy) *The Almagest* (The Mathematical Composition). As printed in Hutchins, Robert Maynard (ed.) *Great Books of the Western World* Volume 16. Chicago: Encyclopedia Britannica, Inc., 1952. Pages 1-478.

Theory-Ladenness and Theory Comparison

William B. Jones

Old Dominion University

Both the most controversial and the most widely publicized aspect of the "new empiricism" of Feyerabend ([2], [3]), Hanson ([5],[6]), Kuhn [8], and Toulmin ([12],[13]) is the thesis (rightly or wrongly) attributed to them that irrationality and subjectivity play a central role in science. The idea that science is pervaded by apt parallels to religious conversion and political propagadizing has achieved truly surprising currency. The epistemological root of this view is the belief that the theory-laden conception of observation which these philosophers espouse effectively prevents any rational comparison of the observational support commanded by competing theories. It has even been pointed out that in the absence of a neutral observational language in terms of which to identify and describe the objects or processes with which rival theories are (supposedly) concerned, there can be no real basis for calling them "rivals".

It is the intent of this paper to set forth conditions which, if met, would provide a rational basis for believing, even in a context in which observation is theory-laden, (1) that two theories are rivals and (2) that one of them enjoys observational support that is superior to that enjoyed by the other. Feyerabend, Hanson, Kuhn, and Toulmin have prompted interest in these problems, but this paper is not an explication, analysis, or defense of their views. Nor does this paper offer a theory of reference or an analysis either of rivalry or of observational support. The intent is simply to set down conditions under which it would not be possible to dismiss as irrational and subjective the belief that two theories are rivals and that one of them commands superior observational support--even if no rule is available for translating the sentences used to describe the evidence for or against either of them into sentences that might be used to describe evidence for or against the other, even if the two theories make use of incommensurable observational vocabularies, and even if learning and accepting either of the theories does somehow affect in a substantial way what one observes, at least in some cases.

PSA 1978, Volume 1, pp. 83-92

1. Conditions for Rivalry

Consider first the charge that two theories which do not make use of a common observational vocabulary, to use the standard terminology, cannot be said to be rivals because they are not concerned with the same objects or processes. Let it be conceded that if one cannot defend the claim that two theories are concerned with at least some of the same things (objects, events, or processes) then he can have no reason to claim that they are rivals, in the sense that is relevant here. They might be "rivals" for such titles as "most elegant","most significant for human welfare", and the like, but they could not be rivals in the sense that only one could be believed or accepted as (more) "correct" or "true". Here, saying that they must be concerned with some of the same things does not mean that the two theories must postulate the same sort of events or processes but only that (some of) the observable states of affairs to which the theories apply, which the theories explain, or which serve as evidence for or against the theories, must be the same.

Obviously, the two theories need not deal with precisely the same class of things. It is a commonplace that successful theories often address themselves to wider classes of phenomena than their competitors. At the same time, successful theories may not be at all concerned with entire subclasses of the events that preceding approaches were addressed to. For example, only in the last few hundred years have theories that deal with the motion of the planets been entirely free of concern with the menstrual cycle of the human female and with the personal fortunes of individual human beings generally. Even so, it is obvious that there must be a strong overlap between the classes of things dealt with by competing theories--if, indeed, they are to be rivals. It is on just this point that much of the criticism of Kuhn's views hinges. It is argued that because of the incommensurability of what is observed under different Kuhnian "paradigms", they are not really about the same things and thus not truly rivals in this relevant sense ([7], pp. 460, 472-473; [10], pp. 15, 82).

What this argument ignores is that rival theories can be concerned with the same things (objects, events, processes, etc.) even in the absence of a common, neutral, observation language and even if, because of Gestalt-like influences, these things are neither conceived nor experienced in the same way by those who subscribe to different "paradigms", People can disagree profoundly about a thing because they experience it in radically different ways and yet agree as to just which thing it is that they are disagreeing about. Now, if two people do experience things differently because they are committed to different "paradigms" or possessed of different Gestalts, it may, in some sense, be correct to say that they are not even observing the same object. The concern here, however, is not with this question. It is rather to show that two such discordant observers may yet be said to be observing the same things--in another, and perhaps more literal, sense of "same". Specifically, two such disputants can be said to be referring to the same thing in that they can point to or otherwise identify the object(s) about which they are disagreeing. They may accomplish this identification by physical gestures, such as an extended arm or finger, or they may accomplish it verbally, using such expressions as: "This one"; "The one on the left"; "The Figure 4"; "The brilliant yellow-white disc centered between the green and blue expanses"; etc. It is thus possible for two disputants to agree as to just which object it is that they are disagreeing about--even if they do experience it differently.

Quite obviously, agreement of this sort would be impossible if the two observers had absolutely and totally different experiences of the world. Then they would not be able to identify or focus upon anything as a common object of disagreement. It would then be impossible for either of them to point to anything and have the other understand what is intended. Thus, for example, in order for two hypothetical observers to disagree about what is depicted in Hanson's Figure 4 ([5], p. 13), they must first identify "Figure 4" as the common object of dispute. This means that they must have similar experiences of the surroundings of the figure and of its location and boundaries. And it must be possible for each of them to have reasons for believing that his antagonist is referring to the same thing in whatever contrary claims he makes about it. This mutual understanding may be fairly full, as is the case when the two are each able to use the expression "Figure 4" and to have the other understand what is intended. Minimally, however, it need involve no more than each's being able to direct the other's attention in a particular direction by a physical gesture.

Being able to agree upon the identity of an object, however, does not require that there be anything in common in the experiences which the two observers have of the relevant object *per se*. That is, for example, they need not have similar visual (or other) experiences of what lies *there*, *within the boundaries* of Figure 4. To see that this is true, one need only reflect upon the unlikely but conceivable case of a person who, for unknown reasons, sees a pattern totally unlike that which a normal person sees when he looks at Hanson's Figure 4. It would be quite possible for this person and a normal person to have a strong disagreement concerning what the figure represents without their experiences of the figure itself having anything in common and thus without their being able to agree upon a common neutral description of it. Indeed, in a sufficiently bizarre world, one of them might see a tiny window into Hades but still use the same finger-pointing gesture and the words "Figure 4" to direct the other's attention to the source of his wonder and consternation. Whether or not the other sees this tiny window or merely an ink drawing in the rectangle above the words "Figure 4" these words and the gesture would serve the same purpose.

The important point, for present purposes, is that the two people can disagree totally about what is to be found within the rectangular space above the words "Figure 4" and still agree as to which thing it is they are disagreeing about. It is thus conceivable that there should be an entire class of things (objects, events, or processes) with respect to which two persons (or two groups of persons, for that matter) agree as to their identity (which things) but disagree as to how they are to be described. Now, suppose that there exist two groups of people in a bizarre world like that suggested above and that the members of one group experience a certain class of things completely differently than do the members of the other group but that the two groups agree as to just which things it is they are in such disagreement about. Suppose further that each of the two groups has developed a theory that deals with this special class of things. Should one say that these two theories are rivals in the sense that, say, Tycho Brahe's and Kepler's cosmological schemes were rivals?

One certainly hesitates to answer this question in the affirmative, and for good reason: being concerned with the same things may be a necessary condition for rivalry between two theories, but it is not a sufficient condition. Theories may be concerned with the same thing or things without being rivals. For example, a falling brown-pelican egg might be an object of study for both a physicist and a wild-life ecologist--one being concerned with its trajectory, the other with the thickness of its shell; but the two accounts that result from these studies are certainly not rivals in the requisite sense. On the other hand, if the two scientists were both concerned with the _same aspect_ of the situation, the motion of the egg, for example, then their accounts would be rivals. The Galilean and the Aristotelian accounts of the motion of such falling objects are well-known as competitors in the sense here being discussed.

The problem that must be addressed here is: what sense can it possibly make to say that two theories are concerned with the _same aspect_ of a situation if their proponents have different experiences when confronted with the situation and describe it differently? Can one say that the _same aspect_ of the situation is experienced and described differently by the two groups? To take an oversimplified example, it might be argued by some "new empiricists" that an Aristotelian and a Galilean physicist would have different experiences if confronted with a falling bird's egg, and that although they use some of the same terms to describe the situation, such as "falling" and "motion", they are not using the terms in the same sense. Rather, it may be explained, these expressions are like the "theoretical terms" of the standard accounts: in order to grasp their meaning it is necessary to understand each's role in the theoretical system of which it is a part. Since Aristotle's and Galileo's understanding of "motion" and "falling" are so different, they--or their followers--cannot even be said to be concerned with the same aspect of the situation under discussion. Rather, each is concerned with something that the other cannot even see--if the theory-laden conception of observation is correct. How, then, can it be said that the two accounts are rivals?

The answer to this puzzle lies in the realization that one can use criteria for "being concerned with the same aspect of phenomena" other than the requirement, say, that the two competing accounts must use the _same terms_ with the _same meanings_ to describe it. Specifically, as Israel Scheffler has suggested, one can focus upon _extension_ instead of _intension_ ([10], pp. 45-66).[1]

Suppose that the term T_1 occurs in one of two contending theories and that the term T_2 occurs in the other and that the two theories differ to a considerable degree as to number and character of axioms and as to number of primitive terms occurring therein. Suppose further that, at least as far as some (or even most) of the terms that have an observational role are concerned, there is no simple relationship between the terms of one system and those of the other. Some terms may even be supposed to have no counterpart at all in the opposing theory. And suppose: that by pointing and the like (as explained previously), the adherents of the two opposing theories can agree upon the identification of the members of a certain class of things (objects, events, or processes); that neither of the theories is concerned with things not included in this class; and that each of the contending groups uses its own theory-laden observational vocabulary exclusively in describing the properties of those members of this class which interest them--except for those "properties" that are necessary for <u>identifying</u> the members of the class, such as their location, their surroundings, their boundaries, etc. (see the above discussion). The two groups are presumed to be in rough agreement about these latter matters. Of course, this "agreement" may amount to little more than each's being able to be guided by the pointing gestures of the other. Suppose finally that there is an extremely high correlation between the members of the class to which the adherents of the first theory apply the term "T_1" and those to which the adherents of the second theory apply the term "T_2".

The view here being advanced is that if these conditions are met, it would be reasonable (i.e., not arbitrary and not subjective) to believe that the two theories are concerned with the same aspect of the things making up the class described above. Clearly, one is much more likely to find such a correlation if T_1 and T_2 have observational roles. Furthermore, if the two theories do, indeed, have radically different structures, one is more likely to find, rather than simple correlations between pairs of terms, correlations between, say, a single term in one theory and the conjunct or disjunct of two terms in the other and the like. In such cases it would also be reasonable to believe that the two theories are concerned with the same aspect(s) of the class of things to which they are applied and that they are rivals in the requisite sense.

The reasonableness of this proposal is apparent if one considers the hypothetical case of an anthropologist who happens upon two isolated tribes of primitives whose respective languages are in structure very different not only from each other but also from the language of the anthropologist. The anthropologist notices that the members of both tribes regularly engage in a certain kind of verbal behavior when in the presence of a particular class of things (objects, events, or processes), the identities of which they seem to be able to agree upon by means of gestures and/or by verbal expressions commonly understood. Furthermore, he discovers that there is a very high correlation between the members of the class to which the members of one tribe apply the term T_1 and those to which the members of the second tribe apply the term T_2. He surmises that the two tribes may be using the two terms to refer to the same aspect(s) of the class of things in question. Note that in order to reach this conclusion, the anthropologist need not have figured out just what aspect of the things it is that the tribesmen were referring to. Indeed, it may be something that he has not even learned to recognize, just as most of us have not learned to recognize the different kinds of bird calls.

It should be noted that judgments of rivalry of the sort here being discussed do not have to be made by a third party, employing his own separate language in the process of identifying and describing the things to which the two rival theories are addressed. On the contrary, the proponents of either or both of the theories may properly judge that the others' theory is somehow concerned with roughly the same aspects of the same class of things as their own and is a rival to it, provided they have reason to believe that the conditions outlined above are satisfied. In order to do this, neither group need be able to make the sort of theory-laden observations made by the other--nor even know that their opponents', or their own, observations are theory-laden. Thus, to use the example cited just above, the members of one or both of the tribes of primitives may regard the other as subscribing to a rival view of some sort which deals with essentially the same class of commonly identifiable things (objects, events, processes, etc.) as their own account. The members of each tribe may have only a minimal understanding of the others' view, perhaps only enough to realize (a) that in spite of its being concerned with essentially the same things as their own account, the rival view differs radically from it and (b) that the other tribesmen believe in the existence of and even claim to observe things which they themselves neither believe in nor observe.

A second point that should be noted at this juncture is that the proposal outlined above is by no means infallible or definitive, nor is it offered as an analysis or theory of reference or even of rivalry between theories, though it does perhaps make a small contribution toward an analysis of rivalry. Thus it will surely be judged unsatisfactory by those for whom an adequate account of rivalry between theories must provide a basis for answering the following question: with exactly which aspects of what phenomena is each of the theories concerned, and precisely how are they related to those with which the other theory is concerned. Of course, it is this question which cannot in principle have an answer if certain accounts of theory-ladenness in observation are correct, namely those which view theory change as involving something like a change of Gestalt of such a character that it is absolutely impossible to view or think about the phenomena of interest from both points of view at the same time--and hence impossible to answer the question of just how are the observational statements made under the aegis of one of the theories related logically to those made under the aegis of the other. The claim being made here is that even if the thesis of radical incommensurability of theories were true, and even if clear and definitive assessments of rivalry were thus impossible, it would still be reasonable, under the conditions outlined, to regard two theories as rivals.

2. Conditions for Comparison

J. W. Meiland ([9], pp. 180-186) has argued that to be objective, an evaluation of a theory must make use of standards that are external to and logically independent of the theory. This criterion rules out self-justification by theories. As Meiland sees it, insisting that the <u>same</u> external standard be applied to all theories <u>and</u> insisting that one of the external criteria be agreement with observational data would mean that the <u>same data</u> must be used to evaluate competing theories. This requirement would rule out any theory-laden conception of observation, such as Kuhn's, according to which competing theories generate incommensurable sets of observations. Meiland holds that a theory which is in principle incapable of generating unfavorable observational evidence is self-justifying and violates the very notion of objectivity based upon independent standards ([9], pp. 181-182).

One must certainly agree with Meiland that trivial self-fulfillment by scientific theories must be ruled out. Observations made under the aegis of a theory that are favorable to it cannot be regarded as evidence for its acceptance if the theory is in principle incapable of generating unfavorable evidence. Making observations in such a context can in no sense be counted as confronting the theory with experience. Data so obtained cannot serve as an independent check of the theory. Furthermore, unless there is a real possibility that at least some of the observational data gathered under the aegis of a theory will be unfavorable to it, any <u>externally</u> imposed criterion for judging competing theories on the basis of the observations that each generates will be almost, if not entirely, vacuous.

If observational data can serve as an independent check on a theory only if there is a genuine possibility of its being unfavorable to the theory, then any philosophical account of science which includes a theory-laden conception of observation must provide for this possibility. If it does not, it will have failed to provide the basis for any genuinely objective empirical control whatever. That the "new empiricists" need to develop an analysis of theory-ladenness which allows for theory-laden observations running counter to the very theories with which they are "laden" has already been noted ([7], pp. 476-478). In the limited space available here, no attempt can be made to provide such an analysis, nor even to determine if such is really possible. What can be done is to provide some reason for thinking that such disconfirming theory-laden observational data just might be possible. Many of the basic theories which are involved in the examples of theory-laden observations most often cited by the "new empiricists" do or did manage to generate contravening evidence. Consequently, if the observations made under the aegis of these theories was theory-laden, then theory-ladenness is no bar to the generation of disconfirming evidence. Of course, this only pushes the argument back a step to the question of whether the relevant observations were theory-laden, a question that cannot be settled here. Given the amount of discussion this topic has received, however, it would seem reasonable and appropriate to explore the question of what would follow <u>if</u> these observations were theory-laden and thus <u>if</u> it were possible for theory-laden observational data to prove unfavorable to the very theory with which it is laden.

It will now be argued that if such theory-laden but disconfirming observations are possible, at least in principle, then under some conditions one can have reasonable grounds for believing that one of two theories enjoys greater observational support than the other--even if the observational data supporting each is theory-laden with that theory. Suppose that there are two theories with separate, theory-laden observational languages like the two discussed in "Part 1." and that they are believed to be rivals for reasons like those outlined (there being a very high correlation between the things to which the adherents of one theory apply the term T_1 and those to which the adherents of the other theory apply the term T_2). Now suppose that the adherents of the second theory have made at least as many observations under the aegis of their theory as the first group have made under theirs. Here determining what truth-value to assign to an observational sentence (in Carnap's sense, [1], pp. 454-455) counts as "making an observation". Note that no obvious or simple connection is assumed to exist between the observational data gathered by one group and that amassed by the other. Of course, they must agree upon certain identifying "properties", such as approximate location and boundaries, of the things that the theories are concerned with, but these need <u>not</u> be included among the features referred to in the observational reports of the two groups of practitioners.

Now, if all of the observations made under the aegis of the second theory are favorable while a significant proportion of those made in connection with the first theory are not, then the second theory can reasonably be said to enjoy observational support superior to that accruing to the second.

There are a number of comments that need to be made regarding the suggestion just put forward. First of all, it makes no pretense at being a complete theory and does not provide a means of deciding between theories under all circumstances. Secondly, there is admittedly a certain amount of vagueness and imprecision associated with even some of the central elements of the proposal. Thirdly, if adopted, the procedure suggested would almost certainly lead to the acceptance of theories which would later have to be rejected for one reason or another. Fourthly, there would seem to be no reason to think that a quantitative account of empirical support could ever be erected within a framework like that proposed. Those for whom one or more of these aspects of the problem are crucial will, of course, find the present proposal unacceptable, but this is irrelevant to the primary intent of this paper, which is to show that, under some circumstances at least, theory-ladenness in observation does not inevitably relegate theory-choice to the arena of irrationality and subjective caprice.

Notes

[1]It is to be noted that considerable difficulty has been encountered by those who have tried to develop a complete account of rivalry based upon extensional considerations, so much so that some ([4], pp. 17-32) have proposed that the project in effect be abandoned. However, the alternative offered in place of the extensional approach is as yet only embryonic ([11], pp. 1-12). The outcome of this matter is unlikely to affect the claim made herein that extensional considerations can provide a basis for believing that rivalry exists.

References

[1] Carnap, R. "Testability and Meaning." *Philosophy of Science* 3-4 (1936-37): 419-471, 1-40.

[2] Feyerabend, P. "Against Method: Outline of an Anarchistic Theory of Knowledge." In *Analysis of Theories and Methods of Physics and Psychology* (*Minnesota Studies in the Philosophy of Science, Volume IV*). Edited by Michael Radner and Stephen Winokur. Minneapolis: University of Minnesota Press, 1970. Pages 17-130.

[3] -------------. "Problems of Empiricism." In *Beyond the Edge of Certainty* (*University of Pittsburgh Series in the Philosophy of Science, Volume 2*). Edited by Robert G. Colodny. Englewood Cliffs, NJ: Prenctice-Hall, 1965. Pages 145-260.

[4] Fine, A. "How to Compare Theories: Reference and Change." *Nous* 9 (1975): 17-32.

[5] Hanson, N.R. *Patterns of Discovery*. Cambridge: Cambridge University Press, 1958.

[6] -----------. *Perceptions and Discovery* (ed.) Willard C. Humphreys. San Francisco: Freeman, Cooper & Company, 1969.

[7] Kördig, C.R. "The Theory-Ladenness of Observation." *The Review of Metaphysics* 24 (1971); 448-484.

[8] Kuhn, Thomas S. *The Structure of Scientific Revolutions* 2nd ed. Chicago: University of Chicago Press, 1970.

[9] Meiland, J.W. "Discussion: Kuhn, Scheffler, and Objectivity in Science." *Philosophy of Science* 41 (1974): 179-187.

[10] Scheffler, I. *Science and Subjectivity*. Indianpolis: Bobbs-Merrill, 1967.

[11] Shapere, D. "Fine on Reference and Theory Change." (Unpublished typescript, 1975).

[12] Toulmin, S. *Foresight and Understanding*. New York: Harper and Row, 1961.

[13] ----------. *The Philosophy of Science*. New York: Harper and Row, 1953.

Observationality: Quine and the Epistemological Nihilists

P. William Bechtel and Eric Stiffler[1]

Northern Kentucky University — Western Illinois University

For several years now Quine has been proclaiming a solution to problems concerning observation sentences and evidence: "The dislodging of epistemology from its old status of first philosophy has loosed a wave . . . of epistemological nihilism. This mood is reflected somewhat in the recent tendency of Polanyi, Kuhn, and the late Russell Hanson to belittle the role of evidence and to accentuate cultural relativism ... . It is ironical that philosophers, finding the old epistemology untenable as a whole, should react by repudiating a part which has only now [with Quine's account of observation sentences] moved into clear focus." ([15], pp. 87-88).[2] Attacks on the notion of observationality by the critics of positivism have been viewed by many (for example Scheffler [21], Chapter 1) as seriously undermining the objectivity of science. Quine's claim to have solved the problems, therefore, deserves careful consideration and evaluation.

Before we can evaluate Quine's claim we need to examine the major objections that have been raised against traditional accounts of observation sentences and consider how Quine's account differs from traditional ones. These tasks occupy the first two sections of the paper. In the last section we argue that Quine avoids two major objections but it is less clear that he has solved all problems stemming from a third objection--the charge that all language, even that used to report observations, is theory-laden. On balance, our view is that Quine's criterion for observation sentences marks a clear advance over previous formulations. Whether it meets the most significant objection raised by the critics of positivism is a question we take up at the end.

1. Criticisms of Traditional Accounts of Observation Sentences

Three basic criticisms have been leveled against the traditional accounts of observation sentences--observation sentences no longer have a significant function in the wake of the downfall of positivism, adequate criteria have not been given for distinguishing observation sentences from theoretical sentences, and observation sentences are

PSA 1978, Volume 1, pp. 93-108

theory-laden and hence not objective. We shall briefly develop each of these in turn.

Suppe ([23], pp. 67-80) makes a general attack on positivism which he claims has the consequence of showing that observation sentences are not relevant to science. Traditionally observation sentences have been seen as serving the dual functions of providing the evidence for and the semantical foundation to our cognitive scheme. Two of the major tasks of philosophy of science, according to positivism, are to explain the meaningfulness of scientific discourse and to show how evidence accrues to theories. For Carnap these two tasks are closely connected: "If we knew what it would be for a given sentence to be found true then we would know what its meaning is." ([4], p. 47). Only observation sentences, though, can be directly tested. Other sentences, according to positivism, are only testable in light of their logical relation to observation sentences via correspondence rules. Hence, observation sentences play a fundamental role in explaining both the meaning and the evidence for scientific theories.

Suppe attacks this picture of the function of observation sentences. He employs Quine's own attack on the analytic-synthetic distinction and reductionism in attacking the view that theoretical statements are connected to observation sentences via correspondence rules.[3] This view is what Quine labeled "the dogma of reductionism", the dogma "that each statement, taken in isolation from its fellows, can admit of confirmation or infirmation." ([19], p. 41). Quine's criticism of reductionism is based on an alternative view of scientific theories which treats them as networks of sentences which only confront experience as a whole. In this model there is no possibility of reducing theoretical statements to observation sentences. Once the possibility of reducing theoretical statements to observation sentences as a means of establishing their meaning and verifying their truth is forsaken, Suppe argues that the importance of identifying the set of observation sentences is also eliminated. Thus, the first objection against accounts of observation sentences is that such accounts are unnecessary. (There is irony in this argument, if it is valid, for it employs one Quinean program, the attack on reductionism, to undermine another Quinean program, the articulation of a criterion for observation sentences.)

The second objection to traditional accounts of observation sentences is the charge that none of the criteria that have been proposed for observation sentences suffices for distinguishing observation sentences from theoretical sentences. The distinction between observation sentences and theoretical sentences has traditionally been presented in terms of the vocabulary used in the sentences--observation sentences employ only observational terms. Carnap acknowledged that the actual distinction between theoretical and observational terms is ultimately arbitrary and vague, but maintained that it can be meaningfully made and is necessary for explaining scientific theories.[4] He suggested two criteria for distinguishing observational terms from theoretical terms--one in terms of whether the properties and relations referred to are observable ([2], p. 41) and the other in terms of the ability of language speakers to render verdicts on predications "with the help of a few observations." ([4], p. 63).

Achinstein ([1], pp. 160-165), while acknowledging an intuitive difference between terms like "red" and "warm" and others like "electron" and "temperature," argues that no proposed criteria, Carnap's included, capture this distinction. He attacks Carnap's observability criterion by arguing, in the spirit of ordinary language analysis, that we can correctly speak of observing things which are not immediately present to the observing subject but whose effects are. For instance, we can speak of observing a forest fire although we only have visual contact with the smoke and we can speak of observing an electron although we actually only see the track in a cloud chamber. Further, we can speak of observing something (e.g., a molecule) when we actually only see an image of it (e.g., with the aid of an electron microscope). Achinstein further criticizes this criterion by noting that even paradigmatic instances of observable events may be reported using what are generally taken to be theoretical terms (e.g., we can report the chemical reaction in a test tube in terms of chemical theory as well as in terms of the color changes that occur). Hence, one cannot use a criterion of observability to distinguish observational vocabulary and hence a set of observation sentences.

Achinstein also criticizes the criterion that observational predications are those on which a subject can render a verdict "with the help of a few observations." A trained scientist can give a theoretical account of some phenomena after only a quick observation. On the other hand, there are cases where several observations may be needed to decide on the correctness of what seems to be a very observational report, e.g., whether something is navy blue or black ([1], pp. 175-6). The conclusion Achinstein reaches after examining the criteria of observationality offered by Carnap and other positivists is that there is no adequate criterion for distinguishing observation sentences from theoretical sentences ([1], p. 177).

The third objection to traditional accounts of observation sentences is that observation is theory-laden. This charge has been urged in a bewildering variety of ways. For the purposes of assessing Quine's criterion for observation sentences we shall consider two broad forms of the objection. One view is that observation in general is theory-laden because an observer's knowledge, beliefs, and language all exert an influence on his observational reports. The objection is urged in a second, more specific, way when it is claimed that acceptance of a particular scientific theory influences the observations that can be adduced to count for or against the specific theory. Hence, holders of different theories cannot turn to neutral observation sentences to test their theories.

In Patterns of Discovery Hanson states a comprehensive case for the first version of the theory-ladenness objection. The basic insight Hanson presses is that observation is not simply a matter of vision but is the link to our cognitive scheme. He argues in particular: "Observation of x is shaped by prior knowledge of x. Another influence on observation rests in the language or notation used to express what we know, and without which there would be little we could recognize as knowledge." ([7], p. 19). Central to Hanson's argument is the claim that seeing that such and such is true of an object is an important element involved in seeing the object. For instance, "to see an X-ray tube is at least to see that, were it dropped on stone, it would break." ([7], p. 21). If this is the case, an observer without this knowledge could not see the X-ray tube. Thus, Hanson claims that a layman cannot see what a physicist sees looking at the same X-ray tube because "the layman must learn physics before he can see what the physicist sees." ([7], p. 26). Feyerabend and Kuhn also press this version of the objection. Feyerabend writes: "On closer analysis we even find that there are no 'bare facts' at all but that the facts that enter our knowledge are viewed in a certain way and are therefore essentially ideational." ([5], p. 20.) Kuhn holds that paradigms or disciplinary matrices, which include background knowledge and beliefs over and above the theory proper, determine what we see ([11], p. 111f. and [10]).

Hanson, Feyerabend, and Kuhn further agree that observation is theory-laden in a second way. Each cites a number of examples intended to show that if different observers adopt competing theories, they will observe different things. Thus, Kuhn contends that "Lavosier . . . saw oxygen where Priestley had seen dephlogisticated air and where others had seen nothing at all." ([11], p. 118). In his famous example of Kepler and Tycho observing the sun at dawn, Hanson argues that there is an important sense in which they do not observe the same thing ([7], Chapter I and [8], Part II). Although their retinas are irradiated in much the same way, and they have much the same phenomenal experience, the fact that Kepler holds the heliocentric theory while Tycho the geocentric theory is responsible for a difference in what they see. Feyerabend supports this version of the objection with two theses: "the description of every fact is dependent on some theory" and "the meaning of every term we use depends upon the theoretical context in which it occurs." ([6], p. 175 and p. 180). Hence, he too repudiates the notion of a theory neutral observation language.

The two versions of the theory-ladenness objection complement the two previously considered objections. The general charge that what is seen is influenced by one's knowledge and beliefs supplements the second objection that observation sentences cannot be distinguished from theoretical ones. The more specific charge that competing theorists will see different things and hence not have any neutral observation reports to turn to complements Suppe's claim that observation sentences have no important role to play in science. Certainly, if our theories determine what we observe, observation sentences cannot be appealed to in deciding which of a number of competing theories should be adopted.

2. Quine's Alternative Criterion for Observation Sentences

Quine appeals to observation sentences for the traditional purposes of securing cognitive significance and empirical evidence for theoretical beliefs. To meet these objectives we need, Quine claims, "sentences which, as we learn the language, are most strongly conditioned to concurrent sensory stimulation rather than to stored collateral information." ([15], p. 85). The task in realizing this objective is to formulate a criterion for observation sentences that excludes sentences whose use depends on "stored collateral information." The task here closely parallels the problem raised by Achinstein: how to discount theoretical descriptions of observations from the class of observation sentences.

Quine recognizes that one does not want to eliminate all reliance on stored information for using observation sentences, since "the very fact of our having learned the language evinces much storing of information." ([15], pp. 85-86).[5] The most one can aspire to is to rule out stored information which exceeds that which is involved in knowing how to speak the language. Quine suggests that the information which is involved in knowing how to speak a language is information shared by all speakers of the language. Information that is idiosyncratic to particular speakers is more likely to be theoretical.[6] Hence, speakers of the language should agree on sentences which invoke only that information needed to speak the language and will more likely disagree on sentences which present theoretical information.

Quine is thus led to the following criterion for observation sentences: An observation sentence is an occasion sentence on which all speakers of the language who are present and attentive when asked for their verdict will concur.[7] The rationale for this criterion is, first, that observation sentences must be occasion sentences, where these are just sentences to which speakers' verdicts change with varying stimulatory circumstances, in order for them to be conditioned to concurrent stimulation. Second, requiring that all speakers should concur when asked for their verdicts ensures that the use of an observation sentence involves only information shared by the linguistic community as a whole. This rules out the possibility of verdicts to observation sentences turning on special knowledge possessed by some but not all speakers. The criterion thus has the consequence that a sentence will count as an observation sentence only in so far as acceptance or rejection of it depends on the present stimulatory situation and the information needed to use the sentence.

Refinements and qualifications will be needed to make this criterion fully adequate and precise. For instance, to avoid the problem that some speakers may be obstinate and purposely disagree with their fellow speakers' verdicts, the degree of concurrence we require to count a sentence as observational may need to be reduced. A number of problems also arise in explicating the notion of an occasion sentence. How frequently, for example, must verdicts change if we are going to identify a sentence as tied to concurrent stimulation? There may, as Siegel

[22] suggests, not be an absolute answer to this question for the frequency may change depending on the kinds of observations being made. In addition, there is the question of how to define the membership of the linguistic community which must agree on the verdict. Quine suggests the ability to carry on fluent dialogue as the test for membership in the language community ([18], p. 39). Despite the need for further specification on these points, Quine's criterion is still rather clear and easily applicable.

One aspect of our formulation of Quine's criterion requires special consideration. Quine's objective in identifying observation sentences is to distinguish the sentences "most strongly conditioned to concurrent sensory stimulation." In actually formulating the criterion, however, we made no mention of the stimulations which actually condition verdicts to sentences. Rather, we have identified sentences in terms of the agreement to the verdict of those present when the sentence was queried. This enables us to avoid specifying at the beginning the classes of stimulations to which verdicts to observation sentences are to be conditioned.[8] After the observation sentences are identified, however, we can empirically discover what classes of stimulations control the verdicts to particular observation sentences. There will be two such classes: the class which prompts an affirmative verdict Quine identifies as the affirmative stimulus meaning, and that which prompts a negative verdict he calls the negative stimulus meaning.

The two classes of affirmative and negative stimulus meanings constitute a unique type of meaning. These classes are composed simply of the stimulations which causally prompt verdicts to sentences and are identified in an empirical manner. Hence, the notion of stimulus meaning escapes the kinds of difficulties that most motions of meaning have encountered (particularly at the hands of Quine).

Quine's behavioral criterion for observation sentences and his empirical analysis of their meanings significantly differentiates his treatment of observation sentences from earlier statements. Quine's criterion is not formulated in terms of any special epistemic awareness of observables. Hence, he does not need to enter into discussion of the primitive data of which we are aware--no sensory core or myth of the given is needed. Rather, Quine is able to treat the relation between the world and observation sentences as simply a causal relation. In "Problems of Empiricism" Feyerabend also argues for a causal treatment of the production of observation sentences ([6]), p. 212), and so here Quine and Feyerabend are in agreement in their break from the tradition.

By distinguishing observation sentences behaviorally and by analyzing their meaning in terms of stimulus meaning Quine has developed an account which does not require knowledge of the semantics of a language for discovering observation sentences. All that is needed is identification of a language's signs of assent and dissent, which Quine indicates can be distinguished (allowing for some indeterminacy) in a behavioral fashion ([17], p. 312). Hence, problems of meaning can be avoided when considering observation sentences. This is one reason Quine calls such sentences "anontological." ([16], p. 16). This will become significant when we consider Quine's response to the theory ladenness objection in Section 3.

Quine, in a more recent paper, "On Empirically Equivalent Systems of the World," has wavered on his acceptance of the behavioral criterion for identifying observation sentences. He notes that some sentences, which intuitively report observations, may neither command community wide concurrence in verdicts nor be reducible to sentences which do command such assent--e.g., sentences which are prompted by a particular wine taste or a particular musical sound. To circumvent these problems, Quine proposes that for "the really distinctive trait of observation terms and sentences" we must turn to how the sentences are learned: "Observational expressions are expressions that can be learned ostensively." ([14], p. 316).

This approach to distinguishing observation sentences, however, presents serious difficulties. How, for instance, are we to distinguish cases where one is able to learn merely on the basis of ostension from cases where one must learn a sentence by developing his own theory? The only way out of this difficulty would seem to be to distinguish types of objects which one can learn to label ostensively. This leads back to the notion of a special epistemic awareness of observables which is precisely what Quine's behavioral criterion allows us to avoid. (Or, at least, we must be able to identify special types of stimulus meanings to which verdicts to observation sentences can be conditioned. This will present its own problems.) Although Quine's behavioral criterion may not precisely pick out what intuitively we take to be observation sentences, it is more precise than his ostensive learning criterion and avoids the difficulties encountered by older epistemologies. Moreover, we shall see shortly that it does pick out the class of sentences we want as observation sentences.

3. Evaluation of Quine's criterion

Part of the test of Quine's proposed alternative criterion of observationality is how well it meets the objections that have been raised against other criteria for observation sentences. We shall now return to the objections considered in Section 1. to see whether Quine can answer them.

The first objection was Suppe's claim that observation sentences will no longer be seen to play a special role in science once positivism and its dogma of reductionism are rejected. Hence, there is no point in distinguishing observation sentences. Quine, however, claims the two traditional important epistemological roles for his observation sentences--they provide a basis of evidence and meaning for our theoretical discourse. Quine's observation sentences are the kind of sentences a community of scientists would turn to when they are seeking evidence. Since they are occasion sentences, i.e., verdicts to them vary in different stimulatory circumstances, it may be that these verdicts are governed by some part of the stimulatory situation. The fact that all speakers of the language present in any given stimulatory situation concur on the verdict makes it even more reasonable to expect that something in the stimulatory situation is regulating the verdicts subjects give to the sentences. It is possible, of course, that continual concurrence on verdicts is purely fortuitious or can be explained in some other fashion, but this seems less likely than the explanation that something in the environment is causing the concurrence on verdicts. Hence, Quine's observation sentences are linguistic devices which can reasonably be assumed to record some aspect of the world.

Moreover, these sentences can play an evidentiary role even if they belong to a Quinean theoretical network and are not treated as primitives to which theoretical sentences are expected to be reduced (à la positivism). Even if different scientists connect sentences into different theoretical networks, that will not affect the evidentiary function of observation sentences. When scientists are seeking evidence about the world, what they are looking for are statements on which they concur. Observation sentences are, by definition, sentences on which members of the community concur in their verdicts. Scientists have no need, moreover, to go beyond that on which they agree. Hence, the alternative networks into which the scientists embed the observation sentences will not undermine their use. The fact that they agree on the verdicts to the sentences in all stimulatory circumstances is sufficient to make these sentences touchstones to the physical world and consequently evidentiary reports.

Quine can thus avoid Suppe's attack that a criterion for observation sentences serves no function once the analytic-synthetic distinction and reductionism are abandoned as the basis for the philosophy of science. Observation sentences can still play an evidentiary role. Hence, the possible irony of Quine undercutting his own enterprise is avoided.

Quine's observation sentences can further be viewed as providing the semantical foundation of the language. Since all speakers agree on the verdicts to these sentences in all stimulatory circumstances, one reasonable expectation on someone learning the language is that he learn to match the community's use of these statements. This common use of observation sentences is one of the few things which associates speakers as speaking a common language. Moreover, in so far as the verdicts to these sentences are responses to stimulations belonging to the stimulus meaning, observation sentences will be sentences capable of being learned simply by behavioral conditioning. Hence, they should be the easiest sentences to learn, the entering wedge into a language. Quine [18] has further tried to trace the process by which theoretical discourse can be learned on the basis of ability to use observation sentences.

Quine's enterprise of giving a criterion for observation sentences, thus, remains an important activity for epistemology and philosophy of science. Just as for Carnap, for Quine observation sentences serve the functions of giving the evidentiary and semantical foundations for theoretical discourse.

The second objection we considered was the charge that previous attempts to distinguish observation sentences failed to distinguish appropriately paradigmatic cases of observational reports. The previous attempts all depended on distinguishing observational terms from theoretical ones. One of the major differences between Quine's criterion and more traditional ones, though, is that Quine does not try to classify terms as observational. Rather, he considers whole sentences. He allows the same vocabulary to figure in both observation sentences and theoretical sentences. Quine therefore escapes the objections that were directed at the problem of differentiating observational terms from theoretical ones.

Quine also avoids some of the more general features of Achinstein's objections. Achinstein posed difficulties for determining what is observable and for specifying the vocabulary observations are to be reported in. Quine's criterion, though, is not in terms of what is observable. The relation to the observable world is analyzed causally after the sentences that meet Quine's criterion are identified. Further, for Quine the community's ability to agree on their verdicts to the sentence sets the limits on the types of reports that count as observational. Reports of observations are simply ruled out from the class of observation sentences unless all members of the relevant community can agree on the report. This gives a sharp distinction, for the relevant community, between observation sentences and theoretical sentences.

It is important to recognize, though, that Quine's criterion does not pick out an absolute set of observation sentences (as, for instance, Carnap's criterion in terms of what is observable sought to do). Rather, Quine's criterion is relative to particular language communities. Moreover, what sentences count as observation sentences may change over time. If some members of the community change their pattern of use of a sentence (perhaps because of theoretical commitments), there will no longer be unanimity in verdict and the sentence will no longer count as an observation sentence. Yet, we have already indicated the utility of the class of sentences meeting Quine's criterion. Even if it does not pick out pre-analytic paradigmatic cases of observation sentences and the class it picks out changes, the criterion remains useful. Quine thus seems to have avoided the problems Achinstein raised to earlier criteria for observation sentences.

In assessing Quine's criterion with respect to the theory-ladenness objection we will continue to distinguish two forms of the objection as we did above. Let us first consider the claim that observation sentences in general are theory-laden due to the fact that an observer's knowledge, beliefs, and language influence what he sees. Quine will happily admit part of this claim. As we saw above (Section 2.), his criterion for observation sentences allows for the fact that some knowledge is needed to use any sentence, including observation sentences. It merely seeks to differentiate knowledge needed to speak the language from more idiosyncratic knowledge. Quine further admits that certain beliefs over and above those involved in learning the language may affect verdicts to observation sentences if these beliefs are shared by all members of the community ([20], pp. 37-38 and [15], p. 86). However, the influence of theory on observation is restricted to just these two domains on Quine's account. Individual differences in knowledge, beliefs, and language among the members of a linguistic community are not held to affect any particular member's use of observation sentences.

The manner in which Quine excludes the possible influence of individual differences on the use of observation sentences is made clear by his reply to Hanson's claim that an object like an X-ray tube can only be seen by someone with sufficient knowledge: "Now my answer to the X-ray example was already hinted at a little while back: what counts as an observation sentence varies with the width of community considered. But we can also always get an absolute standard by taking in all speakers of the language, or most." ([15], p. 88). Quine denies that any significant disagreement over observation sentences occurs here. Although the physicist is able to identify the object as an X-ray tube, we need not conclude with Hanson that the physicist accepts an observation sentence which the layman fails to accept. A sentence counts and functions as observational only relative to a particular linguistic community. Within the community of physicists the sentence "X-ray tube" may suffice as an observation sentence and serve evidentiary and semantical functions. But, according to Quine's criterion, "X-ray tube" will not count as an observation sentence for the broader community which includes non-physicists.

The main difference between Quine's view of the effect of theory on observation and the positions defended by Hanson, Kuhn, and Feyerabend, is that information and patterns of linguistic usage peculiar to individual speakers are not permitted to influence what counts as an observation sentence for Quine. The criterion's requirement of community wide agreement in the various members' dispositions to assent to and dissent from observation sentences rules out the possibility of any such influence. Two justifications can be given for restricting the intrusion of theory on observation in this way. First, the purposes for which observation sentences are needed--giving evidence and a semantical foundation--arise in community situations. We are concerned with what the community takes as evidence and what information they need to speak the language. Second, Quine does not employ observation sentences to report what is observable to an individual. As we have noted before, Quine, along with Hanson, Kuhn, and Feyerabend, repudiates the idea of a sensory core in favor of a pragmatic treatment of observation. Hence, Quine allows for the basic claim of the first version of the theory-ladenness objection--that knowledge, beliefs, and language affect observation--without going to the extreme of allowing idiosyncratic information or usage to play a role in deciding which sentences count as observational reports.

It is more difficult to frame an adequate reply to the second version of the theory-ladenness objection, which claims that scientists holding different theories will see different things and so there are no neutral observational reports on which to judge rival theories. This charge is more serious because it lends support to the view that there is no rational procedure for reaching decisions between theories. So far as we can tell, Quine has not addressed this problem directly. Let us briefly examine a reply that seems open to him.

Quine's response to this objection depends on whether the disagreeing theories imply different verdicts to sentences which satisfy his criterion for observation sentences.[9] In cases where they do imply different verdicts, a convincing answer to the objection is available to Quine. He can argue that the verdicts actually given in stimulatory circumstances provide a way of testing the theories. Suppose theory T implies 0 and theory T' implies not-0. If 0 is an observation sentence, then the community agrees in verdicts on 0 in all stimulatory circumstances. Consequently, the disagreeing scientists need only put themselves in the appropriate circumstances and note whether they are prompted to assent to or dissent from 0. The fact that they hold different theories cannot inhibit their verdicts so long as 0 qualifies as an observation sentence. This is where the benefit of not introducing semantical criteria for observation sentences makes itself obvious. In treating a sentence as an observation sentence, we are only concerned with the linguistic entity and the verdicts individuals give to it, not the meaning they attach to it. (cf., [12]). Quine's claim that observation sentences serve as arbiters between rival theories is clearly vindicated in such cases.

If there are no sentences meeting the criterion on which the opposing theories imply different verdicts, then Quine might respond that there is no observational disagreement between the parties. He would then be forced to admit that unless the disputants work out their theories' implications in terms of common observation sentences, observation sentences do not generate a rational means of resolving the dispute.[10] In such circumstances Kuhn and Feyerabend lead us to believe that the dispute can only be resolved when one side capitulates and adopts the other's theory. The appeal to observation sentences apparently decides nothing, although it may serve as an effective psychological ploy in winning adherents to one's own theory.

This reply raises a crucial question for both defenders and opponents of observation sentences. How frequently is it possible for disagreeing scientists to test their theories by sentences meeting Quine's criterion for observation sentences? Quine's criterion suggests two conditions which have to be met for testing to occur: certain sentences must qualify as observational for the community jointly composed of the disagreeing parties and these theories must imply contrary verdicts to one or more of these sentences in at least some stimulatory situation. The force of the theory-ladenness objection remains fully intact if either of these conditions fails. Of course, in such cases, the conflicting theories may delimit their own classes of observation sentences.

These will be the sentences which satisfy the criterion with respect to the narrower communities consisting of the proponents of each of the different theories. There will not, however, be disagreement on any common observation sentence. In such circumstances Quine's criterion fails to solve the serious problem although it may still prove useful in comparing the theories. For example, it may be useful to compare what sentences meeting Quine's criterion the theories agree on. These sentences provide a means of connecting the theories as theories about a common domain of phenomena. Further, they provide the bridge for competing theorists to learn each other's theory. Thus, the scientists may be able to figure out where their disagreement enters. With this as a basis they may be able to evaluate the scope and explanatory power of the alternative theories. Nevertheless, Quine's criterion provides no ultimate foundation for testing such radically opposed theories.

Quine's position seems to be that this situation is not typical and that disagreeing scientists are often able to appeal to a common observation language to resolve disputes.[11] It is beyond the scope of the present paper to decide whether this view is correct. Only a case by case study of theoretical conflicts can settle the issue. Unfortunately, most of the studies done to date adopt a traditional definition of observation sentences. They therefore tell us little or nothing about whether sentences meeting Quine's quite different criterion are generally accessible to disagreeing scientists. In the absence of detailed case studies it is no more plausible to assume with Quine that such observation sentences will generally be available than to suppose with the proponents of the theory-ladenness objection that they will not be.

In this paper we have measured Quine's criterion for observation sentences against objections that have been made to earlier definitions of observation sentences. It has successfully met most of these objections, but it remains a question for further study whether sentences satisfying the criterion play a significant role in resolving theoretical conflicts. It is on the answer to this question that the utility of Quine's criterion must ultimately be judged.

Notes

[1]We dedicate this paper to the late Donald J. Lipkind, our close friend and fellow graduate student at the University of Chicago.

[2]Compare: "Ironically enough . . . some iconoclastic philosophers of science have taken to questioning the notion of observationality only now that it ceases to present a problem." ([16], pp. 4-5).

[3]Suppe actually bases most of his argument on Putnam's criticism of the comprehensiveness of the analytic-synthetic distinction, but does acknowledge and consider the Quinean form of the attack. At the

end of his remarks he draws the explicit connection with Quine's attack on reductionism.

[4]Carnap states: "This explanation is necessarily vague. There is no sharp line between observable and non-observable predicates... ." ([4], p. 63). Elsewhere he says: "There is a continuum which starts with direct sensory observations and proceeds to enormously complex, indirect methods of observation. Obviously no sharp line can be drawn across this continuum . . . the line separating observable from non-observable is highly arbitrary." ([3], p. 226).

[5]Mary Hesse likewise argues that "no predicates, not even those of the observation language, can function by means of direct empirical association alone." ([9], p. 38).

[6]Quine states this point as follows: " . . . a sentence that is true by mere meanings of words should be expected, at least if it is simple, to be subscribed to by all fluent speakers in the community." ([15], p. 86). In his comment of Davidson's "Belief and the Basis of Meaning" Quine suggests the criterion of community agreement may provide a general distinction between the meaning component and the belief component in a person's use of language ([13], p. 325).

[7]This statement of Quine's criterion is drawn from his various formulations of the criterion. It comes closest to his formulation in "Grades of Theoreticity": " . . . the behavioral criterion that qualifies a sentence as an observation sentence . . . is just that all speakers of the language who are present and attentive when such a sentence is affirmed are apt to agree in assenting to the sentence or to agree in dissenting from it. Someone may want to qualify the definition so as to exclude sentences everyone would assent to, or everyone would dissent from, independently of concurrent stimulation; but that is an easy emendation." ([16], p. 4).

[8]In other places Quine identifies observation sentences as sentences "on which speakers of the language will give the same verdict when given the same concurrent stimulation." ([15], p. 87). This does raise the problem of what constitutes the same stimulation. In [18] Quine proposes mechanisms for classifying stimulations as the same or similar. Although the human nervous system undoubtedly employs some such mechanisms, the reasonable guide to follow in locating such mechanisms would be the actual behavior of giving verdicts to observation sentences. Since the criterion as we have formulated it (drawing from [16]) allows one to directly distinguish observation sentences, it seems preferable to the version Quine sometimes gives in terms of same stimulation.

[9]As Quine has pointed out, theories do not actually imply observation sentences, but observation conditionals, that is, conditional sentences composed of two observation sentences ([14], p. 318). For simplicity, though, we have continued to speak of observation sentences themselves being implied.

[10] Quine might respond that if opposing scientists agree on verdicts to all observation sentences common to them (i.e., all sentences meeting Quine's criterion for their composite language community), then their theories are really empirically underdetermined. Quine holds that all theory is ultimately underdetermined by empirical evidence in the sense that alternative theories can always be devised to account for the same evidence. The situation we are supposing, though, is not a case of conflict between underdetermined theories, but rather a conflict between theories making different and contrary empirical claims. It may be a difficulty for Quine's criterion that it counts only sentences upon which the whole community agrees on use as observational. Consequent it cannot distinguish what we intuitively recognize as observational disputes between people who use sentences in different ways. How such cases of empirical dispute are to be recognized requires independent exploration, but it is cases where competing theories intuitively make conflicting empirical claims (even if not in the same language) that we have in mind.

[11] Quine speaks as if there were always a set of observation sentences to which a community could fall back: "The specialist can always be driven back to the less technical levels of evidence, though in practice he cites only what is needed for reassuring his trained colleagues." ([18], p. 41). Quine does not explain in what sense we can be "driven back to the less technical levels of evidence." Given his attack on reductionism, it certainly should not be in terms of a reduction. The more serious problem, though, is that Quine has not established that there will be any commonly employed, less technical idioms. Moreover, this is not something that can be established *a priori* but only through detailed linguistic study of scientific practice. Some support, though, is given to this view by Wimsatt's analysis of how intra-level reduction is possible over periods of theoretical change. Between a theory and its immediate predecessor a reduction will generally be possible, but it will be less feasible with more remote predecessors. Complete incomparability is the result of multiple changes in theory ([24], pp. 216-218). This accords well with Quine's favorite analogy of theoretical change to the rebuilding of Neurath's ship while at sea. According to this picture, only part of the linguistic network is changed at any one time while most remains intact. If historical cases do fit this pattern, then Quine's observation sentences will have a major function in comparing theories.

References

[1] Achinstein, Peter. Concepts of Science: A Philosophical Analysis. Baltimore: The Johns Hopkins Press, 1958.

[2] Carnap, R. "The Methodological Character of Theoretical Concepts." In The Foundations of Science and The Concepts of Psychology and Psychoanalysis. (Minnesota Studies in the Philosophy of Science Vol. 1.) Edited by H. Feigl and M. Scriven. Minneapolis: University of Minnesota Press, 1956. Pages 33-76.

[3] ---------. Philosophical Foundations of Physics. New York: Basic Books, 1966.

[4] ---------. "Testability and Meaning." Philosophy of Science 3 (1936): 420-468 and 4 (1937): 1-40, (As reprinted in Feigl, H. and May Brodbeck (eds.). Readings in the Philosophy of Science. New York: Appleton-Century-Crofts, 1953. Pages 47-92).

[5] Feyerabend, P. "Against Method: Outline of an Anarchistic Theory of Knowledge." An Analysis of Theories and Methods of Physics and Psychology (Minnesota Studies in the Philosophy of Science, Volume IV). Edited by Michael Radner and Stephen Winokur. Minneapolis: University of Minnesota Press, 1970. Pages 17-130.

[6] -------------. "Problems of Empiricism." In Beyond the Edge of Certainty (University of Pittsburgh Series in the Philosophy of Science, Volume 2). Edited by Robert G. Colodny. Englewood Cliffs, NJ: Prentice-Hall, 1965. Pages 145-260.

[7] Hanson, N.R. Patterns of Discovery. New York and London: Cambridge University Press, 1958.

[8] -----------. Perception and Discovery. (ed.) Willard C. Humphreys. San Francisco: Freeman, Cooper, and Company, 1969.

[9] Hesse, M. "Is There an Independent Observation Language." In The Nature and Function of Scientific Theories. (University of Pittsburgh Series in the Philosophy of Science, Volume 4). Edited by R. Colodny. Pittsburgh: The University of Pittsburgh Press, 1970. Pages 35-105.

[10] Kuhn, T. "Second Thoughts on Paradigms." In The Structure of Scientific Theories. Edited by Frederick Suppe. Urbana: The University of Illinois Press, 1974. Pages 459-482.

[11] -------. The Structure of Scientific Revolutions. 2nd ed. Chicago: The University of Chicago Press, 1970.

[12] Ostien, Philip A. "Observationality and the Comparability of Theories." In PSA 1974. (Boston Studies in the Philosophy of Science Vol. XXXII.) Edited by Robert S. Cohen, C.A. Hooker, A.C.

Michalos, and J.W. Van Evra. Dordrecht, Holland: D. Reidel Publishing Company, 1976. Pages 271-289.

[13] Quine, W.V. "Comment on Davidson." Synthese 27 (1964): 325-329.

[14] ----------. "On Empirically Equivalent Systems of the World." Erkenntnis 9 (1975): 313-328.

[15] ----------. "Epistemology Naturalized." In Ontological Relativity and Other Essays. New York: Columbia University Press, 1969. Pages 69-90

[16] ----------. "Grades of Theoreticity." In Experience and Theory. Edited by L. Foster and J.W. Swanson. Amherst: The University of Massachusetts Press, 1970. Pages 1-17.

[17] ----------. "Reply to Hintikka." In Words and Objections: Essays on the Work of W.V. Quine. Edited by D. Davidson and J. Hintikka. Dordrecht, Holland: D. Reidel Publishing Co., 1969. Pages 312-315.

[18] ----------. The Roots of Reference. LaSalle, Ill.: Open Court Publishing Co., 1974.

[19] ----------. "Two Dogmas of Empiricism." In From a Logical Point of View: Nine Logico-Philosophical Essays. 2nd ed. rev. New York and Evanston: Harper and Row, Publishers, Inc., 1963. Pages 20-46.

[20] ----------. Word and Object. Cambridge, Mass.: M.I.T. Press, 1960.

[21] Scheffler, Israel. Science and Subjectivity. New York: Bobbs-Merrill, 1967.

[22] Siegel, Judith. "Logic Without Mentalism: A Study of Quine's Views." Unpublished Ph.D. Dissertation: The University of Chicago, 1970. Xerox University Microfilm Number unavailable.

[23] Suppe, Frederick. "The Search for Philosophical Understanding of Scientific Theories." In The Structure of Scientific Theories. Edited by Frederick Suppe. Urbana: The University of Illinois Press, 1974. Pages 3-241.

[24] Wimsatt, William C. "Reductionism, Levels of Organization, and the Mind-Body Problem." In Consciousness and the Brain: A Scientific and Philosophical Inquiry. Edited by Gordon Globus, Grover Maxwell, and Irwin Savodnik. New York: Plenum Press, 1976, Pages 205-267.

Analyticity Without an Observational-Theoretical Distinction

Mark Tamthai

Chulalongkorn University
Bangkok, Thailand

The problem we are concerned with in this paper is that of separating out, among the consequences of an empirical theory, those that play a purely linguistic role (the "analytic" consequences) from the rest. This problem we shall refer to as the problem of theoretical analyticity. In a sense, the problem was solved by Carnap's proposal ([1], p. 270) to consider the analytic consequences to be exactly the consequences of a special sentence associated with each theory, along with results of Winnie [10], which gave further justification to this proposal. Besides Carnap's proposal there have been other attempts to solve the problem of theoretical analyticity (e.g., [7]). However, all such attempts (including Carnap's) have relied upon a distinction between observation and theoretical terms in the language of empirical theories. Since there has been much debate concerning this distinction, and since there do not seem to be any good reasons why "analyticity" should be tied so closely to it, we shall attempt to attack the problem of theoretical analyticity without assuming an observation-theory distinction among the vocabulary of empirical theories.

1. Carnap's Proposal

Let T be an empirical theory whose non-logical vocabulary consists of observational and theoretical terms. We denote the Ramsey sentence of T by R(T). It is well known that the class of observational consequences of T and the class of consequences of R(T) are identical. Carnap proposed as the analytic consequences of T, the consequences of $R(T) \rightarrow T$. (We denote this sentence by C(T), the Carnap sentence of T.) Since the conjunction of R(T) and C(T) is equivalent to T, and since the only shared consequences of R(T) and C(T) are the logically valid consequences, this proposal seems quite reasonable. Winnie [10] has shown that the consequences of C(T) are exactly those consequences of the theory T which are observationally vacuous in the following sense:

PSA 1978, Volume 1, pp. 109-115

<u>Definition 1</u>: A consequence S of T is observationally vacuous iff for all consequences S' of T, if an observation sentence O is a consequence of S&S', then O is a consequence of S'.

This seems to correspond to the intuitive notion of an analytic consequence of a theory, and thus lends further justification to Carnap's proposal.

2. The Evidential Base of Empirical Theories

Carnap's approach to theoretical analyticity is based upon a distinction between observation and theoretical terms in the language of a theory. It considers the observational term to be the basic carrier of empirical import. Another alternative would be to consider sentences rather than non-logical terms as the carrier. The first problem is to decide what kinds of sentence would perform this function. In explications of analyticity observation terms have played a significant role in the context of <u>observation sentences</u>. This suggests that what is needed is an analog of the concept of an observation sentence, but definable independently of the idea of an observation term. Now, the observation sentences are the means by which evidence for and against a theory is assessed; they are used in the process of confirming and disconfirming theories. Therefore, we may conjecture that the basic empirical units we are searching for are those sentences which are used for just this purpose. What must be done is to characterize such sentences without recourse to machinery involving a reference to "observation" terms. Three attempts at such a characterization are Karl Popper's basic statements [6], Quine's observation sentences ([8], p. 42), and Feyerabend's quickly decidable sentences [3].

The two main conditions which Popper imposes on statements in order for them to qualify as "basic" statements are, first, that they cannot be deduced from a universal statement without initial conditions, and secondly, that they can contradict a universal statement. Popper gives the following rule concerning basic statements which guarantees that these two conditions will be satisfied: basic statements have the form of singular existential statements. Besides these formal requirements for basic statements, Popper also imposes a material requirement on them the event which a basic statement tells us is occurring at place k must be an "observable" event. The purpose for this requirement is to insure us that basic statements are "testable, inter-subjectively, by observation." Thus, basic statements are those statements which assert that an observable event is occurring in a certain region of space and time.

Quine's observation sentences are defined in terms of "occasion sentences" and "stimulus meanings of sentences". Occasion sentences are sentences which "command assent or dissent only if queried after an appropriate prompting stimulation." The "stimulus meaning" of a sentence S is defined as the ordered pair consisting of the affirmative stimulus meaning of S and the negative stimulus meaning of S. The affirmative stimulus meaning of S, for a speaker, is the class of all stimulation that would prompt his assent. The negative stimulus meaning of S, for a speaker, is the class of all stimulations that would prompt his dissent. Quine then defines an observation sentence as follows: observation sentences are occasion sentences whose stimulus meaning do not vary under the influence of collateral information. Thus, the observation sentences of a theory are those sentences "on which a scientist will tend to fall back on when pressed by doubting colleagues." They are the datum sentences of science.

Feyerabend's quickly decidable sentences share with Quine's observation sentences their characterization in terms of subjective and pragmatic notions. According to Feyerabend, observation sentences are not characterized by their content but by the fact that their production conforms to certain behavioral patterns [4]. He gives five conditions which a sentence S must satisfy in order to be considered to be an observation sentence:

(1) S must be an atomic sentence;
(2) There is a situation s such that every class of observers C, when presented with S in s, will run through a series of states and operations which terminates either in the acceptance or rejection of S by C;
(3) The associated series mentioned above must be passed through quickly;
(4) If in a particular situation s, S is accepted (or rejected) by some C, then it will be accepted (or rejected) by nearly every C in that same situation s; and
(5) The decision made by the members of C, concerning S, is causally dependent on the situation s and not only upon S or the internal state of C.

These three examples indicate that there is no unanimity concerning the specific nature of the sentences which might be considered as the basic empirical units - the test sentences - of a theory. Nevertheless, there are some general formal features which it seems reasonable to impose on test sentences, and which are in the spirit of the above views. Letting t be the class of test sentences for a theory T, and L(T) the language of T, we propose the following as requirements to be satisfied by t:

(a) t is a recursive set of 1st-order sentences of L(T); and
(b) If k is a consistent set of test sentences, then all consequences of k are test sentences.

We now turn to explicating the concept of an "analytic component" of an empirical theory making use of the concept of a test sentence and assuming only that the class of such sentences satisfies (a) and (b).

3. Analyticity as Immunity from Confirmation and Disconfirmation

Before developing the details of this explication we state three theorems which will be applied to construct the desired explicatum. The first theorem is an elementary result of Recursion Theory. The second and third theorems are results of W. Craig [2] and S. Kleene [5] respectively.

<u>Theorem 1</u>: If A is a recursive set of sentences and B is a recursively enumerable set of sentences, then the set of sentences $A \cap B$ is recursively enumerable.

<u>Theorem 2</u>: Every recursively enumerable set of sentences which is closed under logical consequence is recursively axiomatizable.

<u>Theorem 3</u>: Every 1st-order theory T has a finitely-axiomatizable conservative extension T', such that the number of non-logical terms of L(T') which are not part of L(T) is finite.

We will now characterize a particular sentence R*(T) associated with each 1st-order theory T.

(1) Let t be the set of test sentences of T, and Con(T) the set of logical consequences of T.
(2) Let $t^* = t \cap \mathrm{Con}(T)$. Since t is a recursive set of sentences, and Con (T) a recursively enumerable set, by Theorem 1 we know that t* is recursively enumerable. Since we assume T to be consistent, t* must be satisfiable, and thus by (b), it is closed under logical consequence.
(3) By Theorem 2 we also know that t* is recursively axiomatizable. Call this axiomatization t*a.
(4) Since t is a set of 1st-order sentences, t*a is a 1st-order theory. Thus, by Theorem 3, a conservative extension of t*a is finitely axiomatizable. Let t*a,f be the conjunction of some finite set of axioms for this theory, and let $V_1,\ldots,V_n$ be the additional predicates.
(5) The R*(T) is the sentence:

$$\exists X_1 \ldots \exists X_n \left(t^*a,f \begin{bmatrix} V_1 \ldots V_n \\ X_1 \ldots X_n \end{bmatrix} \right)$$

This sentence R*(T) which is associated with every theory T will be referred to as "the Ramsey sentence analog". As we recall, the fundamental property of the Ramsey sentence of a theory T is that it has the same observational consequences as T. Since the test sentence has now taken the place of the observation sentence, it is natural to demand of the Ramsey sentence analog of a theory T that it should have the same test-sentence consequences as T. The following theorem indicates that this is indeed the case.

Theorem 4: If t(i) is a test sentence of a theory T, then t(i) is a logical consequence of T iff t(i) is a logical consequence of R*(T).

<u>Proof</u>: Let t(i) be a test-sentence consequence of T. Therefore, t(i) is a logical consequence of the theory t*a. Since the

theory axiomatized by t*a,f is an extension of t*a, t(i) must be a logical consequence of t*a,f. Now t(i) is a sentence in the language of t*a, and the predicates $V_1 \ldots V_n$ do not occur in t(i), therefore, t(i) is a consequence of R*(T). Conversely, if t(i) is a logical consequence of R*(T) then since $V_1 \ldots V_n$ do not occur in t(i), it must be a logical consequence of t*a,f. But t*a,f axiomatizes a conservative extension of t*a and t(i) is in the language of t*a, therefore, t(i) is a consequence of t*a, and hence of T as well.

Being in possession of a proper Ramsey sentence analog for a theory, we turn next to the problem of analyticity. Our search is once more for an analog, this time an analog to the Carnap sentence C(T). We consider the following obvious candidate: R*(T) → T, and denote this sentence by C*(T). We are now in a position to present our proposal to identify the analytic component of a theory without utilizing any observational-theoretical distinction.

<u>Definition</u> <u>2</u>: A sentence S is an analytic component of a theory T iff S is a logical consequence of C*(T).

What must be examined next is the adequacy of the explication of analyticity given by the above definition. What conditions are there which all proposed explicata of this concept must satisfy? One such condition seems to be the following:

(A) If A is an analytic component of T, then A is immune from confirmation and disconfirmation within T.

Before considering whether or not the given definition satisfies condition (A), we must make the notion of "immunity from confirmation and disconfirmation" more precise. To say of a sentence that it is confirmed or disconfirmed is to say that it is supported by, or clashes with, experience. Thus, it seems reasonable to equate "immunity from confirmation and disconfirmation" with "not occurring essentially in the deduction of any test sentence", or, to be more precise, with "testable vacuity", which we define as follows:

<u>Definition</u> <u>3</u>: S is testably vacuous in T iff S is a logical consequence of T and for all consequences S' of T and test sentences t(i) of T, if t(i) is a consequence of S&S', then t(i) is a consequence of S'.

The precise question which can now be posed is: "Are all analytic components of a theory T testably vacuous in T?" The following theorem answers this question affirmatively.

<u>Theorem</u> <u>5</u>: If S is an analytic component of a theory T, then S is testably vacuous in T.

<u>Proof</u>: Let S,S' be consequences of T, and t(i) be a test sentence of T which is a consequence of S&S'. Since S is analytic in T,

it is a consequence of $C^*(T)$ and hence of $R^*(T)$. By Theorem 4 we know that $t(i)$ must be a consequence of $R^*(T)$, and hence $S' \to t(i)$ is a consequence of $R^*(T)$ as well. But $S' \to t(i)$ is a consequence of S, and hence of $R^*(T)$. Therefore, $S' \to t(i)$ must be logically true, or in other words, $t(i)$ is a consequence of S'.

Thus S is testably vacuous in T.

We now have an explication of analyticity which does not rely on any observation-theory distinction, and for which a justifying theorem - Theorem 5 - has been proven. A proof of the converse of Theorem 5 would lend further justification to this proposal.

References

[1] Carnap, R. *Philosophical Foundations of Physics* (ed.) Martin Gardner. New York: Basic Books, 1966.

[2] Craig, W. "On Axiomatizability Within a System." *Journal of Symbolic Logic* 18(1953): 30-32.

[3] Feyerabend, P.K. "An Attempt at a Realistic Interpretation of Experience." *Proceedings of the Aristotelian Society* 58(1957-58): 144-170.

[4] ---------------. "Explanation, Reduction, and Empiricism." In *Scientific Explanation, Space, and Time*. (*Minnesota Studies in Philosophy of Science*, Volume III.) Edited by H. Feigl and G. Maxwell. Minneapolis: University of Minnesota Press, 1970. Pages 28-97.

[5] Kleene, S.C. "Finite Axiomatizability of Theories in the Predicate Calculus Using Additional Predicate Symbols." *Memoirs of the American Mathematical Society*, 10(1960): 27-68.

[6] Popper, K. *The Logic of Scientific Discovery*. New York: Basic Books, 1959. (Originally published as *Logik der Forschung*. Wien: J. Springer, 1935.)

[7] Przelecki, M. and Wojcicki, R. "The Problem of Analyticity." *Synthese* 19(1969): 374-399.

[8] Quine, W.V. *Word and Object*. Cambridge, Ma.: M.I.T. Press, 1960.

[9] Rogers, H. *Theory of Recursive Functions and Effective Computability*. New York: McGraw-Hill, 1967.

[10] Winnie, J. "Theoretical Analyticity." In *PSA 1970*. (*Boston Studies in the Philosophy of Science*, Volume 8.) Edited by R.C. Buck and R.S. Cohen. Boston: D. Reidel, 1970. Pages 289-365.

Part IV

Case Studies with Applications to the Philosophy of Science

The Role of Psychology in Functional Localization Research

Barbara V.E. Klein[1]

Yale University

It takes no more than a passing acquaintance with functional localization research in the neurosciences to notice that psychological notions often come into play in the characterization of human brain functions. In the past, such notions were taken primarily from one form or another of associationism; currently, they are adopted from the language of information processing. A rather typical example can be found in a recent article by A.R. Luria (see also [3]). He describes the human brain as made up of three main blocks incorporating basic functions. These functions are described as follows:

> The first block regulates the energy level and tone of the cortex, providing it with a stable basis for the organization of its various processes ... . The second block ... plays a decisive role in the analysis, coding and storage of information ... . We can easily identify areas in the second block that are respectively responsible for the analysis of optic, acoustic, cutaneous and kinesthetic stimuli. Each of these cortical areas has a hierarchical organization: a primary zone that sorts and records the sensory information, a secondary zone that organizes the information further and codes it and a tertiary zone where the data from different sources overlap and are combined to lay the groundwork for the organization of behavior ... . The third block ... is involved in the formation of intentions and programs for behavior. ([5], p. 66-68).

That psychology plays some role in attributions of function to the human brain and its parts is evident. But what precisely this role is and whether it is legitimate is far from clear. It is to these questions that this paper is addressed.

I shall begin by attempting to answer three preliminary questions:

(1) What is a functional localization hypothesis?

PSA 1978, Volume 1 , pp. 119-133

(2) To what sort of neural model is it relevant?

(3) What is the relationship between a functional localization hypothesis and such a neural model?

The use of the term 'function' in attributions of functional localization seems to be related to its use in biology. Cummins [1] has suggested that the biological notion of function can be explicated, in part, by the notion of capacity. For some component of a system or organism to have, say, digesting as one of its functions is for that component to be capable of digesting.[2] But more is needed by way of definition, because not everything that a component of a system is capable of doing ought to count as one of its functions, as Cummins points out. The heart, for example, is capable of producing sound; yet we would not regard producing sound as one of the functions of the heart. It has frequently been noted that what distinguishes the capacities of system components that are _functions_ of components from those that are not, has something to do with the relationship of the component capacity to the functioning of the whole system or organism. Precisely what this relationship is has been a matter of philosophic controversy.[3] Cummins' view is that what is relevant is whether the component capacity figures in an _explanation_ of some capacity of the entire organism.

Presumably what we want to explain is how the organism is able to do something, for example, how Mr. Jones is able to name objects upon confrontation. Explaining this capacity is to be contrasted both with explaining why Mr. Jones engages in object naming upon some specific occasion (the psychologist asked him to) and explaining how Mr. Jones came to be able to have this capacity (his mother taught him). Cummins suggests that a system's (organism's) capacity to do something can be explained by a _functional analysis_ of that capacity. By a functional analysis of some capacity C, he means, roughly, a breakdown of the complex capacity into a set of constituent capacities (c_1, c_2 ... c_n), with a specification of the sequence in which those constituent capacities must be exercised for the result to be a manifestation of the complex capacity.

Let us consider an example. A person's capacity to cook an omelet can be analyzed into a set of constituent capacities such as the capacity to break eggs into a bowl, the capacity to stir the eggs, the capacity to pour the mixture into a pan, etc. But simply possessing all the requisite skills will not do the trick. The order in which they are to be exercised must also be specified. Thus it is essential that one stir the eggs before pouring the mixture into the pan; otherwise, the result will be scrambled eggs. In effect what is required is an algorithm or program for exercising C in terms of c_1, c_2, ... c_n. Such algorithms can be displayed graphically by a flow chart such as Figure 1. Note that a capacity like omelet-making differs in one important respect from the sorts of capacities of concern to an aphasiologist. The constituent capacities of omelet-making are normally exercised intentionally by a person. In contrast, the constituent capacities of, say, object-naming, must be assigned to something inside the organism. Suppose that naming an object involves perceiving the object, computing a conceptual representation, looking up the appropriate lexical entry for that representa-

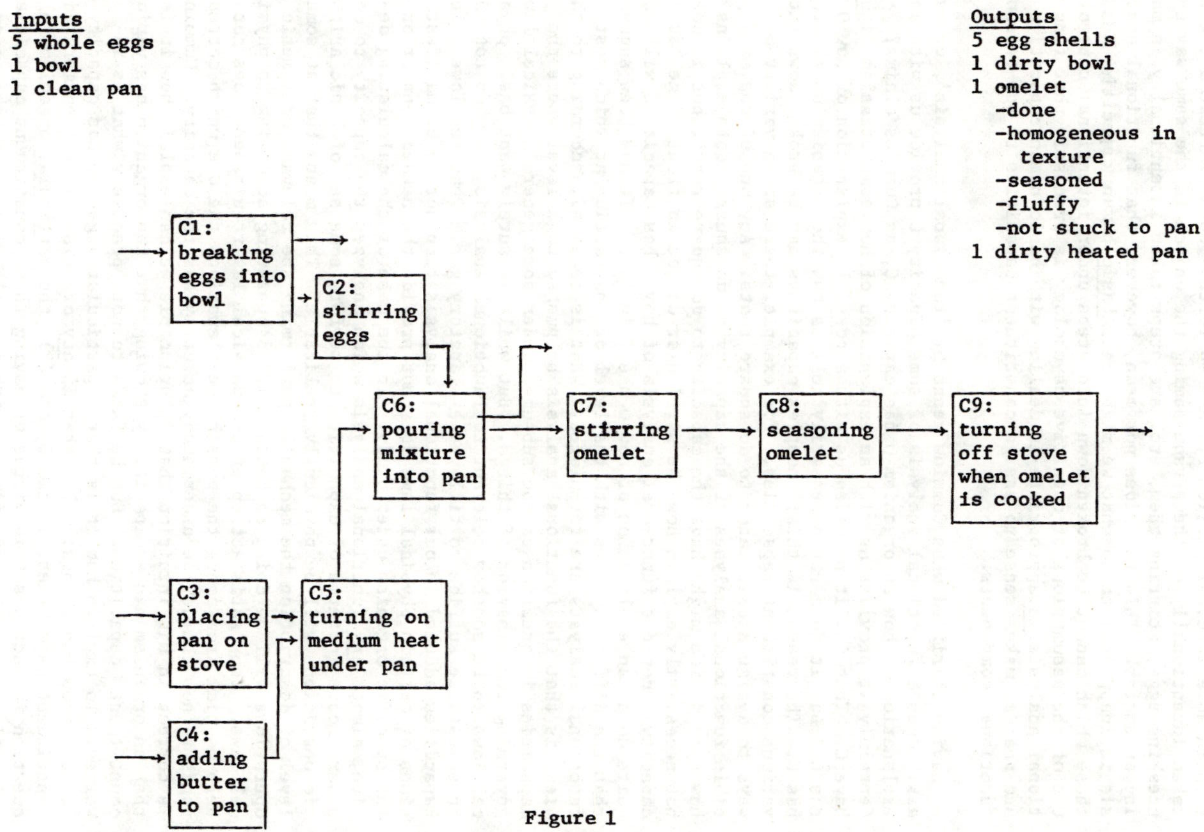

Figure 1

A Functional Analysis of Omelet Making

tion in some internal dictionary, and so forth. None of these are steps taken intentionally by the person naming the object in the same way that breaking eggs, stirring them, etc. are steps taken intentionally in making an omelet. This fact does not mean, however, that functional analysis is inappropriate to what we might call *basic* human capacities, i.e., those which cannot be broken down into steps under intentional control. One of the assumptions of cognitive psychology is precisely that functional analysis *is* appropriate in dealing with basic human capacities and that it makes sense to assign constituent capacities to sub-personal functional components.

Having clarified what Cummins means by 'functional analysis', we must ask: does a functional analysis of some capacity C provide us with an explanation of how an organism O has capacity C, as Cummins claims? Certainly it provides us with an explanation of how some organism *might* have C. That is, it provides us with a *possible* explanation of how O has C. But it does not necessarily tell us how the organism under study has C. The reason is this: complex capacities may be broken down into various constituent capacities. For example, there are a variety of ways to design a calculator to do square roots. Any one of the many possible functional analyses of the capacity to do square roots explains how some device might have the capacity to do square roots, but it does not necessarily explain how the particular device in front of me has that capacity. Hence a functional analysis of how O has capacity C will not quite do as an explanation of how O has C in fact. It has been suggested that the difficulty lies with the level of abstraction at which most functional analyses are formulated. What is required, according to this view, is that the functional analysis be pushed to a level where the capacities in question can be subsumed under some general physical law governing the behavior of things. But while we surely want basic physical laws to be somehow relevant to functional analysis, they do not in this instance suffice to fill the explanatory gap. That the lowest level capacities adduced in our functional analysis turn out to be manifestations of certain physical laws does not provide the needed connection to the specific organism under study. In the case of the calculating device, almost *any* functional analysis which analyzes the capacity to do square roots in terms of capacities to perform some set of basic arithmetic operations will be physically realizable. This means that at some level of description the mechanism performing the relevant arithmetic operations will be doing so because it is operating according to physical laws. Thus satisfaction of the physical law requirement does not help us pick from among the possible explanations the one which corresponds to how the device under study actually has the capacity. Cummins is correct in his intuition that in principle we can explain how it is that an organism has a capacity by showing that the organism has various constituent capacities. But he fails to note precisely what is involved. To explain how it is that a particular organism has capacity C, we must show two things: (1) that if *any* organism has the enumerated constituent capacities, it has C, and (2) the *particular* organism in question in fact has C in virtue of having those constituent capacities. A functional analysis provides us with only (1). Hence we have an explanatory gap.

To fill this gap we must supply (2), that is, we must tie the analysis to the organism under study. For reasons that will become clear shortly, we will call a functional analysis for which (2) is the case a <u>structurally adequate functional analysis</u>. The notion of a structurally adequate functional analysis is important for the following reason. We have defined a function to be a constituent capacity which can figure in an explanation of a capacity of the system as a whole. But since only a structurally adequate functional analysis can explain how an organism has some capacity C, what counts as a function can be determined only relative to a structurally adequate functional analysis.

What is required for a functional analysis to be structurally adequate? A functional analysis A of how an organism O has capacity C is structurally adequate if there exists a <u>structural decomposition</u> of O into component (physical) parts such that:

(1) operation of those parts results in O's exercising C;
(2) for each constituent capacity of A, there exists at least one of those structural components which has that capacity; and
(3) the order in which those component parts operate when O is exercising C mirrors the algorithm specified in A.

Condition (1) guarantees the existence of a physical mechanism underlying the capacity C. Condition (2) insures that some part of the mechanism corresponds to each constituent capacity specified by the functional analysis. Condition (3) guarantees further that the sequence of operation of these physical parts corresponds to that specified by the functional analysis. Condition (1) alone will not insure the structural adequacy of A because (1) alone is compatible with O's exercising C in a way which does not involve the exercise of any of the constituent capacities of A. Conditions (1) and (2) alone are not enough because they are compatible with O's exercising C by exercising the constituent capacities in a sequence different than that specified by the functional analysis.

Consider now a model which describes a possible structural decomposition of O such that if it were true of O, conditions (1)-(3) would be met. We shall call such a model a <u>functional component model of how O has C</u>. Such a model <u>incorporates</u> A.

To give an example, a functional component model incorporating our functional analysis for making omelets might consist of a component for cracking, a component for stirring, a component for adding ingredients to a pan, a heating unit, a heat switch, and so forth. In other words, it would be a description of an omelet-making machine in terms of parts which can exercise the constituent capacities specified in the analysis.

It is important to realize that for any given functional analysis of some capacity C, more than one functional component model incorporating A is possible. Suppose, for example, that capacity C is analyzed as involving steps P and Q, and these in turn, are analyzed into the primitive operations s, t, and u as follows:

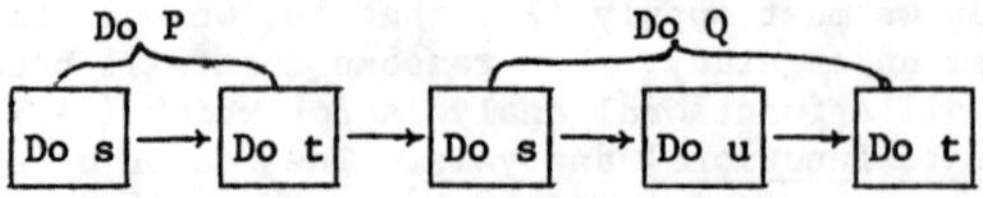

Figure 2

Functional Analysis of C

At least two structural decompositions of a system which has C are possible: one which maps capacities onto structural components only at the level of primitive operations, (Model 1), and one which reflects the hierarchy of capacity analysis and, hence, includes a device corresponding to each intermediate capacity P and Q (Model 2).

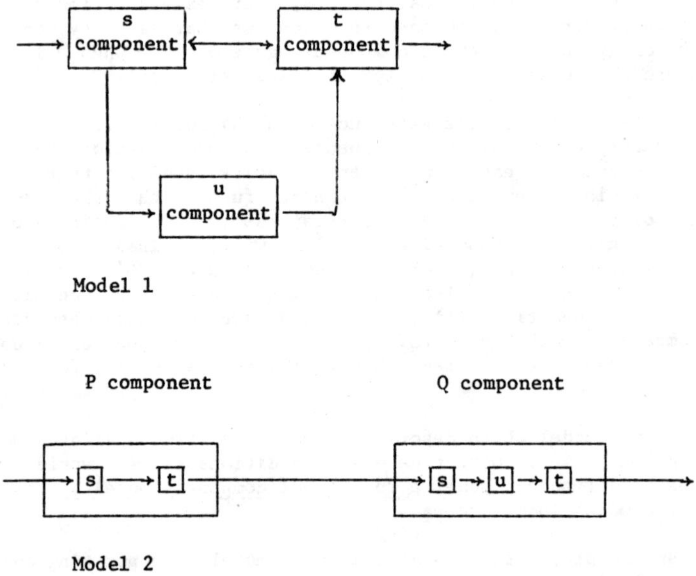

Figure 3

The existence of a functional component model incorporating a functional analysis of how O has C does not yet suffice to bestow structural adequacy on the analysis. For just as a functional analysis provides a description only of how the organism might exercise C, so a functional component model provides a description only of a possible structural decomposition of O. Imagine for example that a system exists for making omelets by piercing eggs, sucking the insides out of the shell, whipping the batter in a blender and then pouring the batter into an appropriately shaped heated slot much like that in a toaster. Neither the functional analysis we described nor the functional component model we sketched would be true of this system. Hence neither would provide an explanation for how this system makes omelets. What is required for there to be an explanation is that the functional component model be true. We shall say that such a model is true of an organism just in case there exists a part of the organism which is the physical realization of each functional component specified by the model and these parts are connected in such a way as to constitute the input-output network described by the model.

The conditions of structural adequacy for a functional analysis can now be restated. A functional analysis A of how O has C is structurally adequate if there exists a functional component model which both incorporates A and is true of O.

We can now use the account we have developed to make Cummins' suggestion about what a function is more precise. Recall that Cummins suggested that a function of some part of an organism is a capacity of that part if that capacity figures in an explanation of how the organism as a whole has some complex capacity. To be more precise:

F is a function of some part of O, where O has capacity C, just in case :

(i) there exists a functional analysis A of how O has C in terms of constituent capacities c_1, c_2, ... c_n;

(ii) A is structurally adequate; and

(iii) $F = c_i$ where $1 \leq i \leq n$.

We introduced the notion of a functional component model in order to satisfy the requirement that the functional analysis of how O has C counts as an explanation of how O has C. Note that such a model is important for another reason. Without a functional component model, we have no way of determining which particular parts of the organism have the functions in question. While attributing a function like pumping blood to the organism as a whole seems legitimate, it is hardly illuminating. The value of functional analysis derives from the value of understanding properties of a whole in terms of properties of its parts. To do so, it is essential to be able to identify the relevant parts.

Let us now return to our three preliminary questions, viz., what is functional localization hypothesis? to what sort of neural model is it relevant? and, what is the relationship between functional localization hypotheses and such neural models?

A functional localization hypothesis is a hypothesis of the form: brain structure S in O has constituent capacity c, where c is a function of some part of O. The sort of neural model to which such a hypothesis is relevant is a functional component model. The functional localization hypothesis specifies one aspect of the structural realization of the functional component model. Thus it is not itself part of a functional component model; rather, it is part of the determination that such a model is true of O, in other words, part of a theory of neural realization for the model.

Returning to our main theme, we are now in a position to see why attributions of function to the brain and its parts are often couched in psychological terms. If Cummins is correct in his account of function, what constitutes a brain function can only be determined relative to a functional analysis of the complex capacities of the organism. And presumably, in the case of cognitive capacities, it is the business of psychology to provide such functional analyses.

But the role of psychology in functional localization research is not exhausted by psychology's contribution to what constitutes a brain function. This can best be brought out by considering the question of the legitimacy of functional localization hypotheses. To do so, we must narrow our focus a bit, and consider the justification of such hypotheses relative to a certain evidential base and a certain method of inference. We shall consider the classical method of inferring functional localization in patients with brain lesions.

The "method of deficit analysis" (as I shall call it) is a method of inferring localization of function in the normal human brain, based on two kinds of evidence from neurologically impaired patients: evidence of abnormal behavior in the patient and evidence concerning the site of brain damage. What is inferred are hypotheses of the form: brain structure S in organism O has constituent capacity c, where c is a function of some part of O, or alternatively, brain structure S is identical with the functional component which has constituent capacity c. Such hypotheses are inferred in two steps: first, a functional deficit in the patient is hypothesized from the kind of abnormal behavior which the patient exhibits; second, localization of function in the normal brain is inferred on the basis of the functional deficit hypothesis conjoined with evidence concerning the site of the brain damage.

Breaking the problem into two parts corresponding to the two inferential steps of the method, we can ask: under the what conditions are we justified in inferring a functional deficit hypothesis from behavioral data? And then: under what conditions can localization of (normal) function be inferred, given a functional deficit hypothesis plus lesion data?

Considerations of psychology are relevant primarily to the first step. The second shall thus be ignored for present purposes.

Confronted with two patients exhibiting a certain pattern of symptoms--they could not understand spoken language, they could produce meaningful sentences, they could handle objects correctly, etc.--the nineteenth century aphasiologist Wernicke [6] inferred that both suffered from a deficit in the sound image center. Based on his writings, it seems plausible to reconstruct his reasoning thus:

> Given my model of the neural components relevant for language capacity, positing a deficit in the speech center allows us to *explain* the kinds of symptoms manifested by my two patients. Hence, given the existence of such symptomatology, it is reasonable to infer the existence of such a deficit.

I would like to suggest that the pattern of reasoning attributed here to Wernicke is typical of that employed to infer a functional deficit hypothesis from behavioral data. In fact it is an instance of a rather widespread species of scientific reasoning that philosophers have called "inference to the best explanation." The general form of the inference is this: we are warranted in inferring some proposition T on the basis of some proposition U if the truth of T provides us with the best available explanation for the truth of U. In the aphasic case, inferring the existence of a functional deficit on the basis of some behavioral pathology is justified *if* positing the existence of that functional deficit provides us with the best available explanation for the existence of the aphasic symptoms.

To see how positing the existence of a functional deficit can provide such an explanation, we must extend the account of functional explanation which we developed in the context of normal behavior. We have seen how a structurally adequate functional analysis can provide an explanation of a capacity or set of capacities. Given such a functional analysis, one important way in which a set of *incapacities* can be explained is by positing that one of the normally operating constituent capacities can no longer be exercised. Reconsidering our functional analysis for omelet making (see figure 4) will make this point clear. Given a deficit in any of the functional components underlying a constituent capacity, it is easy to predict exactly what will happen to the output. This is so because each of the constituent capacities makes its own particular contribution to the eventual output. Thus if a constituent capacity cannot be exercised, its particular contribution to the output will be missing.

Sometimes this contribution is direct, as in the case of stirring the omelet (results in the omelet being fluffy), seasoning the omelet (results in the omelet being seasoned), adding butter to the pan (results in the omelet not sticking), etc. Sometimes, however, the contribution is indirect; the exercise of a constituent capacity may not be reflected directly in the output but instead may be a necessary condition for the subsequent normal exercise of another constituent capacity. Thus breaking the eggs into the bowl is a necessary condition for stirring the eggs to produce an omelet which is homogenous in texture.

What makes a functional system work at bottom is the existence of structures and causal mechanisms which realize that system. Similarly, what makes a functional system break down is that some of these structures and causal mechanisms fail to operate normally. When this happens, there are three possible results. First, one of the functional components of the system may simply fail to operate, resulting in inability to exercise its constituent capacity. This case is exemplified in Figure 4.

Functional deficit	Output Result (normal unless otherwise specified)			
C1	no shells	bowl clean	5 slightly battered whole eggs with butter and seasoning and shells	pan clean
C2			omelet non-homogeneous in texture	
C3			omelet not done, not fluffy	pan not hot, with liquid residue
C4			omelet sticks to pan	pan with more than normal egg particles on it
C5	(same as C3)	....	....	....
C6		bowl filled with egg mixture	no omelet	clean pan
C7			omelet not fluffy	
C8			omelet not seasoned	
C9				burnt pan

Figure 4

Possible Functional Deficits of Omelet-making

of how O has C. It is unclear precisely what is involved in meeting this stronger requirement. Roughly speaking, what we want is an explanation which is not only adequate "locally"; that is, for the capacity under study, but also provides the best possible "fit" with available explanations of other capacities of the organism. Apparently the structural adequacy of a functional analysis does not insure that the analysis constitutes the best available explanation for how O has C. A functional component model incorporating A may map onto the structure of the organism (and hence be true of the organism according to our definition) and yet still be unfelicitous. Again the omelet example helps to make the point. Suppose we have a machine capable of cooking a variety of things in a skillet, including omelets. Suppose further that our functional analysis of omelet making treats salting the omelet and peppering the omelet as separate constituent capacities; likewise, the functional component model incorporating this analysis includes a component for salting and a component for peppering. Now such a functional component model might be true of our skillet cooking machine (that is, the machine might have a part which salts and a part which peppers) without providing the simplest functional description of the machine, given its other capacities. The simplest description might be that the machine has as one of its parts a seasoner, namely, a device which sprinkles whatever it is given as input onto whatever is in the skillet. To reflect this fact, our original functional analysis of omelet-making must be changed. Where it posited originally two constituent capacities, namely, the capacity to salt and the capacity to pepper, it now will have only one, namely, the capacity to season.

Conditions like (1)-(5) have an important role to play in scientific practice, although it is often left to the philosopher to make both the conditions and their role explicit. In debating the relative merits and demerits of competing functional deficit hypotheses, it is such conditions which are ultimately appealed to when a case is made for accepting or rejecting a candidate hypothesis. The appeal can be made in two ways. If there is strong evidence that a candidate hypothesis fails to meet one of the conditions, this failure constitutes grounds for rejecting the hypothesis. If we have neither sufficient evidence to claim that all the conditions have been met, nor sufficient evidence to rule out any of the candidate hypotheses on the grounds that it fails to meet some condition the case is more complicated. In the early stages of research on the neural realization of language, this situation will be typical for functional deficit hypotheses. This is because in the early stages of research, little or nothing will be known bearing on the question of structural adequacy implicitly required by condition (3). Even in this situation, however, the conditions for regarding a functional deficit hypothesis as established have a role to play. While we will not be in a position to claim that any candidate hypothesis meets all the conditions, we may be in a position to choose hypothesis H_1 over hypothesis H_2 on the grounds that, given the total available relevant evidence, the probability that H_1 meets the conditions is higher than the probability that H_2 does. The functional deficit hypothesis to which we assign the highest probability of meeting conditions (1)-(5) might be regarded as the <u>best available working hypothesis</u>.

Second, the underlying structural or causal change may transform one of the components into what is in effect a different functional component with a different constituent capacity. Or third, what amounts to a new functional component may be added to the system. The existence of a functional analysis of how a system or organism has some capacity C, which provides an explanation of the normal range of output resulting from exercise of C, is sufficient to provide an explanation of pathological output only in the first case. In the second and third cases, we must resort to information outside the scope of the functional analysis of normal behavior to account for the deviant output. Either a functional analysis specific to the abnormal organism must be provided, or the explanation of deviant output must occur at a non-functional level, i.e., at the level of causal mechanism. In our omelet-making system, for example, severe lack of attention by the human element of the system might result in C9 being modified to sprinkling detergent on the omelet. The result would be a clearly deviant output, but the nature of the deviance in this case can only be accounted for by knowing about the character of the change, in particular, about the properties of detergent in combination with egg mixture in combination with the human palate. Knowing about the normal functional system does not help us in this case to explain the output. Likewise, if between steps C6 and C7 a severe earthquake occurred, accounting for the nature of the resulting output (e.g., a gooey mess on the floor) would require knowing something about earthquakes and how they affect objects like pans and stoves.

It should be clear that when faced with a pathological or deviant output of a system or organism, we cannot always explain that deviance in terms of a deficit in one of the functional components required for normal operation of the system. However, what is important for our purposes is that although positing a deficiency in a normal functional component is not the only possible explanation of pathology, it is often the best available explanation. Where it does constitute the best available explanation, we are justified in inferring the existence of the functional deficit on the basis of the pathological evidence. To be more precise: given a functional deficit hypothesis H that the functional component in O with constituent capacity c is deficient, we are justified in regarding H as established on the basis of evidence of pathological behavior P just in case:

(1) P is correctly interpreted as the failure to exercise some capacity C;
(2) there exists a functional analysis A of how it is that O normally has C which includes c as a constituent capacity;
(3) A provides the best available explanation of how O has C;
(4) operation of the program described in A, minus the operation of the functional component with constituent capacity c, results in P as an output; and
(5) there is no better available explanation for the existence of P.

Note that (3) calls for the functional analysis to provide not only an explanation of how O has C, but also the best available explanation

How we ought to determine such probabilities is an important topic for further research. One possibility is that given two functional deficit hypotheses H_1 and H_2, each of which meets conditions (1), (2), and (4) and presupposes a functional analysis that constitutes a possible explanation of how O has C, we ought to assign the highest probability to the one which (a) constitutes the best available possible explanation of how O has C, and (b) is such that there is no better available possible explanation explanation for the existence of P.

The role of psychology, at least for the first step of the functional localization inference, should now be clear. The method of functional deficit analysis for inferring localization of function relies on the following sort of argument:

Localization Argument

H: There exists a deficit in O in functional component F_c.

L: There exists a lesion in O in brain structure S.

FLH: Therefore, under normal conditions, brain structure S would be the structural realization of F_c in O; or, brain structure S would have constituent capacity c in O.

To regard the functional deficit hypothesis H as established, or even the best available working hypothesis, requires the identification of the relevant functional components of the system as well as the explanation of the behavioral pathology in terms of a deficit in one of those components. Where the constituent capacity performed by the functional component is one required for cognitive skills, the existence of an adequate psychological theory will be required for both such identification and such explanation. Whether psychological considerations must be taken into account, in addition, in justifying the second step of the inference or in generalizing what has been learned about localization of function in one individual to all normal individuals must be left as a topic for another time.

Notes

[1]This paper is a shorter version of a chapter to appear in Carey, S., Chomsky, N., Forster, K.I., Kean, M., Klein, B.V.E., Resenfield, R., and Walker, E. Explorations in the Biology of Language. Montgomery, Vt.: Bradford Books, Publishers, 1978. I would like to acknowledge George Smith, Ned Block, Jeffrey Poland for helpful criticisms.

[2]It might be argued that such an explication is not helpful because the notion of capacity is just as obscure as the notion of function. This is not correct, however. For while Cummins does not do so, the notion of a capacity can be precisely defined in terms of an input-output description of the organism having that capacity.

Let us assume that there exists an optimal description of the behavioral repetoire of a system within a domain of the form:

$$f: (S \times I) \longrightarrow (S \times O),$$

i.e.,as an input-output device, where S represents the set of relevant internal states of the system (i.e.,those states that make a difference to the behavior), I, the set of relevant inputs, and O, the set of relevant outputs. The function f is then a mapping from all possible state-input pairs into all possible state-output pairs.

Given such a description we can define a capacity C of *S* in *D* as a subset of the set of all possible input-output pairs I x O which satisfies some naturalness condition. The point of the naturalness conditions is that most subsets of I x O will not correspond to what we ordinarily think of as a capacity.

S then has C provided that for each input-output pair <i,o> where <i,o> ε C there exists an internal state s, where s ε S such that f(s,i)= o.

[3] Cummins' account of functional explanation is by no means the classical account. For the classical account, see Hempel [4]. For an argument that the classical account is not correct, see Cummins [1].

References

[1] Cummins, R. "Functional Analysis." *Journal of Philosophy* 72(1975): 741-765.

[2] Cohen, Robert S. and Wartofsky, Marx W. (eds.). *Boston Studies in the Philosophy of Science*, Volume IV. Dordrecht, Holland: D. Reidel, 1969.

[3] Geschwind, N. "Anatomy and the Higher Functions of the Brain." In [2]. Pages 98-136.

[4] Hempel, C.G. "The Logic of Functional Analysis." In *Symposium on Sociological Theory*. Edited by Llewellyn Gross. New York: Harper & Row, 1959. Pages 271-307. (As reprinted in Hempel, C.G. *Aspects of Scientific Explanation*. New York: The Free Press, 1965. Pages 297-330.)

[5] Luria, A.R. "The Functional Organization of the Brain." *Scientific American* 222(1970): 66-79.

[6] Wernicke, Carl *Der aphasische Symptomencomplex: Eine psychologische Studie auf anatomischer Basis*. Bresland: Max Cohen and Weigert, 1874. (As reprinted as "The Symptom Complex of Aphasia: A Psychological Study on an Anatomical Basis." In [2]. Pages 34-97.)

Scientific Problems and Constraints

Thomas Nickles[1]

University of Nevada, Reno

What is the relation between a scientific problem and the constraints on its satisfactory solution? To what extent can problems be identified with (or how far are they determined by) such constraints? What types of constraints are there, and how do the various types function to determine problems and their solutions? I wish to consider these questions in the context of a single historical case, the blackbody radiation problem, 1859-1900. Since only the most condensed treatment of even one historical problem is possible, I select one with a familiar and accessible history (see [15],[16],[20],[22] for details). Despite the historical attention it has received, I know of no philosophical treatment of the blackbody case from a problem-oriented viewpoint.

More attention to constraints on problem solutions can help philosophers of science in several ways. Among them are: sufficient attention to constraints, their many varieties and functions, would have quickly exploded what became a nearly universal misconception of scientific problems as "empirical"--problems of theoretically explaining the data--rather than the highly conceptual puzzles many of them are. We shall find that most of the constraints defining the blackbody problem were not requirements of factual agreement with the data, or were far more than that. Once the misconception that problems are merely a matter of explaining the data is eliminated, it becomes apparent that problems themselves, and not just the theories which solve them, can be conceptually deep and interesting, and not all alike. This realization can help remove the theory-oriented bias that has plagued philosophy of science. Problems, their structures, relations, and varieties, have received negligible attention in comparison with the enormous effort devoted to the analysis of theories. This neglect has harmed the discussion of theories also, e.g., by making plausible the view that theories are all *essentially* explanations of observable data. Finally, attention to constraints will be central to the analysis and classifi-

---

PSA 1978, Volume 1, pp. 134-148

cation of scientific problems. Can scientific problems be fully analyzed in terms of constraints?

1. Historical Development of the Blackbody Problem

The history of the blackbody problem is a history of problem transformations and reductions. Qualitatively, the blackbody problem is the problem how and why bodies change colors as they are heated to ever higher temperatures. Somewhat more precisely, it can be resolved into a cluster of three related problems: (a) At various times, early and late, between 1859 and 1900, one empirical problem was simply to obtain data, or more accurate data, on the relationship of the energy radiated by "black" bodies to the frequency ν of the radiation and the temperature T, particularly in regions of high and low ν/T; (b) A second empirical problem was that of finding a simple formula to fit this data on the distribution of energy over frequencies and temperatures; (c) A third problem, partly empirical but largely conceptual, was to provide a sound theoretical derivation for the best available distribution formula. For simplicity, I often follow the practice of lumping together these (and the host of specific problems which they encompass) as "the" blackbody problem. This policy begs questions about the individuation of problems which ought to be discussed in a longer treatment of my subject.

Gustav Kirchhoff, who was interested in the chemical constitution of the sun and stars, first sharply formulated the blackbody problem and established its importance. In works of 1859, 1860, and 1862 ([17],[18],[19]) Kirchhoff established via equilibrium arguments that for each given ν and T, the ratio of emissive to absorptive power (the ratio of the rate at which a body emits energy to the fraction of the incident energy absorbed) is the same for all bodies ("Kirchhoff's law"). He then introduced the concept of a perfectly absorbing "black" body (absorptivity = 1), and deduced that: (1) the emissive power of blackbodies depends on ν and T alone and is independent of the material constitution of the bodies and of their environment, i.e., the energy density ρ of the emitted radiation is a function of ν and T alone; (2) blackbody radiation at temperature T is equivalent to "cavity radiation", the radiation field inside a black walled cavity, at T; and (3) cavity radiation depends only on ν and T and is independent of both the material nature and the shape of the cavity.

In establishing (1), Kirchhoff in effect "reduced" the blackbody distribution problem to that of determining the explicit form of a function of just two variables, $\rho(\nu,T)$, expressing the energy density of the emitted radiation as a function of frequency and temperature. More than twenty years later, (2) and (3) provided the theoretical basis for transforming both the experimental and the theoretical blackbody problems. Experimentally, a radiating cavity with a small opening is a much better source of blackbody radiation than a hot surface. Theoretically, (2) and (3) transformed the problem to one of cavity radiation and permitted simplifying assumptions about the structure of the walls

(Planck, see below) or obviated matter theoretic assumptions altogether (the approach of Rayleigh [30], Ehrenfest([6],[9]), Debye [4]).

In 1879 Stefan [34] conjectured from available data that the total radiation density is proportional to the fourth power of T, a result which Boltzmann [2] in 1884 derived theoretically from the Second Law of Thermodynamics. When extended explicitly to the entire frequency distribution, this Stefan-Boltzmann law imposed on the determination of $\rho(\nu,T)$ the condition that $\int_0^\infty \rho(\nu,T)d\nu \propto T^4$. Then in 1894 Wien [35] established another, very important condition when he showed that $\rho(\nu,T) = \nu^3 f(\nu/T)$. This was Wien's "displacement law" (which implies the Stefan-Boltzmann law). As Einstein ([11],p. 200) and many others have noted, the displacement law "reduced" the blackbody problem to that of determining the form of a function of a single variable, ν/T. The justification for speaking of problem <u>reduction</u> is obvious.

By linking the blackbody distribution for radiation to the Maxwell velocity distribution for molecules, Wien was able to propose a specific distribution law in 1896. Wien's [36] bold conjecture fit the available data, and so it tentatively solved the empirical blackbody problem of finding a formula. However, contrary to what one might expect on the classical Popperian view, a bold conjecture without serious competitors which "proves its mettle" against the data is not thereby adjudged satisfactory, pending the emergence of empirical anomalies. From the first Wien's distribution law was considered merely a conjecture, a guess, without theoretical foundation. Its very problem was that it was <u>too</u> conjectural. The distribution law cried out for the theoretical grounding accorded the displacement law (and even that grounding was refined by later work). The problem Max Planck set for himself at this point, in the 1890's, was precisely to furnish the theoretical justification which Wien's law lacked, to derive it from thermodynamics and electromagnetic theory, and to do so by developing a theory of passage to equilibrium for radiation analogous to Boltzmann's work on material gases. This aspect of Planck's problem was highly conceptual rather than narrowly empirical.

Kirchhoff had shown the theoretical equivalence of the blackbody problem to the cavity radiation problem, permitting the transformation or reduction of the former to its more tractable, cavity form. Planck realized that the initial distribution of the radiation field would not settle down to equilibrium (i.e., blackbody radiation) of its own accord, since distinct modes of vibration do not interact. Such interaction can be achieved only by introducing resonators which both absorb and emit radiation, he concluded in 1899 [28]; and so, appealing to Kirchhoff's result (3), which licensed arbitrary assumptions about the material nature of the cavity walls, Planck made the highly simplifying assumption that the walls contained an array of harmonic oscillators.[2] Technically, this transformed the problem into one of an idealized matter theory, but Planck, unlike most investigators, viewed the problem as one of theoretical principle and not one concerning the structure of matter.[3] In any case, he thereby was able to demonstrate that (4) $\rho(\nu,T) = 8\pi\nu^2/c^3 \cdot \bar{E}_\nu(T)$, where $\bar{E}_\nu(T)$, the only function remaining to be explicitly determined, is the average energy of a simple harmonic oscillator of frequency ν, as a function of T. This result further reduced the theoretical blackbody problem to that of determining how the mean energy of an oscillator depends on T. As is well known, the latter problem was "solved" by Rayleigh in 1900 [30], using the equipartition theorem; but plugging his result into (4) would have led directly to the ultraviolet catastrophe. Klein [20] has corrected the standard, textbook account of Planck's problem of 1900 as being the "catastrophe" problem of a conceptual blowup in classical physics. The problem as Planck perceived it apparently did not involve this conceptual dilemma at all; anyhow, the blackbody problem soon received an empirical twist. Experimental results of Lummer and Pringsheim [25] and (especially) of Rubens and Kurlbaum [32] now definitely refuted Wien's distribution law! Hence Planck's program to establish this "law" theoretically had to be redirected, if not abandoned, and the empirical blackbody problem itself was now both unsolved once again and more urgent than ever. Rather than being discarded, however, Wien's law continued to function as an important constraint on both the empirical and the theoretical blackbody problems, since it remained valid in the limit of high ν/T. It became a _limit condition_ or _limit constraint_. Rayleigh's equipartition paper of 1900 provided a limit constraint for the opposite end of the spectrum, but we know that Planck was unaware of (or at least ignored) Rayleigh's paper. Fortunately for Planck, the data of Rubens provided precisely the same constraint for the low frequency limit: $\bar{E}_\nu(T)$ becomes proportional to T.

By considering how his energy-entropy function must behave in the limits, and employing simplicity as an additional, heuristic constraint, Planck [27] arrived at his _Interpolationsformel_, his empirical distribution law, which he then, over a period of weeks, succeeded in grounding in a quasi-Boltzmannian statistical argument. Here begins the most fascinating phase of Planck's work, but here I must stop.

2. Toward a Typology of Constraints

What kinds of constraints are there, and how do they function? Let us take some first steps toward a typology, using the suggestive blackbody history as a guide. Obviously, I can make no pretension of completeness or finality on the basis of a single case. I shall term _reductive constraints_ those which "reduce" the problem in some way or are otherwise related to reduction. Wien's displacement law imposed a strong condition on any solution of the blackbody problem and itself reduced the problem task by partially solving it. In this case the partial solution was a partial determination of the form of the function , reducing the problem to that of determining a function of a smaller number of independent variables. Kirchhoff already had reduced the problem to a two variable problem when he established (1).

There are many ways in which to simplify an equation or in which to partially determine a function, and it seems natural (to me) to say that all of these "reduce" the problem in some way while defining it more sharply. Not all of these partial determinations reduce the number of independent variables; for instance, Planck's result (4), which reduced the problem to that of determining the mean energy of a harmonic oscillator did not. And the method of Lagrangian multipliers actually increases the number of unknowns and the number of simultaneous equations, while reducing (eliminating) the auxiliary conditions on a variational problem (see, e.g., [23], p. 47).

A second broad type of reductive constraint need not function as a constraint so much as a transformation of the problem (either the physical problem or the purely mathematical problem, or both) to another, more tractable form.[4] Reductive relationships are stronger than the mere modelling of one problem on another. Thus Kirchhoff's proofs (2) and (3) enabled Planck, Rayleigh, and others, in different ways, to reduce the blackbody problem to that of cavity radiation, more or less independently of matter theoretic assumptions. The problem was then more obviously (for Planck and a very few others) an "absolute" problem of theoretical principle rather than a problem in matter theory. It is natural to speak of such a case as one of problem *reduction*, however, only so long as the transformed problem appears more easily solvable, or is already solved. An example of the latter sort of case, during 1913-16, is Debye's [5] conjecture and Ehrenfest's [6] theoretical reduction of the quantization problem for an asymmetric oscillator to that for the simple harmonic oscillator, which Planck already had solved. It is important to note that problem reductions of this second type may occur without being part of a theory reduction. This type of problem reduction is not parasitic on (the concept of) theory reduction (see my [26] for details).

A third important type of reductive constraint is the *limit constraint*, which requires that an adequate problem solution reduce, in the appropriate limit or approximation, to an available *limit theory*--a theory or law which holds in the limit but is otherwise defective. Wien's distribution law and Planck's theory of it functioned as a limit law and theory, respectively, in Planck's thinking after they were refuted by Rubens and Kurlbaum. Rayleigh's result was a limit theory for the low end of the spectrum, although Planck did not employ it. Indeed, the data of Rubens and Kurlbaum, in the form in which Planck used them, functioned more as a limit law than merely as a set of brute facts to be explained, for it was the "limit lawlike" extrapolation from the data ($\bar{E}_\nu$ proportional to T) which functioned as a limit condition in Planck's thinking. Empirical constraints usually are more useful when given conceptual form. Indeed, often data are nearly useless, their relevance not apparent, until they are cast in conceptual form, such as a theoretical "model" which relates and accounts for them. Thus Einstein's [10] famous quantum theory paper of 1905 (the "photoelectric" paper) developed a particle theory of radiation, based on Wien's law, which Einstein knew to be defective. I think we can consider this the development of a limit theory (the solution of a limit problem) which would impose a powerful conceptual condition on any adequate theory of radiation and at the same time bring to bear empirical information about the production and transformation of light in a manner that was inconceivable before. Most important scientific problems are highly conceptual and not merely empirical.

There are several kinds of nonreductive constraints or conditions that may be imposed on solutions to problems of various sorts. There are the obvious logical and semantical constraints of consistency and clarity. There are general methodological constraints such as simplicity (which played both a heuristic and a justificatory role in Planck's work, especially in the final phase) as well as methodological constraints more specific to a field and to a particular research program. There may also be "metaphysical" constraints, such as Planck's concern with the "absolute," which play a role not only in selecting research problems but also in evaluating the initial plausibility or at least the interest of problem solutions (witness Planck's attitude toward special relativity). Metaphysical constraints sometimes find a more scientific expression--as in the universal, "absolute" nature of Kirchhoff's laws. The principles of thermodynamics and electromagnetic theory were also constraints on Planck's work. These functioned as more than _consistency constraints_--conditions with which the problem solution must be consistent. We might call them _foundational constraints_, since any fully adequate problem solution had to be well founded, i.e., derivable from them as a foundation. They contrast with _derivability constraints_: results such as the Stefan-Boltzmann law which must be derivable from any admissible problem solution. The former constraints tell you where the new theory (problem solution) must come from, and the latter tell you what results it must lead to (cf. [31] on problem vectors). Among derivability constraints are limit constraints (derivability in the limit, e.g., Wien's distribution law), some other kinds of reductive constraints (e.g., the displacement law), some consistency constraints (displacement law again, Stefan-Boltzmann), and empirical constraints. A given condition may fit two or more classifications. Also, we must distinguish strict logical derivability from approximate (or limit) derivability. This done, it is obvious that every strict derivability constraint is a consistency constraint, while no merely approximate derivability constraint is.

The constraints on a problem solution are flexible in varying degrees and must be if deep conceptual change is to be possible. Foundational constraints may change if the responsibility for handling the problem is shifted from one branch of science to another, and sometimes the very problem will be to find new foundations. Historically, even the requirement of logical consistency has been violated fruitfully many times (although inconsistency remains a defect); and a major problem solution which satisfies _all_ its constraints is a rarity indeed. Although no one recognized that fact immediately, Planck's quantum theory of 1900 violated both of his principal foundational constraints: the Second Law and electromagnetic theory. Whether and how it is possible to work _rationally_ from a set of constraints to a theory in violation of some of them is an interesting question--_the_ interesting question in this area. In short, there are no absolutely inflexible constraints in scientific research (although certain conditions may remain unbending in particular historical cases), and the set of consistency constraints on a problem may itself turn out to be inconsistent. In Planck's case, we can now see that there was no way in which to meet in full the demands of classical thermodynamics, electromagnetic theory, and the empirical data. My point should not be confused with Feyerabend's attack on "the consistency condition" ([12] and elsewhere). If Feyerabend's point is that it is not methodologically defensible to employ consistency constraints, then I must reject it. In the absence of consistency constraints (which include foundational and strict derivability constraints, _inter alia_), one can hardly formulate a highly conceptual problem. Despite his emphasis on the theoretical determination of observational language, Feyerabend too treats problems as primarily a matter of explaining empirical data.

3. Problems and Constraints

What then is the relation between scientific problems and the constraints on their solutions? Constraints obviously do much to determine a problem and its formulations: "... the problem confronting Planck could not have been defined in such detail much before 1895." ([13], p. 92). Do they determine the problem completely? My initial feeling that they do, at least up to the type of query made (the 'Why', 'How', or 'What' of the matter) has given way to the view that until several muddy issues are clarified, we cannot even say whether the thesis is obviously false or trivially true.

Could a problem be simply a set of _known_ constraints on an adequate solution? No, because knowledge may vary among persons working on the same problem. Every change of knowledge would be a change of problem; progress in solving a problem would change the problem. If we consider a problem to be a set of constraints, known _or_ unknown, then we can intelligibly speak of discovering a problem, of progress in formulating it more exactly, and of progress toward solving it. This conception of problems still fails, however, if constraints are taken to be linguistic entities, since every reformulation of a problem would change the problem. The corresponding view of theories as sets of sentences is now widely rejected for implying too fine-grained conditions of individuation. Like theories, it is more fruitful to think of problems as abstract structures than as linguistic entities. Of course, we may easily think of constraints as nonlinguistic, conceptual conditions expressible in more than one way, whence this difficulty does not arise. We can then say that an appropriate list of linguistic items expressing constraints is not the problem itself but a (partial) _formulation_ or representation of the problem. Like theories, problems may be reformulated in different ways, incompletely formulated, etc.

Some other difficulties in identifying problems with sets of constraints are: (i) distinct problems seemingly may be bounded by the same set of constraints; (ii) not all constraints are equally important; nor is simply weighting them enough, since (iii) the various constraints in the set do not all serve the same function. Thus a still better way of analyzing problems is as ordered or structured sets, in which the constraints are classified according to function and weighted as to importance and degree of flexibility. This suggests a matrix form of representation. A structured set or problem matrix would distinguish limit constraints from foundational and other consistency constraints, and so on. By examining the structure one could determine <u>from</u> what theory or principles (if any) a problem solution must be derived, <u>to</u> which results it must lead, and by which methods or techniques. This matrix would exhibit the <u>structure of the problem</u>. Since the set of constraints is the basis for constructive reasoning to a problem solution, we can expect that the structure of the solution (the structure of the resulting theory) will in some measure reflect the structure of the problem, and not just in its "starting" and "end" points. The requirement of correspondence in the limit, for example, may impose significant structural conditions on the new theory. Indeed, since the problem solution "must" satisfy the constraints, there is some point to saying that the structure of the theoretical solution is identical with the structure of the problem. This does not mean that the structure of the problem can be determined readily by analyzing the structure of the theory which solves it in a manner independent of the problem context, as did positivistic, logical analyses of theory structure. On the contrary, the problem was there first, and not just any "structural" feature of a theory is relevant to what we might call its "problem structure". (Of course, both problems and problem structures can be studied from either a logical or an historical point of view.)

An interesting objection to identifying the structure of a scientific problem and the (problem) structure of its solution and to analyzing a problem as a set of constraints is that many scientific problem solutions, especially the more innovative ones, violate some of their constraints. It follows that such solutions do not, strictly speaking, solve the problems they were designed to solve; instead, they solve "new" problems more or less closely related to the latter. I do not think this objection is fatal to the analysis, for is there not much truth in speaking this way? In a real sense, the classical blackbody problem was unsolvable, and it follows that its eventual "solution" really solved a replacement problem constituted by a revised set of constraints. Here again we run up against the matter of how to individuate and identify problems. Should we view the blackbody problem as a single problem which underwent quite radical transformations, or should we conceive it as a geneology (Toulmin's term) of distinct but closely related problems? It would be fruitless to individuate problems so finely that two investigators (or the same researcher at different stages of his research) rarely address the same problem, but we also want to avoid the opposite extreme, which is insensitive to both logical and historical distinctions. Here I must confine myself to a single point, *viz*., that the term 'problem' is ambiguous in just the ways 'theory' is: either term may designate a subject matter, field, or general approach, as well as something quite specific and concrete. I am inclined to think that 'the blackbody problem' more comfortably fits a subject matter than a *single* concrete problem spanning the years from 1859 to 1912.

Even if we consider solutions which seriously violate certain kinds of constraints to change the problem, however, the view of problems as structured sets of constraints continues to cause individuation embarrassments. A structured set is still a set, and altering a single element changes it to another set. We do not want *every* change in the importance or flexibility of a constraint to be a change of problem, for example. Yet conceiving problems to be some other type of abstract conceptual structure runs into trouble when the historical set of constraints is inconsistent, for logically impossible structures do not exist. Again, we meet the familiar but touchy difficulty of relating abstract, logical considerations to historical inquiry. While treating problems as abstract structures unrelated to human knowledge does avoid certain difficulties, it creates others. Are not scientific problems problems of human knowledge and cognition, after all? Another difficulty facing the general strategy of analyzing problems solely in terms of constraints is that while this strategy gives real point to the observation that adequately formulating a problem is half the solution, we might wonder why (on the analysis) a fully adequate formulation is not the *full* solution. If a partial solution (e.g., Wien's displacement law) further defines the problem, why not say that same of the full solution? But we obviously do not want to identify the problem with its solution. If we allow the problem and its solution to have the same structure, in some sense, a structure given by the constraint matrix, then it appears that the matrix or structured set of constraints alone cannot distinguish the problem from its solution and hence cannot fully capture the nature of the problem as a *problem*. A problem raises a question; it postulates something that needs resolution. The problem solution is the resolution and does not need it. Even if a question can be considered the set of admissible answers (as erotetic logicians do), how could it be identified with a single answer?

It is tempting and perhaps illuminating to relate my inquiry to the work of erotetic logicians. "Knowing what counts as an answer is equivalent to knowing the question," postulates Hamblin ([14],p. 162), who resists the reduction of questions to statements. Belnap and Steel elaborate: "The meaning of a question addressed to a query system is not to be identified with how the system processes the query ..., but rather it is to be identified with the range of answers that the question permits ... [with] what counts as an answer to the question, regardless of how, or if, any answer is produced." ([1], pp. 2f). Now in real scientific cases, the set of possible answers to a question or solutions to a problem are not laid out before you in advance, like *hors d'oeuvres* on a platter ([1], pp. 12, 17), so it may appear that the precise formal apparatus of erotetic logic is utterly inapplicable to real science. Inapplicable, I think, but not utterly! For in a qualitative sense, it is just the structured set of constraints, or a subset, which delimits the range of solutions and thereby gives meaning to a problem. Indeed, the set of known constraints does more than this, for it restricts the class of admissible solutions to those which are essentially correct, so far as the scientific community can then determine correctness.

Roughly speaking, the _presupposition_ of a scientific question (that information which logically must be true for the question to have an answer) is furnished by the relevant theoretical background, including foundational and certain other constraints. I say 'roughly' for several reasons. (a) Scientific theories and constraints cannot be known to be true and in any case do not logically guarantee that the problem has a solution. The guarantee is of the weaker sort that Kuhn has in mind when he says that a paradigm guarantees the solvability of puzzles. (b) As previously noted, there are interesting scientific cases, including the blackbody problem, in which the set of constraints turns out to be inconsistent, although no one may realize it at the time. Such scientific problems are both scientifically meaningful and important, yet not genuine questions in the formal, erotetic sense. Here we must go beyond available formal analyses of simple questions into the area of corrective answers--answers which correct the questions. (c) Because scientific problem solutions frequently are corrective or even "super corrective" (when they introduce new conceptual frameworks and do not merely correct items of misinformation within the old) and thereby violate one or more constraints, and for other reasons, there is no effective test of whether a proposed theory adequately solves a given problem. In the terminology of Reitman [31] and Simon [33], scientific problems are "ill defined". (d) The simple identification of scientific problems with questions in the erotetic sense requires examination. Kuhn ([21], Ch. 4) distinguishes problems and "puzzles," and Bromberger ([3], p. 61) makes a question a proper part of the problem situation he terms a "p-predicament". Moreover, a scientific "problem situation" may include a number of related but distinct questions or problems.

One final matter: Can distinct problems have the same relevantly structured set of constraints? If so, then there is more to a problem than can be expressed in terms of constraints. Although constraints do much to determine a scientific problem, it may seem that there are at least some cases in which they do not determine the question asked completely. To take the simplest, empirical example, let the primary constraints be an observational datum (description of a phenomenon) and some foundational constraints. We still may not know what the question is. Are we to explain _why_ the phenomenon occurred, explain _how it was possible_, given the foundational information, explain _what_ the phenomenon is by redescribing it in the prescribed theoretical terminology, or what? The structured set of constraints apparently fails to fully determine the question. The remaining component is roughly the W- part (Why, How, What, etc.). Let us call it the _query_.[5] In general a matrix or structured set of constraints does at least partly determine the query, since not just any query may be made of a given matrix. In our example the presence of foundational constraints probably excludes the 'What?' query, since they represent starting points of a derivation and not merely a basis for redescription. Does the fully specified and fully structured set of constraints determine completely the query and thus the entire question? Does it determine even what is left when we subtract the query from the problem situation?

Answers to these questions await the resolution of four complex issues in addition to difficulties mentioned above: (A) What kinds of things can be included in the set of constraints, so as not to trivialize the identification of a problem with a set of constraints? (B) How can these constraints be structurally organized so as to best display the structure of the problem? Answering this question will require a reasonably complete and precise taxonomy of constraints and a typology of scientific problems as well (see [24] and [31] for a beginning). (C) Can any clear sense be given to the idea of a "fully specified and fully structured set of constraints"? Since fuller specification of a problem (constraint structure) amounts to partial solution, will the fully specified constraints differ from a full solution? Observe that the full solution trivially constitutes the constraint: "Any adequate problem solution must be equivalent to me." (D) What are the relations of scientific 'why', 'how', 'how possibly', etc. questions? Constraints can be expected to distinguish some kinds of questions, at least, and a great many 'why', 'how', and 'how possibly' questions can be put in the form: "What is an example of a theory or model satisfying this constraint matrix?" (cf. [1], p. 84). But surely not all of them can be; moreover, there are other types of questions.

While I have not been able to determine how far scientific problems may be analyzed in terms of constraints, it is clear that further attention to this issue can illuminate neglected areas of scientific inquiry.

Notes

[1]Much of the research for this paper was done three years ago under a National Science Foundation grant. The paper was written with the aid of a grant from the Research Advisory Board, University of Nevada. I am grateful to both organizations for support and to Maurice Finocchiaro and a PSA referee for helpful, if mutually incompatible, criticisms.

[2]Ehrenfest [8] showed that Planck's mechanism still was inadequate to alter the initial radiation distribution, leading Planck to conclude, at the end of [29] that more complex matter theoretic assumptions were necessary.

[3]See [13] for details. I can no longer pretend that the blackbody problem(s) were the same for all interested parties; however, I shall continue to speak as if there were one objective problem, or cluster of problems, viewed somewhat differently from the perspectives of different

research programs, knowledge, and interests. In this paper I must gloss over the fact that the blackbody problem was a problem cluster treated quite differently by different people, and I ignore the question of how physical problems are related to mathematical problems.

[4]These problem equivalences and other transformations do function as constraints in the sense that a fully adequate solution must solve all forms of the problem in the same way (a kind of invariance under the transformation). In the history of relativity theory, where the transformations have a precise quantitative expression, this became a strong constraint indeed.

[5]What I label the 'query' overlaps but is not identical with Belnap's 'request'.

References

[1] Belnap, Nuel and Thomas Steel. *The Logic of Questions and Answers*. New Haven: Yale University Press, 1976.

[2] Boltzmann, L. "Ableitung des Stefan'schen Gesetzes betreffend die Abhängigkeit der Wärmestrahlung von der Temperatur aus der electromagnetischen Licht theorie." *Wiedemannsche Annalen der* Physik 22(1884): 291-294.

[3] Bromberger, Sylvain. "Science and the Forms of Ignorance." In *Observation and Theory in Science*. Edited by Ernest Nagel et. al. Baltimore: Johns Hopkins University Press, 1971. Pages 45-67.

[4] Debye, P. "Der Wahrscheinlichkeitsbegriff in der Theorie der Strahlung." *Annalen der Physik* Ser. 4, 33(1910): 1427-1434.

[5] --------. "Zustandgleichung and Quantenhypothese mit einem Anhang über Wärmeleitung." In *Vorträge über die Kinetische Theorie der Materie und der Elektrizität*. Leipzig: Teubner, 1914. Pages 19-60.

[6] Ehrenfest, Paul. "Over adiabatische veranderingen van een stelsel in verband met de theorie der quanta." *Verslag Koninklijke Akademie von Wetenschappen te Amsterdam* 25(1916): 412-433. (Reprinted in [7] as "On Adiabatic Changes of a System in Connection with Quantum Theory." Pages 378-399).

[7] ---------------. *Collected Scientific Papers* (ed.) Martin J. Klein. Amsterdam: North Holland Publishing Co., 1959.

[8] ---------------. "Über die physikalischen Voraussetzungen der Planck'schen Theorie der irreversiblen Strahlungsvorgänge." *Sitzungsberichte der Kaiserlichen Akademie der Wissenschaften, Wien, Mathematische - Naturwissenschaftliche Klasse*, 114(1905): 1301-1314. (Reprinted in [7]. Pages 88-101).

[9] ---------------. "Welche Züge der Lichtquanteshypothese spielen in der Theorie der Wärmestrahlung eine wesenthliche Rolle?" *Annalen der Physik* Ser. 4, 36(1911): 91-118. (Reprinted in [7]. Pages 185-212.)

[10] Einstein, A. "Über einen die Erzeugang und Verwandlung des Lichtes betreffenden heuristischen Gesichtspunkt." *Annalen der Physik* Ser. 4, 17(1905): 132-148.

[11] -----------. "Zur Theorie der Lichterzeugung und Lichtabsorption." *Annalen der Physik* Ser. 4, 20(1906): 199-206.

[12] Feyerabend, Paul. *Against Method*. London: New Left Books, 1975.

[13] Garber, Elizabeth. "Some Reactions to Planck's Law, 1900-1914." Studies in History and Philosophy of Science 7(1976): 89-126.

[14] Hamblin, C.L. "Questions." Australasian Journal of Philosophy 36(1958): 159-168.

[15] Jammer, Max. The Conceptual Development of Quantum Mechanics. New York: McGraw-Hill, 1966.

[16] Kangro, Hans. Vorgeschichte des Planckschen Strahlungsgesetzes. Wiesbaden: Franz Steiner Verlag, 1970. (Translated as Early History of Planck's Radiation Law. London: Taylor & Francis Ltd., 1976).

[17] Kirchhoff, Gustav Robert. "Über den Zusammenhang zwischen Emission und Absorption von Licht und Wärme." Monatsberichte der Akademie der Wissenschaften zu Berlin. December(1859): 783-787.

[18] ------------------------. "Über das Verhältnis zwischen dem Emissionsvermögen und dem Absorptionsvermögen der Körpe für Wärme und Licht." Poggendorffs Annalen der Physik und Chemie. 109(1860): 275-301, translated in Philosophical Magazine 20(1860): 1-21.

[19] ------------------------. Untersuchung über das Sonnenspectrum und die Spectren der chemische Elemente. Berlin: Dummler, 1862.

[20] Klein, Martin. "Max Planck and the Beginnings of the Quantum Theory." Archive for History of Exact Sciences 1(1962): 459-479.

[21] Kuhn, Thomas. The Structure of Scientific Revolutions. Chicago: University of Chicago Press, 1962.

[22] ------------. The Black-Body Problem and the Quantum Discontinuity, 1894-1912. Oxford: Oxford University Press, in press.

[23] Lanczos, Cornelius. The Variational Principles of Mechanics. Fourth Edition. Toronto: University of Toronto Press, 1970.

[24] Laudan, Larry. Progress and its Problems. Berkeley: University of California Press, 1977.

[25] Lummer, O. and Pringsheim, E. "Über die Strahlung des schwarzen Körpers für lange Wellen." Verhandlungen der Deutschen Physikalischen Gesellschaft 2(1900): 163-180.

[26] Nickles, Thomas. "Theory Generalization, Problem Reduction, and the Unity of Science." In PSA 1974. (In Boston Studies in the Philosophy of Science, Volume XXXII). Edited by R.S. Cohen et. al. Dordrecht: D. Reidel, 1976. Pages 33-75.

[27] Planck, Max. "Über eine Verbesserung der Wienschen Spektralgleichung." <u>Verhandlungen der Deutschen Physikalischen Gesellschaft</u> 2(1900): 202-204.

[28] -----------. "Über irreversible Strahlungsvorgänge." <u>(Monatsberichte) Sitzungsberichte der Könglich - Preussischen Akademie der Wissenschaften, Berlin, Mathematik - Naturwissenschaftiche Klasse</u>. (1897): 57-68; 715-717; 1121-1145. (1898): 449-476. (1899): 440-480.

[29] -----------. <u>Vorlesungen über die Theorie der Wärmestrahlung.</u> Leipzig: Johann Barth Verlag, 1906.

[30] Rayleigh [Lord]. "Remarks upon the Law of Complete Radition." <u>Philosophical Magazine</u> 49(1900): 539-540.

[31] Reitman, Walter. <u>Cognition and Thought</u>. New York: John Wiley, 1965.

[32] Rubens, H. and Kurlbaum, F. "Über die Emission langwelliger Wärmestrahlen durch den schwarzen Körper bei verschiedenen Temperaturen." <u>(Monatsberichte) Sitzungsberichte der Könglich - Preussischen Akademie der Wissenschaften, Berlin, Mathematik - Naturwissenschaftiche Klasse</u>. 1900: 929-941.

[33] Simon, Herbert A. <u>Models of Discovery and Other Topics in the Methods of Science</u>. Dordrecht: D. Reidel, 1977.

[34] Stefan, Josef. "Über die Beziehung zwischen der Wärmestrahlung und die Temperatur." <u>Sitzungsberichte der Kaiserlichen Akademie der Wissenschaften, Wien, Mathematische - Naturwissenschaftliche Klasse</u>, 79(1879): 391-428.

[35] Wien, Willy. "Temperatur und Entropie der Strahlung." <u>Wiedemannsche Annalen der Physik</u> 52(1894): 132-165.

[36] -----------. "Über die Energievertheilung im Emissionsspectrum eines schwarzen Körpers." <u>Wiedemannsche Annalen der Physik</u> 58 (1896): 662-669.

Discoveries and The Emergence of New Fields in Science

Lindley Darden [1]

University of Maryland, College Park

Understanding the growth of scientific knowledge is a central problem in philosophy of science today, a part of which takes the form of understanding how new knowledge arises and how it becomes unified with existing knowledge. As a way of tackling this problem, one may ask: how do new fields arise in science? And how is the knowledge in a new field related to that in other fields?

In recent work I characterized a scientific *field* as an area of science consisting of the following elements: a central problem, a domain of items taken to be facts related to that problem [18], general explanatory factors and goals providing expectations as to how the problem is to be solved, techniques and methods, and concepts, laws and theories related to the problem which attempt to realize the explanatory goals ([6], p. 44). Also in that work, my coauthor and I discussed *interfield theories* which serve to link two fields by postulating specific kinds of relations between them.

The present paper will extend that analysis by examining features of the emergence of new fields in science. Case studies of the emergence of cytology (the study of cells) and biochemistry (the study of chemical processes in biological systems) will provide data about features of emergence. These cases will show that the first step was the discovery of a new entity in a model system, namely a system which, it is claimed, provides information that may be generalized to other systems. In each case the generalization was made that a new *kind* of entity had been discovered; the generalization constituted a new theory. With investigations of the nature of the entity and tests of the generalization, a new domain was formed, eventuating in the emergence of a new field. An important reason for making a general claim about the discovery existed: the generalization provided a solution to an old problem.

PSA 1978, Volume 1, pp. 149-160

1. Cytology.

Although Hooke first observed pores in cork and called them "cells" in his *Micrographia* of 1665, cytology did not emerge as a field until the 1830s. This emergence was a result of Schwann's proposal of the cell theory in which he claimed that cells constitute the elementary parts of all animals and plants and, further, all cells have a similar mode of development involving the appearance of, first, a nucleus, and then a cell wall. The claim that all cells have a nucleus and cell wall provided new and reliable criteria for identifying cells, thereby distinguishing them from smaller globules within the cells or larger fibers formed by groups of cells. A new kind of entity could thus be clearly differentiated from other parts of organisms. An old problem -- what is the basic unit of organization of living things? -- found a proposed solution in the cell theory. Testing the generality of the theory became the central problem for the new field of cytology.

Schwann's statement of the cell theory in 1839 was made possible by several previous developments. Microscopes were improved in the 1830s: effects due to spherical aberration were eliminated, thereby eliminating many of the circular structures microscopists had thought they were observing ([10], ch. 1 and [1], p. 117). Also, interest in the microscopic structures of plants increased in the early nineteenth century ([15], p. 256); plant material proved to be an excellent source for observation of cells.

Prior to the work of the botanist Schleiden, who had a significant influence on the zoologist Schwann, numerous studies of plants included discussion of their basic structures. Since plants have cell walls, their cellular structure is easily observable. However, early proposals had been that cells were not separate units but were merely holes in the network produced by linked walls; "lacework" was one analogy used. Among the first botanists to demonstrate that plant cells are separable units was Treviranus (in work between 1805 and 1811) who succeeded in pulling cells apart after sectioning plant tissue ([3], pp. 161-162). Dutrochet in 1837 improved the technique by use of nitric acid ([3], p. 162). Robert Brown, investigating orchids in 1833, proposed that the nucleus was a structure found in most cells of that species and in those of several other species ([2], p. 101).

Thus, when Schleiden came to his investigations of embryonic development in plants, separable cells had been observed and nuclei within them noted in several species of plants. He extended previous work by showing that the nucleus is found in the developing cells of numerous flowering plants; he then generalized that it is found in all. Furthermore, he stressed the presence of the nucleus as an essential stage in the development of cells. Schleiden's view of development was to have an important influence on Schwann and subsequent workers in the field of cytology for decades. Studying development in the embryonic sac of plants after fertilization (the model system), Schleiden observed a process which came to be called "free-cell formation." In his words: In a previously homogeneous gum within the embryonal sac (a large cell), "sharply defined granules" appear and "soon afterwards the cytoblasts [nuclei] appear, looking like granulous coagulations around the granules. ... so soon as the cytoblasts have attained their full size, a delicate transparent vesicle rises upon their surface. This is the young cell." ([16], p. 238). He explicitly claimed that this process is found universally in developing plants ([16], p. 247).

Schleiden also discussed cell formation in more mature parts of the plant. Before presenting his observations, he outlined the possible ways in which mature cell formation might occur: "Where are the new cells formed? Here three instances comprise all possible replies. Namely, the new cells are either formed outside on the surface of the entire previous mass, or in its interior; and in that case again either in the intercellular spaces or in the cells themselves; quartum non datur." ([16], p. 252).[2] The latter possibility, namely that new cells form within old ones, was the process which Schleiden claimed to have observed.

When Schleiden communicated his findings about cells and their development in plants to Schwann, who was investigating development in tadpoles, Schwann was struck, he says, by the similarity of Schleiden's results and his own observations ([17], p. xiv). Although some zoologists had previously noted cell-like structures in animals, especially in the blood where they are easily observed, and some had even compared observations of animals to those made of cell-like structures in plants, the diversity of animal tissues and the lack of clearly differentiated cells in most of them had made a general form of organization in animals and plants appear unlikely. Schwann, however, did not base his claim of similarity on the forms of adult animal tissues but on the mode of development. Thus, he stated the cell theory: "there is one universal principle of development for the elementary parts of organisms, however different, and...this principle is the formation of cells." ([17], p. 165). He examined numerous different tissues and referred to the work of others in different species to show that, even in tissues that later lose their separate cellular structure, their early development goes through a cell stage. He, with Schleiden, emphasized the importance of the nucleus as a stage in cell formation and claimed to observe a cell wall in animals similar to that in plants. However, he differed from Schleiden by extending "free cell formation": Schwann claimed that cells formed in animals not only in the fluid within cells but also in the fluid between cells (one of Schleiden's rejected possibilities) ([17], p. 165).

Schwann explicitly says that the cell theory is a general claim uniting the animal and plant kingdoms: "The object of the present treatise [in which he states the cell theory] is to prove the most intimate connexion of the two kingdoms of organic nature, from the similarity in the laws of development of the elementary parts of animals and plants." ([17], p. ix). This generalization had not been made by his predecessors; Schwann was thus the first to state the cell theory.

The new field of cytology emerged as a result of Schleiden's and

Schwann's work. They recognized that a single nucleus ([2], pp. 102-3) and the cell wall were essential components of the cell (although Schwann did concede that a few non-nucleated cells are found). Thus cytologists could reliably recognize cells and distinguish them from smaller and larger structures that had been lumped together by previous workers. A new domain was thus formed consisting of information about cells. Furthermore, the cell theory provided a general claim that all cells develop in the same way, making cell development a problem for cytology to investigate. The process of free cell formation in the embryonic sac which Schleiden observed proved to be unusual; it was not a model for understanding cell formation in general. Even though free cell formation was proved wrong in favor of cell division within two decades, the original claim served the important function of unifying studies of cell formation, that is, it created expectations that a single mode of development for all cells existed. Thus, when cell division was discovered in one-celled organisms and in some tissues of animals and plants, it was plausible to claim that that represented the general mode of development of new cells.

Not only did claims about the development of cells change as the field progressed, claims about the nature of the cell also underwent alteration. The distinction between cell walls and cell membranes was discovered. Although the visible cellulose cell walls of plants were shown not to be present in animal cells, the presence of a functional membrane proved to be a common feature of all cells. Further work also showed that protoplasm, a fluid within cells, is another important component of cells, along with the nucleus and membrane. Investigation of microscopic organisms showed that many were one-celled, thus extending the applicability of the cell theory to the microscopic level.

The new field of cytology, at the outset, unified botany and zoology, and later encompassed microbiology also. The unification of botany and zoology which the emergence of cytology in the 1830s accomplished was not the result of uniting items in previously established domains of those two branches of science (as an interfield theory does [6]). Instead, a new domain was formed consisting of data about cells that was based on new observations in animals and plants. Cytology thus became a field at a new level of organization common to animals and plants (and later all living things). Since the organization of tissues and gross morphological structures in animals and plants is quite different, the studies of these structures larger than the cell remained within zoology and botany, although the study of different types of cells in different tissues affected those separate studies. The old problem -- what is the basic unit of organization of living things? -- had found a new solution -- cells.

2. Biochemistry

Biochemistry emerged as a field around 1900 in part as a result of the proposal of the enzyme theory, namely the general claim that life processes are a result of chemical changes catalyzed by specific intracellular enzymes. Eduard Buchner's discovery of "zymase," an enzyme within yeast cells that causes fermentation in cell-free extracts, was a discovery of a new type of entity -- an intracellular enzyme. The generalization that all life processes can be traced to other enzymes within cells provided a program for the new biochemistry. The enzyme theory and the new field of biochemistry replaced the old protoplasm theory of life processes proposed by the physiological chemists of the nineteenth century. No longer was the study of life processes the study of a living substance; instead it became the study of chemical entities and reactions. Investigations to test the enzyme theory thus generated a new domain about intracellular enzymes; chemical techniques were used to study reactions in biological systems. A new chemical answer was provided for the old problem, what is the nature of life processes?

Robert Kohler, in "The Enzyme Theory and the Origin of Biochemistry," [13] argues that the enzyme theory was the single most important development leading to the emergence of biochemistry as a distinct discipline.[3] Before turning to that development, which is the most important for our purposes here, a related line of development occurring at the same time should be mentioned. In addition to the study of enzymes, other biological molecules were being investigated using chemical techniques. Although chemical analyses of parts of organisms had been done since the days of animal chemistry in the early part of the nineteenth century ([8], p. 7), the early analyses had produced such amorphous results that little could be made of them. The development of organic chemistry in the mid-nineteenth century provided important techniques of analysis and synthesis of carbon compounds. Although many organic chemists dissociated themselves from the study of complicated biological phenomena, Emil Fischer and a few others successfully applied organic techniques to the study of fats, carbohydrates and proteins as well as the action of enzymes [7]. Thus, alongside the development of the enzyme theory was development of an understanding of the chemical nature of other biological molecules, many of which were reactants in enzymatically catalyzed reactions.[4]

Enzymes, or "ferments" as they were often called, were known to nineteenth century biologists and chemists. However, prior to Buchner's discovery in 1897, the known enzymes were extracellular, *e.g.*, digestive enzymes such as pepsin. Intracellular processes such as fermentation were known to take place only in the presence of living cells, as Pasteur had so ably demonstrated. Attempts to find intracellular enzymes had failed. The prevailing theory in the late nineteenth century was that life processes were carried out by a living substance within cells, protoplasm. Some speculation about its chemical nature included the proposal that it was a large molecule with side chains responsible for specific processes ([13], pp. 184-185).

Eduard Buchner's discovery of zymase was a somewhat fortuitous one: although he had studied chemical aspects of fermentation in earlier work, that was not the focus of the work that led to the discovery of zymase. His brother Hans Buchner was studying the immunological properties of proteins within cells and wanted a gentle technique for extracting them. Hans and an assistant developed a method using sand to break open the cells; they then added a separating agent to which the cell juices adherred. In order to stablize the mixture, various preservatives were tried, including glucose. Eduard, who had been following the progress of experiments, went to visit the lab and noticed the bubbles coming from the extract and glucose. He immediately recognized the process as fermentation and began a study of its cause, thus finding zymase, an enzyme within yeast cells which causes fermentation [11].

Although there was controversy at first about the nature of this agent (was it perhaps bits of living protoplasm?) and the repeatability of the results [12], it was hailed by such biologists as Emile Duclaux at the Pasteur Institute as opening up "a rich field of research." (Quoted in [12], p. 338). Buchner's result, coupled with the findings of others in the 1890s that enzymes were not only destructive agents but also functioned as oxidases and synthetases, provided an empirical basis for the enzyme theory by 1900 [13]. A new domain of investigation of these new intracellular enzymes was thus begun. Testing the general claim that all life processes are the result of enzyme activity and finding the chemical nature of enzymes occupied biochemists for the decade following 1900. The enzyme theory was eclipsed in the period from 1913 to 1929 with the rise of colloid chemistry ([14], [9]). This approach emphasized amorphous suspensions rather than specific substances undergoing specific reactions. However, in the 1930s, with the crystallization of specific enzymes and their identification as proteins, the enzyme theory and the study of enzymatic pathways came to dominate the field.

The emergence of biochemistry served to link chemistry and biology with the formation of a new domain consisting of information about specific biological molecules and reactions. The important techniques were chemical with the addition of new methods such as that of the Buchners to handle the fragile molecules more gently. A new level of organization between molecules of organic chemistry, on the one hand, and cells and cell organelles of cytology, on the other, was established by the enzyme theory. It was closer to the chemical level than either the amorphous breakdown products studied by the early animal chemists or the living protoplasm postulated by the physiological chemists. Consequently, chemical techniques and characterizations were more the mark of the new biochemistry than they had been of its nineteenth century predecessors. Biochemistry forged a tighter link between biology and chemistry, as the enzyme theory provided a chemical answer to the old problem of the nature of life processes.

3. Conclusions

The two cases of emergence of new fields have a number of features in common which I will first enumerate and then discuss in more detail.

1. The discovery of a new entity is made by someone using a new instrument or technique who recognizes the important aspects of the phenomenon being investigated.
2. A claim is made that this new entity is of a kind found more generally; this generalization is a new theory.

3. The theory proposes a new solution to an old problem.
4. This solution is recognized as a promising one and a new line of research using the new instruments or techniques begins to test the extent of the generalization and determine more about the new entity.
5. A new field emerges as a new domain is generated by this new line of research.

Discoveries of new entities (Step 1) are not simply a matter of looking and seeing, even in such an observational field as cytology. The person making the discovery must pick out certain features as essential, or at least important, and improved instruments or techniques may be important. Although others had observed the cell nucleus, Schleiden was the first to make it an essential part of the developing cell and Schwann was the first to claim that the nucleus and cell wall were essential to all developing cells. Their work was aided by improved microscopes. Buchner, as a result of past work, was able to recognize the process of fermentation; his brother was attempting to study intracellular proteins and thus developed a technique for finding them, although enzymes were not what he was searching for. Thus, asking certain questions requiring new techniques to answer them or focusing on certain features as essential may be necessary for an "observation" to be made.

In these cases, the step of generalization (Step 2) is the step of constructing a new theory. The discovery of a new entity in one system may provide the basis for generalizing that is of a kind found more generally, if the system is an adequate model for others. The choice of model organisms is an important part of biological research, but it is outside the scope of this paper to discuss criteria by which good models may be distinguished from bad ones. Unfortunately, hindsight is often the best way to judge whether the findings in one case are applicable to others. But, any finding is potentially generalizable. Plausibility is given to the generalization since at least one instance has been found. Further testing is, however, immediately in order.

It is sometimes remarked that discoverers do not see the importance of their discovery. That remark, within the framework of this analysis, indicates that the discoverer did not make the transition from Step 1 to Step 2. If a scientist does not believe that the entity discovered in one system may be of a more general type, then he or she is the empirical scientist who is unable or unwilling to be the theoretician. On the other hand, if a speculative generalization is proposed without a discovery's having been made (Step 1), the theory may be less plausible since it will lack the empirical base that a discovery provides. The scientist may then be the theoretician generalizing on the basis of little evidence; the theory may not be plausible enough to warrant testing or no techniques may be available to do so.

These cases of theory construction do not follow either of two patterns which philosophers of science have discussed as means of "discovering" a new theory. ('Discovery' thus far in the paper has been used to refer to new empirical findings; I prefer "theory construction" to "discovery of a new theory.") They are not instances of either induction by simple enumeration or merely guessing a hypothesis. Some inductive evidence existed for both generalizations, but neither the number nor the number of different kinds of observations was large at the time the theories were proposed. Thus, they were not simply inductive generalizations made on the basis of observing numerous instances. On the other hand, one (or more) instances of the kinds of entities postulated by the theory had been discovered. Consequently, the "theoretical" entity was not merely hypothetical, not merely the product of someone's reverie before the fire. Debate did not occur as to whether the entity really existed; at least one instance had already been discovered.

A pattern different from either simple induction or hypothesis-guessing is discernible here. The discovery of a new entity provides a weak empirical basis for the construction of the theory: since one (or a few) has (have) been found, it may be of a kind found more generally. Consequently, the discovery makes construction of the theory plausible. But testing is needed to determine the extent of the generalization and further work is also needed to determine more about the nature of the entity. The theory is developed as this testing and further work proceeds.

The term 'theory' has numerous senses in science, two sharply contrasting of which are (1) a generalization that is not yet proved, i.e., is hypothetical and (2) a well-confirmed set of postulates that adequately explain a domain. The usage of 'theory' in the case of 'cell theory' and 'enzyme theory' is in the former sense. In other words, these theories are, at the emergence of the fields, programs for future work; they are hypothetical claims in need of testing, not well-confirmed, completed explanatory theories.[5] And, as they are elaborated and confirmed, their theoretical (hypothetical) nature is removed. Curiously, in these cases elaboration and confirmation does not produce a theory in the second sense; instead the term 'theory' ceases to be used when referring to knowledge about cells and enzymes. The domains consisting of information about cells and enzymes become accepted bodies of facts and are incorporated into biological knowledge. It is a feature of Shapere's concept of domain [18] that items which were previously theoretical can become, at a later stage, domain items. (Further discussion of the fate of theories within successful fields, ones that have solved their central problem, is outside the scope of the present discussion of their emergence; it will be examined in future work.)

In both the cases examined, the theories proposed new solutions to old problems (Step 3). The recognition of the problem may be important in leading to the generalization. If, for example, Schwann had not been seeking a basic unit of organization for all animals and plants, he would not have generalized that the structures he was observing in the tadpole were examples of such basic units. If the problem of the nature of life processes had not been posed, then the construction of the enzyme theory would have been unlikely. Since the generalizations provided solutions to old problems, it was plausible to make the generalizations. Thus, the theories in these cases satisfied two different plausibility conditions: at least one instance had been found and the generalizations provided solutions to old problems. How these general problems arise and why some scientists are better able to recognize them are important questions to ask but are outside the scope of the present analysis.

A new line of research (Step 4) is made possible by the new technique or methods that resulted in the discovery. There is thus the promise of being able to determine the extent of the generalization. (For a more detailed discussion of promise in the emergence of another new field see [5].) Furthermore, the new techniques also allow more information about the new entity to be found. This new line of research generates a new domain, marking the emergence of a new field (Step 5).

The generalization, that is the new theory, becomes an *intrafield* theory, a theory within a field that proposes a solution to its central problem and is the object of research in that field. Once the field is established, once more is known about the kind of entity under investigation, then questions about the relation of this entity to entities in other fields can be asked. In other words, *interfield* connections can then be sought.

4. Implications of this analysis for other cases

This study has examined two cases of the emergence of new fields that have proved successful in solving their central problems. How general are the conclusions for characterizing aspects of the emergence of other new fields? Some other cases undoubtedly follow the same pattern: molecular biology, although more complicated because of interfield links, emerged as a field in the 1930s as a result of the discovery that important biological molecules are very large macromolecules (the new kind of entity) requiring techniques from physics to investigate their structure. Other cases do not result from the discovery of a new kind of entity but from the discovery of repeated patterns in data; new kinds of entities may be postulated to account for the data. For example, Dalton's atomic theory accounted for the laws of chemical combination. In the emergence of genetics, genes were postulated to account for patterns of inheritance of characteristics. In the atom and gene cases, the entities were *postulated*, not *discovered* experimentally; as a result, debate about their "reality" was a feature of the new fields that emerged to test the general claims made about them. Darwin's theory of the origin of species postulated, not a new kind of entity, but a new kind of process -- selection. The claim that selection is the general cause of the origin of species generated a promising line of research which continues today.

In the above cases of the emergence of new fields, sufficient evidence existed to make the generalization plausible and techniques existed for pursuing promising lines of research. Debate about whether a line of research should be pursued may center on whether there is sufficient evidence for its general claims to make it plausible, whether techniques exist to test the claims, or whether the generalization does provide a solution to a problem. Analysis of the features of the emergence of previously successful fields may thus aid us in assessing the promise of new candidates. At the very least, it helps us to understand aspects of the growth of scientific knowledge in the past.

Notes

[1]This work was supported by NSF Grant SOC 77-23476. I wish to thank Nancy Maull, Frederick Suppe, Stephen Brush, Dudley Shapere, Kevin Mutchler and Jane Voytek for comments on an earlier draft.

[2]For a discussion of the role of enumerating possibilities prior to choosing the most plausible postulate of a theory, see my discussion of theory construction in [4].

[3]Kohler uses the term 'discipline' to refer to both the cognitive aspects of a field (or a discipline in the sense of Toulmin in [19]) and the institutional and sociological aspects of a profession (in Toulmin's sense [19]).

[4]For an alternative to Kohler's analysis that places less emphasis on the development of the enzyme theory and more on these and other lines of development see Fruton [9].

[5]The view that theories are at first only vague ideas in need of much elaboration was suggested to me, prior to examination of these cases, by Dudley Shapere.

References

[1] Barker, John R. "The Cell-theory: a Restatement, History, and Critique, Part I." Quarterly Journal of Microscopical Science 89(1948): 103-125.

[2] --------------. "The Cell-theory: a Restatement, History, and Critique, Part II." Quarterly Journal of Microscopical Science 90(1949): 87-108.

[3] --------------. "The Cell-theory: a Restatement, History, and Critique, Part III." Quarterly Journal of Microscopical Science 93(1952): 157-190.

[4] Darden, Lindley. "Reasoning in Scientific Change: Charles Darwin, Hugo de Vries and The Discovery of Segregation." Studies in History and Philosophy of Science 7(1976): 127-169.

[5] ---------------. "William Bateson and the Promise of Mendelism." Journal of the History of Biology 10(1977): 87-106.

[6] Darden, Lindley and Maull, Nancy. "Interfield Theories." Philosophy of Science 44(1977): 43-64.

[7] Fischer, Emil. "Synthetical Chemistry in its Relation to Biology." Journal of the Chemical Society, London 91(1907): 1749-1765.

[8] Fruton, Joseph. Molecules and Life. New York: Wiley-Interscience, 1972.

[9] --------------. "The Emergence of Biochemistry." Science 192(1976): 327-334.

[10] Hughes, Arthur. A History of Cytology. New York: Abelard-Schuman, 1959.

[11] Kohler, Robert. "The Background to Eduard Buchner's Discovery of Cell-Free Fermentation." Journal of the History of Biology 4(1971): 35-61.

[12] --------------. "The Reception of Eduard Buchner's Discovery of Cell-Free Fermentation." Journal of the History of Biology 5(1972): 327-353.

[13] --------------. "The Enzyme Theory and the Origin of Biochemistry." Isis 64(1973): 181-196.

[14] --------------. "The History of Biochemistry: A Survey." Journal of the History of Biology 8(1975): 275-318.

[15] Sachs, Julian von. History of Botany. Oxford: Clarendon Press, 1890.

[16] Schleiden, M.J. "Beiträge zur Phytogenesis." Müller's Archiv für Anatomie,Physiologie, und Wissenschaftliche Medecin (1838): 137-176. (As reprinted as "Contributions to Phytogenesis" in the appendix to the reprint of [17]. Pages 231-268.)

[17] Schwann, T. Mikroskopische Untersuchungen über die Übereinstimmung in der Struktur und dem Wachstum der Thiere und Pflanzen. Berlin: G.E. Reimer, 1839. (As reprinted as Microscopical Researches into the Accordance in the Structure and Growth of Animals and Plants. (trans.) Henry Smith. London: Sydenham Society, 1847 and New York: Kraus Reprint, 1969.)

[18] Shapere, Dudley, "Scientific Theories and Their Domains," in The Structure of Scientific Theories. 2nd. ed. Edited by F. Suppe. Urbana: University of Illinois Press, 1977. Pages 518-565.

[19] Toulmin, Stephen. Human Understanding. Volume 1. Princeton: Princeton University Press, 1972.

Part V

Explanation

Explanation and Subsumption

Robert Cummins[1]

University of Wisconsin-Milwaukee

1. The Claim

Quine once claimed that modal logic derived from a use-mention mistake ([8], p. 196). In this paper, I will peddle an equally outrageous claim: the thesis that subsumption is an essential goal of fundamental explanatory theorizing rests on a failure to distinguish explaining what made x happen from explaining why we believe what we do about what made x happen. I don't mean to suggest that the distinction wasn't noticed (except perhaps by Hume); only that it wasn't vigorously enforced. There have always been philosophers who assimilated a belief to its justification with malice aforethought.

2. Strategy

Everyone knows that subsumption is not sufficient for explanation: the length of a pendulum is not explained by appeal to its period and the pendulum law, and emission spectra are not explained by Balmer's formula. But the spirit of the deductive-nomological account of explanation lives on in the belief that subsumption is an essential feature of an explanatory theory. Contextualists have argued with some plausibility that subsumption is seldom necessary to particular explanations, and it has been argued by some (myself in [1]) that subsumption is at best incidental to functional explanation. But even when these points are conceded, the majority opinion, I suspect, is still that non-subsumptive explanation is really a stand-in or under-study: if subsumption under law isn't waiting in the wings somewhere, then the play is a farce. Thus, to attack the claim that subsumption is necessary to explanation, it is not enough to provide an explanation that does not involve subsumption--e.g., the car turned over because the left front tire blew out--for defenders of the subsumptive model have always claimed that such cases are cases of incomplete explanation: promissory notes, but not hard cash. This line was taken by Hempel and Oppenheim [5], and it has some plausibility: when we trot out a "because" but do not or cannot

PSA 1978, Volume 1, pp. 163-175

trot out a law, we typically suppose nevertheless that there must be a law of the appropriate sort, even if we don't know what it is. It is therefore open to the subsumptivist to claim that it is only the promise of the missing law and a corresponding subsumption that gives the non-subsumptive explanation point. If we want to block this sort of move, we must set the stage in such a way as to guarantee that there are no unstated laws in the wings. This is what I will try to do: my strategy will be to describe a theory that (1) if true, would explain every event it can describe at all, (2) does not explain by subsumption, (3) is fundamental in that it is understood to be irreducible. If I can do this, I will have shown that the possibility of a thorough-going explanatory theory worthy of the name does not depend for its explanatory import on the possibility of subsumption under law.

3. ANA

I call the fanciful theory I am about to describe ANA because the explanatory strategy it exploits is analytical rather than subsumptive. ANA is very much like Lucretian Atomism without the Swerve or the Downward Tendency. According to ANA, the universe is a collection of elementary parts moving relative to one another. These are of several types, distinguished by shape and inertial mass. Momentum and kinetic energy are transmuted and conserved in the usual way. There is no gravitation, charge or anything else of the inverse-square law variety. Collision is king and is perfectly elastic. No elementary part can change its shape or mass.

According to ANA, every event has a unique micro-analysis: just as every *thing* is a congeries of elementary parts, so every *event* is a congeries of micro-events. Thus, to explain an event is simply to analyze it into micro-events and explain these. In the simplest case, we explain O's moving from P_1 to P_2 by analyzing this event into the corresponding motions of O's elementary parts, and then explaining those by appeal to the stated properties of elementary parts.

ANA recognizes only one kind of micro-event: instantaneous acceleration of an elementary part. Since shape and mass cannot change, the only candidates are change of position and change of velocity. ANA agrees with classical mechanics in refusing to regard *mere* change in position as a genuine change of state. Continued motion is the natural state of things: what requires explanation is a *change* in that state of affairs. The argument here is based on the peripatetic principle that the effect ceases when the cause ceases. Since the only causes are collisions, and since motion continues without continuous impact, it follows that continued motion is not an effect, hence requires no cause to explain it.

According to ANA, then, the only genuine micro-events -- the only real effects -- are accelerations, and these are all caused by collisions. The universe is like a pool table. Thus, the only laws employed by ANA that have a chance of explaining a micro-event are the conservation principles, for they are the only laws that subsume anything that wants explaining at the micro-level. One may suspect this is the Achilles' Heel to be exploited by the wily subsumptivist. Perhaps it is. Before we examine ANA's heel, though, let's take a look at the upper body.

4. Macro-laws

I'm inclined to believe that ANA might be true of a universe that simply did not exhibit any law-like regularities at the macro-level. If so, then the only candidate for subsumptive explanation would be the one just mentioned. But such a universe would be very different from our own, and so perhaps we should expect that explanation of the events in such a universe would be radically different too. Our concept of explanation was evolved for local employment, so we'd best see if we can allow for macro-laws of a sort without allowing them any explanatory force.

I will begin with a very artificial illustration which I call the alphabet phenomenon. Suppose you are gazing out of your office window and notice that the people on the quadrangle periodically form block letter patterns like a marching band. Further observation yields the result that the letters succeed one another in alphabetical order, cyclically, with 'A' following 'Z'. Initially, you suspect design, but subsequent investigation rules this out. The truth is that the path of each person is simply the shortest path between point of off-quad departure and point of off-quad destination, so that the on-quad position of a given person at a given time is fixed by the position, time and initial velocity of departure, these in turn being determined by similar factors independent of the alphabet phenomenon. Thus, every macro-event subsumed by the Alphabet Law -- i.e., every letter change -- has a unique micro-analysis, and each of the analyzing micro-events is independent of the others and independently explicable. Having discovered this, you must conclude that the alphabet phenomenon is just a very striking coincidental artifact of the fact that a lot of people are going their own way. And how striking is it, really? After all, they must exhibit some pattern or other at any given time, and these must succeed one another somehow. What is striking is simply that we happen to have ready to hand, at no charge, a vocabulary for describing it and a corresponding method for predicting it. Generally, adequate taxonomies are not so cheap.

I call things like this artifactual regularities. "Accidental" is misleading: the Alphabet Law supports counterfactuals and licenses predictions. It is no accident that the patterns evolve as they do: "B's" following "A" isn't sheer coincidence. What's coincidental is just that we can describe things in a vocabulary designed for something else. Nor is it law-like in the full sense: knowing what we know, we can see that alphabetical order has no physical significance.

The important point about artifactual regularities is that they have no explanatory force. Consider this: Patterns succeed alphabetically; the last pattern before 2:00 p.m. was A-shaped; therefore, the 2:00 p.m. pattern is B-shaped. As a predictive tool, the Alphabet Law is practically indispensible. As an explanatory tool it is a non-starter. This manifests itself in an interesting way: if we ask why the Alphabet Law is true, almost any answer we give will be misleading. Given the micro-events that are actually happening, we can see why it will continue to hold. So the question comes to this: why do we find these particular causal chains at the micro-level? This is reminiscent of the Cosmological Argument: granted we can explain each event in the chain, but why this whole chain rather than another? This question has no answer (and doesn't need one), and neither does our question about the Alphabet Law.

Could _all_ macro-regularities be artifactual? Well, why not? One reason we might think not is this. If we think that admitting the artifactuality of a regularity is tantamount to conceding that the events falling under it are inexplicable, we will resist artifactuality. But we have just seen that this worry is unfounded: we recognize that the Alphabet Law is artifactual precisely because we see how the events it subsumes _should_ be explained. (I think, by the way, that some philosophers of language and some cognitive psychologists may have been guilty of this mistake: there are good reasons for insisting on explanations of macro-regularities, but among them is not that the only alternative is the inexplicability of the subsumed events.) There is no doubt that many actual macro-regularities are not artifactual. My point is simply that a good scientific theory, a theory capable of explaining everything needing explanation -- would be possible in a world in which all macro-regularities _were_ artifactual.

Another reason for thinking there must be non-artifactual macro-laws is that there appears to be a simple way to construct them: surely there will be laws like this:

(1) If an event having micro-analysis A-1 were to occur, an event having micro-analysis A-2 would occur.

For just consider a macro-event e occurring at t: it has a micro-analysis A(e). Now we use our micro-theory - ANA or whatever -- to find the successor at t' of each of the micro-events specified by A(e). Taken together, these constitute a macro-event e' at t', viz.,the macro-event whose micro-analysis specifies the successors at t' of the events specified by A(e).

This seems okay until we remember that the Alphabet Law is itself of this form: evidently, laws of the form (1) can be artifactual. Indeed, once you think of it, they are bound to be: micro-events don't get together and occur in the way they do in order to constitute a macro-event susceptible of a certain analysis. The succession of events we get by calculating in the manner just rehearsed may well be covered by _no_ macro-law other than those which describe the effect of E as the sum of the micro-effects of the micro-causes constituting E. Macro-laws of that sort are _obviously_ boring: they tell us that the effect of E is whatever follows next in the causal order, and refers us to the micro-theory for details. It might be that this is just the best one could do at the macro-level.

5. More on Macro Laws

In order to tighten our intellectual grip on this possibility, let's consider a non-artifactual macro-regularity, viz., ionic chemical bonding as it is treated in valence theory. Pure valence theory tells us we have a compound - something capable of no further bonding - when the sum of the valence numbers of the elemental constituents is zero. Let's call this the Valence Law. Valence numbers are assigned to elements in such a way as to make this true. At this level, the numbers have no non-dispositional significance: we could multiply them all by 2 or 27 without changing the theory in any important way. What everything having the same valence shares is a certain dispositional property - viz., the disposition to bond with anything having the same valence but for sign. Now it <u>might</u> have turned out that the numbers had a straightforward significance at the micro-level which provided an explanation of the bonds subsumed by the theory. For instance, it might have happened that molecules with a valence of +n had n free hooks, and molecules with a valence of -n had n free eyes. This would provide a straight-forward explanation of why the Valence Law holds true. In particular, it would explain why things having the dispositional property specified by a given valence number have that disposition.

Given this unified account, explanation of bonding by appeal to valence is surely unproblematic: the Valence Law, together with specifications of the relevant numbers, explains why we get NaCl, H_2O, and so on. Actually, of course, there are no hooks and eyes in the picture. But rather than consider the actual case, imagine a rather different fantasy: suppose there is no single explanation of ionic bonding, but several. Sometimes it is hooks and eyes, sometimes something else. So the things that have a disposition to form ionic bonds have it for different reasons. To elaborate the Lucretian fantasy, perhaps some atoms are equipped with one or more clamps, like lobster claws: what can be grabbed depends on the size, shape and number of claws, together with the size and shape of the things to be grabbed.

How does explanation by subsumption under the Valence Law fare under this revision?

Not so well, I think. For now subsumption under the Valence Law is subsumption under a statement that generalizes <u>across</u> explanatory boundaries. Consider this:

> (i) sodium bonds 1-1 with chlorine because (ii) sodium has a valence of +1 and (iii) everything having a valence of +1 bonds 1-1 with chlorine.

(iii) remains true, of course, but it's a put-up job: the uniformity it presents us with is artifactual. This becomes obvious as we increase the heterogeneity of the physical situations underlying (iii): what if there are as many different micro-stories as there are elements? (There aren't, but this is surely a contingent matter.) (iii) would lose all semblance of explanatory force: for it would obviously be nothing more than a convenient recursive way of generating a list of facts like (i). I say "nothing more" because the facts listed would not hold in virtue of anything corresponding to our method of generating the list. Under these conditions the Valence Law would be exactly comparable to the Alphabet Law.

Another way of putting the point is this: (iii) attributes a disposition - the disposition to bond 1-1 with chlorine - to a specific list of elements. But the physical stories underlying this disposition are as numerous as the items on the list. So what (iii) tells us is that sodium is disposed to bond with chlorine for some reason or other which is, in fact, peculiar to the case of sodium. So the "explanation" comes to this: sodium is disposed to bond 1-1 with chlorine for some special reason. Moliere would be pleased.

I want to draw two morals from this story: the first is that the explanatory power of the Valence Law depends on what we know or believe about the facts underlying particular bonds. In the absence of such knowledge or belief, its explanatory force should be zero, for we cannot actually explain anything to ourselves by appeal to a regularity we do not actually know or believe to be non-artifactual. The second is that it is a purely contingent matter that macro-regularities such as the Valence Law are non-artifactual and hence explanatory. This is to be strongly distinguished from the claim that it is purely contingent that ionic bonds can be explained: in the fantasy just rehearsed, we became convinced that the Valence Law is artifactual precisely because of the way we _do_ explain ionic bonds.

6. Micro-laws

If the foregoing is correct, we must admit the possibility of a scientific theory capable of explaining every macro-event without subsuming any since micro-analysis could reveal that all macro-laws are artifactual. If there is to be any subsumptive explanation in the picture, it must be explanation of micro-events. ANA admits of only one sort of micro-event -- accelerations -- and only one sort of explanation of them -- collisions. So our question reduces to the question whether billiard-table physics is explanation by subsumption. We can further simplify matters by supposing that every elementary particle has the same mass -- i.e., that the conservation principles (CV) reduce to conservation of v and v^2.

Is subsumption under CV explanatory? I want to approach this matter via a different but related question: when a particle conforms to CV, does it manifest a disposition? On the face of it, to say the elementary particles satisfy CV is to attribute to them a disposition: CV tells us what a particle would do if ... without telling us why. Furthermore, this disposition is a "brute disposition" there being (given our assumptions) no theoretical prospect of explaining _why_ particles satisfy CV. Brute dispositions are unwelcome for this very reason, a reason made famous (though not clear) by Moliere. Asked why opium puts people to sleep, Moliere's doctor replies that opium has a dormitival virtue. The prospect of having to take seriously at the level of elementary particles what we regard as a joke at any other level is sufficiently unwelcome to make us ask whether we are really dealing with a genuine disposition here. One naturally tends to think of conserving velocity, for instance, as the manifestation of a disposition, viz., the disposition to follow the roads and speed limits allowed by the parallelogram rule. But there are indications that this is a mistake.

FIRST: if conserving velocity and its square -- moving well, as I shall say -- is the manifestation of a disposition, it is peculiar in that the disposition is one that everything has (according to ANA). Thus there would be no point in asking why things have this disposition in the sense in which this means: how do the things that have it differ from the things that do not? This contrasts sharply with uncontroversial dispositions such as flexibility where the need for explanation derives in part from the fact that not everything is subject to the law of flexible things. We want to know about just these things what accounts for the fact that they would flex were they stressed whereas other things would not. No comparable question can be raised concerning the disposition to move well. SECOND: if moving well is a manifestation of a disposition, then it is peculiar in that the disposition is always manifested, never latent. Again, this contrasts with uncontroversial dispositions. Soluble things are sometimes dissolving, sometimes not. This is the point of the familiar subjunctive formulation: we use the conditional to represent the connection between immersions and dissolvings, and the subjunctive because there may be no immersions. A thing may be soluble but not dissolving because conditions are not propitious. The need for explanation derives here from the need to show why the propitious conditions are propitious, i.e., from the need to explain how manifestations are brought about. Once we know this, we can explain, e.g., why sugar dissolves when it does, and why it doesn't when it doesn't. Since the distinction between having the disposition and manifesting it is absent in the case of moving well, the corresponding need for explanation is absent as well. Conservation is not a sometimes thing: there are no special conditions under which things move well, conditions that are sometimes present, sometimes not: hence there is no pressure to explain why some conditions are propitious for moving well while others are not. This is reflected in the fact that a subjunctive formulation is inappropriate. The subjunctive formulation allows for the possibility that the conditions requisite for manifestation may not exist. But there is nothing comparable to allow for in the case of moving well: a particle would move well were it to be the case that ... what? Anything whatever.

Dispositions call for explanation because they are not universal and because they are not always manifested. Brute dispositions cause problems because the required explanations are not and cannot be forthcoming. But CV raises no questions comparable to those raised by flexibility or solubility, and hence we need not be embarrassed for want of comparable answers. It is a mistake, therefore, to think of moving well as the manifestation of a disposition, brute or otherwise. This is not yet to say that it makes no sense at all to ask why particles move-well: it is only to say that asking this is not like asking why soluble things dissolve. This may seem small comfort. If the only possible answer to "Why do particles move well?" is "They just do." doesn't the spirit of Moliere's criticism remain untouched?

I think the way is prepared for a defense of the short answer to the question about moving well, and hence for a negative answer to our question about Moliere's criticism, once we are clear that in giving the short answer we are not acquiescing in a brute disposition. If we think "Why do particles move well?" is analogous to, "Why do soluble things dissolve?" we will think of ourselves as asking after the causes of moving well. For in order to explain why soluble things dissolve we must uncover the etiology of dissolution. Once we construe the question about moving well in this way, the short answer is bound to seem indefensible. We have seen ample reason to resist this construal, however, and this leads me to wonder whether the short answer is defensible after all.

The best way to approach this matter, I think, is by examining another case in which the short answer is unproblematic. Suppose someone were to ask why things move or have sizes or shapes. We might reply that, while it makes perfectly good sense to ask why this or that particular thing has the motion (shape, size) it does, it does not make sense, or anyway not the same sort of sense, to ask why anything at all moves (has a shape or size). This is just the way things are. The question would make sense were we to suppose that there was a time when nothing moved (had a shape or size), for then we might be asking what brought about the onset of motion (shape, size), i.e., what brought about the change in the universe from its supposed previous state to its present one. But, of course, we don't suppose any such thing. Things do <u>change</u> their motion (size, shape), and hence it makes sense to ask how some particular thing came to have the motion (size, shape) it does now have. But there is no comparable question concerning things generally, for the universe didn't change from a motionless state.

This is the stance taken towards inertial motion in classical mechanics: mere change in position is not regarded as a change of mechanical state, and hence it does not require mechanical explanation. Now it seems to me that this stance can and should be taken towards moving well. We may ask why this or that thing is moving well in the way it is, but we cannot in the same way ask why things move well rather than not. Once we cease to see moving well as the manifestation of a disposition, there is no need to see it as involving a change in mechanical state at all, hence no longer any reason to suppose that moving well <u>must</u> be explained. There is, in other words, no reason to suppose that "Why do things move well?" cannot be classed with "Why is there motion (shape, size)?" and given the short answer (as I did in [2]).

If there is no reason why things move well -- why they satisfy CV -- then we cannot, it seems to me, explain a particular case of moving well by appeal to CV. If there's no reason why particles move well, there's no reason why <u>this</u> particle moved well <u>then</u>. A particle doesn't move well _because_ velocity and its square are conserved: its moving well _is_ velocity and its square being conserved.

If CV does not explain why a particular particle moved well on a particular occasion, perhaps it explains why that particle took the particular direction and speed it took. Or is it the initial velocities and positions that explain this? Asking the question in this way is an open invitation to stipulation, or analysis of ordinary language. We'd best find another way of raising the question.

Let's consider a case in which A and B collide "head-on", B initially at rest (relative to the observer), A moving at 2 m/sec. What happens is that A stops and B assumes A's initial velocity. The event to be explained here is B's assuming a velocity of 2 m/sec, or rather this particular change in B's velocity, viz., from zero to two. Let's call this event e. Rather than ask _why_ e occurred, let's ask: (I) what made e happen? This may or may not be a different question, but it's easier, and that's reason enough to shift to it.

Now certainly the laws didn't make it happen. ('Principle' used to (maybe still does) have an ambiguous use in this connection: When a 17th Century writer referred to the principles governing something's behavior, this was ambiguous as between something that makes the behavior happen and a statement subsuming that behavior.) What made it happen was the collision with A, this being, in fact, a collision head-on with something traveling at 2 m/sec due north. Let's call the collision c. The answer to our question then is: _c made e happen_, and this answer involves no appeal to laws at all. Of course we may have to appeal to laws to _justify_ our claim that it is c that made e happen, but that is plainly another matter altogether. We mustn't confuse what made e happen with what justifies us in believing c made e happen.

Is there anything for subsumption to do in connection with answering (I) besides form part of a justification of our answer? I think not. We might go on to ask: Why or how did c make e happen, as opposed to e', say? But this question has no answer at all: there isn't any _way_ in which c made e happen: it just did. Since this question has no answer, subsuming e under CV can have no role in answering it.

What _do_ we do when we subsume an event, or rather a pair of events, under a law? Well, in our illustrative case, CV gives us:

(1) $4 = V_A^2 + V_B^2$

(2) $2 = V_A + V_B$

where V_A and V_B are the final velocities of A and B. Elementary algebra reveals that either $V_A = 2$ and $V_B = 0$, or $V_A = 0$ and $V_B = 2$. Since A cannot pass through B, we have:

(3) $V_A = 0$

(4) $V_B = 2$

The theory and initial values give us a certain mathematical representation of the situation, viz., (1) and (2). A little algebra allows us to substitute (3) and (4) for (1) and (2). What the algebra does is allow us to make it obvious that the required representation is in fact the representation of a situation in which A stops and B continues at a velocity of 2 m/sec. So subsumption of (c, e) under CV comes to this: we construct a mathematical representation by plugging initial values into CV. We then do enough algebra to make it obvious that the representation we have constructed is a representation of e.

Is this "explaining why e happened?". Maybe. But it isn't explaining what made e happen, for it presupposed the answer to that question: we start by characterizing <u>c</u>, the event we suppose made e happen. In the absence of any supposition about what made e happen, subsumption cannot get off the ground.

Let's consolidate:

(1) Subsuming (c, e) under CV is irrelevant to answering (I): "What made e happen?"
(2) The answer to (I) is a singular causal statement.
(3) Justifying singular causal statements often involves appeal to subsuming laws.
(4) Subsumption presupposes an answer to (I), since we begin by describing the event that made e happen.

If subsuming (c, e) under CV explains why e occurred, then explaining why e occurred is different from answering (I). Or rather, since answering (I) is patently part of explaining why e occurred, the contribution of subsumption has to lie in answering whatever question is left over when we take away (I). To borrow an old trick of Wittgenstein's, we want to know what remains when we subtract explaining what made e happen from explaining why e happened. I think the remainder is a justification of the claim that c made e happen.

The right way to view matters, I think, is this: when we explain why e happened we do two very different sorts of things:

(i) we say what made e happen, viz., c.
(ii) we justify the claim that c made e happen by subsuming (c, e) under CV.[2]

How does subsuming (c, e) under CV justify (i)? Well, c certainly occurred. And when we describe c -- write out the equations that depict its mechanical features, -- we find we have depicted e.

Thought of in this way, the CV principles are depiction rules -- rules or schemata for constructing representations of mechanical events, representations which are high in epistemological utility. This sounds bizarre, not to mention instrumentalist, until we recall two points:

(1) there is nothing instrumentalist about saying c made e happen;
(2) causal claims are justified - often enough - by appeal to laws.

What my claim boils down to is this: to explain an event via ANA is to cite its cause, and to justify this citation.

7. Taking It Back

Now, haven't I given up the claim that subsumption is irrelevant to explanation? Yes and no. "Yes" because the complex business we call explanation typically requires specification of causes, these need to be justified, and that's typically achieved via subsumption. "No" because subsumption is given a subsidiary justificatory role which might be played by other actors. Suppose we didn't need to justify singular causal statements by appeal to laws, i.e., suppose we could just tell, without appeal to "constant conjunction" or anything similar, that c caused e. Davidson([3], [4]) has argued that singular causal statements do not entail causal laws. This shows that we can *understand* a singular causal statement without knowing any corresponding law. And once we abandon the Humean claim that causal laws are involved in the analysis of singular causal statements, the doctrine that they are necessarily required for the justification of singular causal statements loses its plausibility. It's rather obvious on the face of it that we often make justified singular causal statements without any serious idea of how to generalize them.

8. Diagnosis

What can*not* be accomplished in the absence of laws is serious, quantitatively accurate, *prediction*, and this, I cannot help but believe, is what's behind the subsumptivist account of explanation. It all goes back to Hume's conflation of two problems:

(i) the analysis and justification of singular causal statements;
(ii) the analysis and justification of predictive inferences.

Whatever Hume *intended* to treat, what he managed to treat was mainly (ii). A review of illustrations of "causal reasoning" in the *Treatise* [7], and *Inquiry* [6] reveals few cases clearly involving singluar causal judgments. And the classic argument -- that you cannot infer the effect from a knowledge of the cause, is a spectacular howler if directed at (i): surely to make a relational judgment requires knowledge of both relata!! (An analogous argument will show that singular size comparisons cannot be justified!!)

Hume has a doctrine about (i): viz., that singular causal statements entail and must be justified by appeal to subsuming laws, but this doctrine is obviously false. It only seemed plausible when (i) and (ii) were not distinguished.

This will seem high-handed to many: there is a long and venerable tradition in philosophy which assimilates truth conditions to evidence conditions on purpose. Philosophers who do this will be unimpressed by a doctrine that depends on distinguishing what makes a singular causal statement true from what justifies it. It is not within my intention nor my ability to take on this tradition. My intention is rather to indicate why subsumptivist accounts of explanation are not likely to find a secure home outside that tradition. My argument took the form of a categorical rejection of subsumptivism only because I cannot find a secure home within that tradition myself.

Notes

[1] I should like to thank John Koethe, Warren Ingber, David Zaret, and Paul Teller for their stimulating attempts to talk me out of most of what follows. I was similarly aided by stimulating audiences at Chicago Circle and The University of Wisconsin where I presented an earlier draft of this paper.

[2] Compare Michael Scriven's views in [9]. I'm inclined to think that one of the things that makes "really" explanatory theories seem so interesting and important is that they provide justifications that "make us understand" an event's occurrence. Hence, we feel cheated when provided a purely inductive or authoritative justification in an explanatory context.

References

[1] Cummins, Robert. "Functional Analysis." The Journal of Philosophy 72(1975): 741-765.

[2] ----------------. "States, Causes and the Law of Inertia." Philosophical Studies 29(1976): 21-36.

[3] Davidson, Donald. "Actions, Reasons, and Causes." The Journal of Philosophy 60(1963): 685-700.

[4] -----------------. "Causal Relations." The Journal of Philosophy 64(1967): 691-703.

[5] Hempel, C. and P. Oppenheim. "Studies in the Logic of Explanation." Philosophy of Science 15(1948): 135-175.

[6] Hume, David. An Inquiry Concerning Human Understanding. ed. Charles W. Hendell. Indianapolis: Bobbs-Merrill, 1955. (Originally published 1748 as Philosophical Essays Concerning Human Nature. Reprint of 1777 edition.)

[7] -----------. A Treatise of Human Nature. ed. L.A. Selby-Bigge. Oxford: Oxford University Press, 1951. (Books I and II originally published 1739 and Book III 1740. Reprint of 1888 edition.)

[8] Quine, W.V. Word and Object. Cambridge, MA: MIT Press, 1960.

[9] Scriven, Michael. "Causation as Explanation." Nous 9(1975): 3-16.

Reasons, Causes, and Empathetic Understanding

J.K. Derden, Jr.

Humboldt State University

The dispute concerning whether reasons are causes and related controversies continue unabated.[1] The conflict manifests itself in various ways. Reasons are not causes: because reasons are not logically independent of actions; because no inductive evidence is required for establishing a reason for why I do something; and because no generalizations are required to explain why someone does something on the basis of reasons. It is also maintained that historical explanation is not explanation by appeal to covering laws and that empathetic understanding is more than an heuristic device for generating hypotheses to explain why people do what they do.[2] It is further maintained that the social sciences are on the wrong track if they see themselves as more akin to the natural sciences, as opposed to philosophy (conceptual analysis) and the humanities.[3]

The amazing fact is that the contradictory of each of the above stated positions has been seriously maintained. It is the contention of this paper that the parties on both sides of the disputes are roughly half-right and half-wrong. It is my belief that the parties involved have both mischaracterized or misdescribed what is involved in acting from a reason. As a result, each party has gotten a part of the picture right and a part wrong, and both parties have presented very confused and fuzzy pictures. This paper attempts to provide the correct description/characterization of acting from a reason, and as a result, provide some explanation for why the above positions and their contradictions have been and still are maintained.

The position of this paper can be briefly and somewhat misleadingly stated as follows: (1) reasons are not causes; (2) there are no reasons for what we do; (3) there are generalizations covering what we do, but they are not the generalizations suggested by Hempel; (4) reasons are not logically connected with actions, but with something else; (5)there is no inductive evidence required for one to know his reasons; and (6) empathetic understanding provides one with a _unique method of discovery_ not

PSA 1978, Volume 1, pp. 176-185

available in the study of non-reasoning creatures. There is much more to the position than the preceding points indicate, but enough has been revealed at this stage to indicate the potential significance of this position, if it is correct.

Assume that an individual does O and has reasons for doing O. In particular, A's reasons for doing O are that he wanted E and believed that doing O was the best means for obtaining E. In these circumstances someone might ask: "Why did A do O?" The answer to this question might vary, but a complete answer might be: A wanted E and believed that doing O was the best means for obtaining E, i.e., A did O for this complex reason. Under these conditions one side of the dispute involving reasons and causes will argue that A's complex reason is not a cause of A's doing O, and further, that without appealing to any covering laws, one understands perfectly well why A does O, given A's reasons for doing O.

The first point is that when one gives A's reasons in response to the question "Why did A do O?", one has _not_ given the reasons for why A did O. But one _has_ given the reasons for something. The question is: For what does one give reasons, when in appearance at least it seems that one gives reasons for actions? The _prima facie_ answer to this question, to be modified subsequently, is that when one gives A's reasons for doing O, one is actually giving A's reasons for _believing_ that he should do O, or A's reasons for _believing_ that he will do or shall do O, or A's reasons for _believing_ that he wants to do O. A better and summary way of putting the matter is that when one gives A's reasons for doing O, one is giving A's reasons for deciding to do O. It is thus my contention that, roughly, deciding to do O and doing O are frequently equated or confused in the papers relevant to the subject. To mention but two papers of this kind, this equation or confusion is much in evidence in Dray [4] and Hempel [5]. Hempel talks about giving reasons for doing something _or_ deciding to do something, as if they were the same ([5], p. 282), while Dray talks about reasons for why someone should do something as if this were the same as doing something ([4], p. 260).

In what follows when I say that A believes that he should do O, I shall mean that A believes that he should do O, or A believes that he will do O, or A believes that he wants to do O, or A believes that he shall do O, and so on. It seems obvious that A's believing that he should do O and his doing O are two distinct "things". It follows that it should be kept clear _what_ it is we give reasons for, when reasons are supplied. It is my contention that we give reasons for what people believe they _should_ do, and not reasons for what they do. If we _can_ give reasons for what people do, then we can do so if and only if we can give reasons for what people believe they should do. In support of this view, I simply point out that one frequently has "reasons for doing O" when O has not occurred and never occurs. In such cases, we do not have reasons for doing O, but rather reasons for believing that we should do O. In short, there are A's reasons, there is A's belief that he should do O, and there is A's doing O. Both parties to the disputes have left out A's belief that he should do O, or A's decision to do O, or they have equated A's decision or belief with A's doing O.[4] It is almost as if both sides have committed a form of post hoc fallacy in that both sides conclude that because an action occurs after the reasons, it occurs because of the reasons. Then both parties argue about the "because" in the expression "because of the reasons". When this occurs, that which interposes itself in some way between the reasons and the action is left out. This "something" is A's belief that he should do O.

There are good reasons for why both parties have either left out A's belief that he should do O or equated this latter with A's doing O. First, the emphasis has been upon explaining why A does O, where it is assumed that the doing of O has occurred. In the case where the doing of O does not or has not occurred, we are more likely to ask what are A's reasons for his decisions, what is the evidence for A's belief, etc. If the emphasis had been upon the case where the doing of O has not occurred, then this equation would have been less likely to take place. Second, and more importantly, under one interpretation of the question "Why did A do O?", the answer to the question is one which most everyone already has. The interpretation that I have in mind is as follows: What is the proximate cause of A's doing O? The answer to this question, given that the doing of O has occurred, is roughly as follows: A believed that he should do O and was able to do O during the time period when he held this belief. We infer from the doing of O that A was able to do O and that A believed that he should do O. We thus do not want to be told what we already know when we ask "Why did A do O?". When we thus ask "Why did A do O?", we are confusedly and obscurely asking the question "Why did A believe that he should do O?". Once again, it is this latter question that we are very likely to ask in the case where the doing of O has not or does not occur. As a first step toward eliminating the confusion, we must realize that in asking the question "Why did A do O?" we are not asking the causal question for which we already have an answer. Unfortunately, our having the answer to this question leads us to neglect the proximate cause which interposes itself somehow between the action and the reasons. Both sides have neglected A's believing that he should do O because it is so obvious and uninteresting. Imagine a psychologist publishing a paper explaining why Ruby shot Oswald by saying that Ruby shot Oswald because he believed that he should and was able to do it while he so believed!

But even if we are aware of the neglect of the proximate cause of A's doing O and are aware that in asking "Why did A do O?" we are implicitly asking why did A believe that he should do O, this is not the end of the matter. I do not know how to put the matter except by saying that the question obscurely being asked when someone asks "Why did A do O?" is doubly ambiguous, i.e., "Why did A believe that he should do O?" is doubly ambiguous. For those who maintain that reasons are causes, to ask the question "Why did A do O?" is to obscurely ask what are the causes for A's being in a belief state or for A's deciding. For those who believe that reasons are not causes, to ask the same question is to obscurely ask what are the reasons for A's belief. Both sides think they are giving a cause or a reason for an action, when in fact they are giving in a very confused manner a cause for a state of believing or a reason for a belief. Both sides have been uninterested in the obvious and already known answer to the question "Why did A do O?" and as a result have mistakenly equated A's belief that he should do O with his doing O or simply have jumped from the action to the reasons, leaving out what is in between in some sense. Both sides have thus been asking a question different from "Why did A do O?", where the question confusedly being asked is itself doubly ambiguous.

The correct characterization of the circumstances involved when one claims to be explaining A's doing O on the basis of A's reasons for doing it is the following: (This characterization is somewhat incomplete, but this does not detract from its essential correctness.)

a. A believes: I want E.
   A believes: The best means for obtaining E is doing O.
   A believes: I can do O.
   A believes: If the preceding is true, then I should do O.
   A infers: So I should do O.

b. Any person's believing any set of premises from which he infers (correctly or incorrectly) a conclusion that he should do something, where there is no other belief which occurs to this person which is in conflict with the conclusion inferred, is frequently causally sufficient for this person's believing what he infers.

c. Any person's believing that he should do something during a time period T when he is able or comes to be able to do it during T is, more frequently than not, causally sufficient for his doing it.

It is (c) above which provides the generalization which gives us the proximate cause for A's doing O. Tell me what a person believes he should do in time period T _and_ that this person is able or comes to be able to do it during T and I will predict that he will do it. Further, if you bet against my predictions, you will lose in the long run - provided that it is _true_ that A believes that he should do O during T and is able or comes to be able to do O during T. My predictions will most frequently fail when A changes his belief or becomes unable to do what is predicted. The first "law of motion" for humans then is the generalization (c). It is so simple that it has been missed. It is uninteresting because people do not want to hear such an explanation for why A does O (everyone already "knows" this); people want to know what A's reasons are for his conclusion that he should do O. Admittedly, there are tremendous problems involved with some of the terms in "the first law", e.g., 'believing or state of belief', 'object of belief', 'being able or coming to be able'. These are, however, problems for anyone interested in the cognitive processes of humans.

I think now that I am in a position to explain why there have been completely contradictory accounts of explaining from reasons.

(i) Those who maintain that reasons are not causes for actions are partially correct; where they are incorrect is in maintaining that reasons "explain or make sense of actions." Reasons are premises in an agent's arguments or deliberational processes, and as such are surely not causes. On the other hand, reasons do not explain or make sense of an action, but rather explain or make sense of an agent's decision to do something or belief that he should do something. Reasons do not explain or make sense of a deciding or a believing, but they do make sense of beliefs or decisions by showing that certain beliefs or decisions are logically connected with reasons as premises in an agent's deliberations.

Those who maintain that reasons are causes are incorrect because they have confused premises in an argument with causes for an action. (Hempel's purported causal generalizations are examined below.) For now, it should be noted that accepting certain premises and inferring a conclusion may be causally sufficient for believing the conclusion, and the believing of the conclusion may be causally sufficient under certain conditions for the doing of something. But the reasons are not causes. It is the believing and the inferring that are causes, if anything is. Reasons in an argument are not causes of a conclusion, where the latter is confusedly equated with an action. A's reasons in A's arguments are not causes of anything, but A's accepting premises, inferring conclusions, and so on may be causes of something.

(ii) Those who argue that reasons are logically or quasi-analytically connected in some way with actions or their descriptions are incorrect, but they are onto something that <u>is</u> logically connected to reasons. In the sketch above, A's conclusion/decision is logically connected with A's premises/reasons. If one confuses A's conclusion/decision with A's action (and this is repeatedly done by both sides of the disputes in the literature)[5], then it is a short but mistaken step to the belief that reasons are logically connected with actions. It follows that those who maintain that reasons are not logically connected with actions are correct, but, as we shall see, their characterization of the role of reasons is incorrect.

(iii) Those who maintain that no inductive evidence is required to know one's reasons for doing 0 are half-right and half-wrong. They are half-right because A does not need inductive evidence to know what the premises of his argument are by which he makes a decision or draws a conclusion. They are half-wrong because they take A's reasons to be connected with A's doing 0 as opposed to being connected with A's conclusion to do 0.

On the other side, the arguments that advocate the need for inductive evidence are most curious.[6] They consist in showing that having reasons for doing 0 cannot establish whether one will do 0 unless one has done something similar to 0 in the past on the basis of similar reasons. (The strange generalizations which result on this account are discussed with the discussion of Hempel's position below.) It suffices to say here that we do not need inductive evidence for knowing what our premises are, and <u>given</u> the generalizations (b) and (c) in the sketch above, we do not need inductive evidence to establish the connection between our believing certain premises and inferring a conclusion to do something, and our doing something. Those who argue for inductive evidence are also half-right and half-wrong. We do need evidence for (b) and (c), but we do not need inductive evidence for what our reasons/premises are for our conclusions/decisions. The advocates of inductive evidence confuse reasons/premises with causes, and conclusions/decisions with actions.

(iv) Those who maintain that we do not need generalizations to explain why A did O are half-right and half-wrong. In the sketch above A's decision to do O follows from A's reasons/premises. Note that there are no generalizations in A's premises. One can thus "see" (logically) how A's decision connects with his reasons without appealing to any generalizations at all. But one cannot see how A's reasons connect with A's action without appealing to generalizations. Once again, the confusion of A's decision to do O with A's doing O is crucial.

Assuming that the covering law model of explanation provides a necessary condition for having a satisfactory explanation, then those who maintain that we need generalizations to explain why A did O are correct. Hempel type generalizations for actions, however, are the wrong generalizations, or perhaps it is best to say that they are not lawlike generalizations at all. Hempel's schema as presented in ([5], pp. 290-291) is as follows:

A is in a situation of type C.
A is rational.
In a situation of type C any rational agent will do X.
Therefore, A does X.

Hempel's covering law mentions a set of particular beliefs that a particular agent holds. For example, in explaining/predicting Derden's going to the Arcata post office, Hempel must invoke the following "law": Whenever a rational agent believes that the best means for mailing a letter in Arcata is going to the Arcata post office and whenever such an agent wants to mail a letter in Arcata, then such an agent will go to the Arcata post office.

It follows from Hempel's account that there will be as many laws of human nature about mailing letters as there are post offices in the world and people who want to mail letters in these post offices. Hempel type accounts are analogous to explaining why particular gas #1 has the particular temperature T=a by using the following "law": Whenever any gas has pressure P=b and a volume V=c, then its temperature T=a. On this account there will be as many laws for gases as there are individual values for pressures and volumes! Needless to say, such purported laws are not acceptable. It is unclear why Hempel presents the generalization he does unless he too missed the proximate and uninteresting cause of A's doing O and attempted to connect specific reasons with specific actions, rather than treat specific reasons as "parts" of _initial conditions_ for the application of a covering law like (b) in the sketch above. If the generalization (b) is approximately correct, then in order to predict what A will believe on the basis of reasons one will need initial conditions like A's believing that his doing O is the best means for obtaining E, etc. As well, in predicting what A will do on the basis of a generalization like (c), one will need initial conditions such as A's believing that he should do O. Hempel is thus correct in insisting upon the need for generalizations, but he has presented the wrong "generalization".

v) Those who maintain as, for example, Nagel does in his Structure of Science ([7], pp. 473-485), that empathetic understanding is merely a useful heuristic device for generating hypotheses are incorrect. There are good reasons why empathetic understanding has been criticized as unobjective, and perhaps as a mysterious seeing on the part of the mind's eye. These attacks are justified for a number of reasons, but from the standpoint of this paper the major criticism of the proponents of such understanding is that they have mischaracterized it. The important point to note is that both parties to the dispute have mischaracterized and subsequently masked what is going on when empathetic understanding takes place.[7]

On the basis of the schema presented, part of what is going on in empathetic understanding (when one "puts oneself inside the head of the other") is that one is appropriating A's premises and seeing (logically) whether A's conclusion/decision follows. One thus sees/understands why A decides to do O or believes that he should do O, and one confuses this seeing/understanding with understanding why A does O. If I can infer A's conclusion from A's premises, then I have empathized with A. There is a logical seeing/understanding of the connection between A's decision/conclusion and A's reasons. This point has been missed by critics of empathetic understanding and has been mischaracterized by proponents as well.

But there is much more to empathetic understanding than this. Certain historians and detectives are better in explaining and predicting respectively than others. Assuming that every historian or detective can have access to an historical figure's or crook's premises, then why would one explain/predict better than another. The answer lies in yet a third generalization to which appeal is made by those who empathetically understand: If a set of premises implies a conclusion, and an agent accepts these premises, and most observers can infer the conclusion from these premises, then it is very likely that the agent will infer the same conclusion. This is a legitimate although unusual generalization. On its basis a detective can infer what is likely to be inferred by a crook. This means that empathetic understanding can be used as a method, not for discovering an hypothesis, but rather as a method for discovering an initial condition to be used in a covering law to predict what the crook is going to do. (Analogous remarks apply in the realm of explanation for the historian.) In short, employing certain generalizations, empathetic understanding can be a method for discovering initial conditions in order that generalization (c) may be applied. If Nagel's hypotheses are generalizations, then one can understand why he and critics have missed the boat, i.e., we already have our generalizations and need to hit upon initial conditions. More importantly, empathetic understanding as a method of discovering initial conditions for predicting an agent's action is unique to the study of reasoning creatures. This is so because we as observers must deduce what the agent deduces, where the conclusion deduced is then needed to set up an initial condition in order that the generalization (c) may be applied. Inanimate objects at least do not hold premises and infer conclusions such that an observer can infer the same conclusion from the same premises and use this conclusion as an initial condition for applying laws. One form of "putting oneself inside the head of the other" is appropriating the other's premises and inferring his conclusion. The mysterious seeing is not mysterious at all; it is a matter of logic. (Note that time does not permit discussion of the case of finding the premises.)

(vi) The comments thus far have been brief, but I now want to finish this paper with even sketchier comments about Winchian social science and rational action. Winch maintains in his The Idea of a Social Science [11] that social science is more like conceptual analysis and philosophy than natural science. This is because Winch is connecting conclusions/decisions with premises/reasons without knowing it, as opposed to his supposed connection of reasons with actions. Conclusions are inferred by the agent in conformity with his rules/norms. The observer's seeing of the agent's supposed conclusions is more like analysis than natural science, i.e., the observer deduces a conclusion from the agent's premises and is thus engaged in logical analysis. Further, the observer can examine how the agent's premises fit in logically with the premises accepted by the majority of the individuals in a given society. This too is more like logical analysis than natural science. We may even need to know the agent's premises in order to discover what the agent is doing, i.e., we deduce the agent's conclusion and use (b) and (c) to get what the agent is doing. Clearly, we need to know the propositional meaning of the agent's premises and conclusions in all these matters. This too involves conceptual analysis. If Winch is maintaining that the social sciences are merely involved in logical/conceptual analysis, then he is mistaken. Many reasons have and can be given for why this is so.[8] One of the major reasons from the perspective of this paper is that Winch is connecting reasons and rules with decisions and not with actions (as he claims). Nevertheless, the social sciences must be involved to a certain extent in the sort of logical/conceptual analysis illustrated above. The point is that there has not been an adequate characterization of the manner in which the logical/conceptual and the causal accounts fit together. The schema presented in this paper is the sort of characterization that is needed.

An action is rational if and only if the belief that one should perform the action is rational. This latter will be rational if and only if the conclusion corresponding to the belief has been "correctly" inferred from a set of premises. In other words, when there are reasons for an action, assessing the rationality of the action is assessing the correctness/soundness of an argument. If the argument an agent constructs is in some sense rational, then in a derivative sense an action connected with agent's deliberations via (b) and (c) is also rational. This point has been missed by a number of writers on the subject, if not all.[9]

Much has been left out of this discussion, e.g., the problem of rationalizations vs. the real reasons for doing something, the nature of action itself, etc. Nevertheless, the significance of the type of analysis proposed should be apparent.

Notes

[1]For a brief discussion of some of the arguments pro and con, see [10], Chapter 5. For additional discussion and further references see [2].

[2]For the positive account see [3] and [4]; for the opposing view see [5] and [7], pp. 473-485.

[3]For the positive account see [11]; for opposition see [6] and [8], Chapters 6 and 7.

[4]Davidson in [2] is a possible exception.

[5]For example, in addition to the pages cited in [5] and [4] above, see [6], p. 19, for "skipping the proximate cause", and [8], p. 146, for equating reasons with causes and decisions with actions.

[6]See [10], pp. 102-104.

[7]Although this is not the paper in which to establish this strong claim, it seems evident to me _at least_ that the characterization of empathetic understanding in this paper is not to be found in the writings of such individuals as Wilhelm Dilthey, Max Weber, R.G. Collingwood, William Dray, Maurice Natanson, Alfred Schutz, Ernest Nagel, Leon J. Goldstein, and others.

[8]See [8], Chapters 6 and 7, and [6].

[9]See [5], p. 297, to illustrate the point. Hempel seems here to be talking about "good" reasoning, i.e., correct or sound arguments, without realizing it.

References

[1] Care, Norman S. and Landesman, C. (eds.). Readings in the Theory of Action. Bloomington: Indiana University Press, 1968.

[2] Davidson, Donald. "Actions, Reasons, and Causes." The Journal of Philosophy LX (1963): 685-700. (As reprinted in [1]. Pages 179-198.)

[3] Dray, William. Laws and Explanation in History. Oxford: Clarendon Press, 1957.

[4] -------------. "The Rationale of Actions." In [1]. Pages 255-280. (This is a slightly revised version of Section 1-5, Chapter V, in [3].)

[5] Hempel, Carl G. "Rational Action." Proceedings and Addresses of the American Philosophical Association XXXV (1962): 5-23. (As reprinted in [1]. Pages 281-305.)

[6] MacIntyre, Alasdair. "The Idea of a Social Science." Proceedings of the Aristotelian Society Supplementary Volume XLI (1967): 95-114. (As reprinted in [9]. Pages 15-32.)

[7] Nagel, Ernest. The Structure of Science. New York: Harcourt, Brace, and World, 1961.

[8] Ryan, Alan. The Philosophy of the Social Sciences. London: MacMillan, 1970.

[9] ----------. (ed.). The Philosophy of Social Explanation. London: Oxford University Press, 1973.

[10] Shaffer, Jerome A. Philosophy of Mind. Englewood Cliffs: Prentice-Hall, 1968.

[11] Winch, Peter. The Idea of a Social Science. London: Routledge and Kegan Paul, 1963.

Approximative Explanation

T. R. Girill

California State University- Fresno

1. The Problem

Sometimes, when an explanandum *E* is a quantitative claim, we are unable to explain why the specific numerical value mentioned in *E* is correct. Sometimes an explanans *X* is available, however, which accounts for a statement *E'* asserting a value close to that mentioned in *E*. Although *X* strictly speaking provides evidence against *E*, it is in some circumstances seen as *approximatively* explaining *E* along with *E'*. Such an approximative explanation may be defined as follows:

> A set of statements *X* constitutes an *approximative explanans* for an explanandum *E* iff (i) there exists a statement *E'* such that *X* adequately explains *E'*, and (ii) *E'* approximates *E*.[1]

Explanans *X* is linked to the desired explanandum *E* only indirectly, by means of *E*'s closeness to *E'*. Hence, a special problem arises about the conditions under which such an approximative explanation is really acceptable. If *E* is a singular statement,[2] when does *X* succeed in "almost" explaining why *E* is true? In other words, when does *E'* "explanatorally approximate" *E* closely enough so that *X* provides an adequate approximative explanation of *E*?

Three alternative answers are possible:

(1) A *single*, universal degree of approximation between *E* and *E'* is required for adequacy (a difference in values of no more than 10%, or 1%, perhaps). This simple approach seems rather farfetched.

(2) "*No general criteria* can be stated for deciding when it is that an explanandum [*E'*] in an explanation argument has approximated the actual case [*E*] enough so that it can be legitimately said that the explanation is acceptable." ([1] , p. 52). This council of despair

---

PSA 1978, Volume 1, pp. 186-196

is perhaps a natural reaction when rejecting alternative (1). But I believe that it too is incorrect. Instead, I develop on the pages below a third option.

(3) The degree of approximation needed (between *E'* and *E*) depends on the *context* in which the explanandum is placed. This context dependence is not so weak as to yield one universal value for the closeness required. However, it is not so strong as to prevent us from describing it in clear, explicit, general terms.

I contend that the *comparisons* which the explanation's audience makes with the quantitative value mentioned in *E* are what influence how closely *E'* must approximate *E* for explanatory adequacy. Section 2. illustrates this claim with a simple example drawn from the literature. Section 3. spells out the principles which underlie these comparisons. Section 4. states the conditions under which these principles operate, i.e., the times at which an audience is willing to accept an approximative explanation at all. Section 5. examines how the contrasts made by the audience influence the degree of approximation they will tolerate, and in turn, which audience traits influence their choice of these contrasts. Thus, the following pages show how the third solution to the problem of approximative explanation is a pragmatic one: it overtly depends on audience beliefs and comparisons for its success.

2. Comparative Case Study

A comparison of two simple approximative explanations, one of which seems adequate while the other does not, suggests the strategy behind solution (3):

> If, for example, one is explaining why a bullet hit a certain point on a target and the explanandum in the [available] explanation argument describes something which is one foot off what actually occurred, this would be considered an unacceptable explanation; on the other hand, if one is explaining why a rocket hit a certain point on the moon and the explanandum in the [available] explanation argument describes something which is one foot off what actually occurred, this would still be an acceptable explanation, ([1] ,p. 52).

What accounts for the intuitive difference in explanatory adequacy between these cases?

My proposal is that when judging the acceptability of such explanations, the audience to which they are addressed *compares* the items mentioned in the desired explanandum (*E*) with other similar ones. For example, here they will tacitly compare the place where the bullet or rocket hit (cited in *E* by coordinates, or distance from some reference point) with other places on the target or on the moon, respectively, where it might have landed but did not. This set of other, possible impact points is the audience's contrast class for explanandum E. It contains items otherwise like the spot mentioned in E, but lacking the explanandum property, i.e., the property of being struck by the bullet or rocket in question.

To be pragmatically adequate, an account of E must have "explanatory power" in the face of this contrast class, in the following sense.[3] Audience beliefs that the projectile could have struck many otherwise similar alternate spots provide evidence by analogy against its striking the one point where it did (against E). So E has a low antecedent subjective probability relative to what the audience believes about the contrast class. To satisfy them, the explanans must therefore increase the subjective probability of E, of hitting the actual impact spot, given this counter-evidence. If the audience is willing to consider an approximative explanation of E, they will settle for an explanans that increases the subjective probability of hitting in the vicinity of the impact point. We can summarize all of this briefly, although somewhat inaccurately, by saying that an acceptable explanans must show why the bullet or rocket hit near the spot it did as opposed to the other places where the audience thinks it might have hit.[4]

The target on a shooting range is small, measured in feet (often no more than one foot square). So the grid of subdivisions which the audience mentally superimposes on the target to generate comparison areas is such that the edge of each sub-area (square inches, for example) will be small compared to one foot. The audience wants to know why (E =) the bullet hit near point p, "instead of" landing in any of the other small subdivisions. If the available explanans X_1 shows that (E' =) the bullet struck one foot from p, this would surely place its impact point in a different neighborhood than (many areas away from) the one containing p. Consequently, X_1 fails to differentiate the vicinity of point p from the members of the audience's contrast class. It appears inadequate as an approximation because a necessary condition for its pragmatic success remains unfulfilled.

The moon, however, is known to be large. The audience's grid of comparison zones will therefore normally consist of units much larger than before. The edge of each subdivision in this contrast class would exceed one foot considerably, perhaps being as great as one mile. Now ex hypothese, explanans X_2 shows that (E' =) the lunar rocket landed one foot from some point q, rather than (E) actually at q. Nevertheless, this still places the impact point in the same (relatively large) sub-area of the moon's surface as the one centered on q. In the context of this comparison, therefore, the approximation of E' to E is close enough so that X_2 still increases the subjective probability of striking in the vicinity of q. So, to the audience, this explanation appears more satisfying than the previous one.

In the first case above, the distance between the places mentioned in E and E' exceeds the edge-length of the typical comparison area (on the target). In the second account, the (same) difference in places between E and E' is much less than the edge-length of the audience's comparative zones (on the moon). This shift in the audience's contrast class reflects a change in their explanatory tolerance. For this reason, the latter approximative explanation was able to achieve pragmatic adequacy where the former could not.

3. Principles Underlying Approximative Explanations

The analysis of the foregoing examples reveals a line of reasoning which I believe is common to assessing all approximative event-explanations. Making this line of reasoning explicit discloses the principles used whenever approximative explanations are accepted, and the conditions under which those principles are thought to hold.

For convenience in generalizing, assume the "desired" explanandum (the one we would ultimately like to see accounted for) has the form

> E: Item i exhibits property P with magnitude m.

The "actual" explanandum (the one literally explained by available explanans X) reads

> E': Item i exhibits property P with a magnitude which differs from m by h.

Of course, regarding the value of P assigned to i in E' as differing from m by some amount (e.g., seeing 5.6 as a "mere departure " of $h = 0.1$ from $m = 5.5$) already betrays a willingness to consider approximations to E. We will also need to discuss what might be called the "tolerance" explanandum E_t, according to which:

> E_t: Item i exhibits property P with a magnitude which differs from m by no more than d.

Accepting any approximative explanation as adequate requires adopting some explanatory principles beyond the usual ones. One possibility is a Principle of Tolerance, which asserts that, pragmatically, there are limits on the accuracy expected of any account of a quantitative claim. A promising formulation is:

> Tolerance Principle
> A request for an explanation of a statement of the form E is pragmatically equivalent to a request for an explanation of a statement of the form E_t (in which $d > 0$).[5]

This principle works as follows. Suppose X adequately explains E'. If $d \geqslant h$, then E' entails E_t. Hence, if $d \geqslant h$, then X together with this entailment adequately explains E_t. We have now accomplished the goal of approximatively explaining E, since, by the Principle of Tolerance, any explanans adequate to explain E_t is "good enough" to satisfy the original demand for an explanation of E.

I believe this analysis is fundamentally correct. But it is open to the objection that people seldom overtly mention surrogate explananda (like $\underline{E}_t$) when assessing approximative explanations. This suggests that a slightly different argument, though equivalent to the preceding one, is actually used.

An Interchange Principle follows from the Principle of Tolerance. It permits explananda (like E and E') mentioning quantitative values within the audience's tolerance interval to be interchanged without loss of pragmatic adequacy. A preliminary, faulty version of such a principle is:

Interchange Principle I
A non-approximative explanation of an explanandum mentioning *any* single value within an audience's tolerance interval is pragmatically adequate as an approximative explanation of *any* other explanandum mentioning a value within that interval.

The diagram below reveals why version I is too generous:

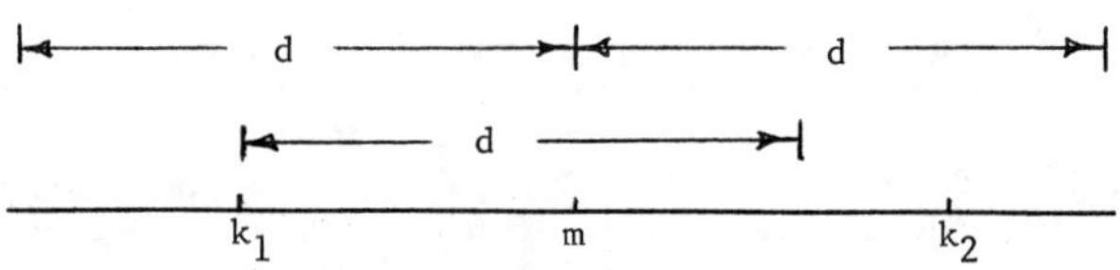

All points inside the tolerance interval are within *d* of *m*, but not within *d* of each other. Interchange Principle I would inadvertently permit an adequate account of E_1 (= *i* has property *P* with magnitude k_1) to approximatively explain E_2 (= *i* has *P* with magnitude k_2), even though k_2 is farther from k_1 than *d*.

A more cautious version of this rule can easily be constructed, however, which escapes this drawback:

Interchange Principle II
A non-approximative explanation of any explanandum attributing to item *i* property *P* with a magnitude within *d* of *m* is pragmatically adequate as an approximative explanation of the explanandum attributing to item *i* property *P* with magnitude *m*.

This principle overtly links an available explanans *X* with an explanandum (*E*) which *X* does not quite cover literally:

1. *X* is a pragmatically adequate non-approximative explanation of *E'* (as above).

2. Interchange Principle II.

/∴ 3. In $d \geqslant h$, then X is a pragmatically adequate approximative explanation of E.

Such an argument makes explicit how approximative explanations are defended. It (conditionally) validates X as adequate to explain E, and does so without mentioning any substitute explanandum E_t. Furthermore, while Interchange Principle II allows tolerance d^t to have just one fixed value, it does not require this.

To stop here with Interchange Principle II, however, is to leave unanswered a pair of key questions: (i) under what conditions is an audience willing to accept the existence of any "tolerance interval" around a quantitative explanandum? and (ii) when it is so willing, how wide is the (variable) interval d which it adopts? Clear responses are vital to further developing solution (3) to the problem of approximative explanation.

4. Accepting a Tolerance Interval

That approximations and idealizations are somehow related is a familiar theme. I suggest that an audience is willing to even consider an explanation of E' as an adequate approximative account of E if and only if they believe that the explanans incorporates an idealization in some form.[6]

The believed presence of an "idealized theory" in the explanans governs the use of the Tolerance or Interchange Principles in the following way. An audience will adopt a tolerance interval d around the value m in E, over which they will seriously entertain an approximative explanation, just in case they believe that:

(1) The explanans X permits the statement of "ideal conditions" such that the more nearly the actual conditions approach the ideal ones, the more closely the explanandum E approaches E' (which is non-approximatively explained by X); and

(2) Whether or not the actual conditions can really be made to approach the ideal conditions, the difference in values between E and E' can be explained in terms of the difference between the actual and ideal conditions.

To illustrate, suppose someone was working with an experimental arrangement as a result of which they desired an account of why

E: The period of this pendulum is 6.27 seconds.

They might use the law of the "simple" pendulum to construct the following explanation:

X: 1. period $P = 2\pi(l/g)^{\frac{1}{2}}$
2. length $l = g$ for this pendulum

$$/\therefore \begin{cases} P = 2\pi(g/g)^{\frac{1}{2}} = 2\pi = 6.28 \text{ seconds} \\ \underline{E'}\text{: the period of this pendulum is 6.28 seconds.} \end{cases}$$

The value cited in E' departs from the one ($m = 6.27$) in E by .01 sec (= h). Nevertheless, X is known to employ an idealized law. This is associated with a pair of ideal conditions (that the pendulum's cord is massless, and that the period is independent of the amplitude of its swing). But if the bob is much more massive than the cord, and the amplitude of each swing is small, these ideal conditions will be very nearly satisfied. The audience will then have good reason to believe that requirements (1) and (2) above are met, and hence will be willing to entertain X as a possible approximative explanation of E.

The case of the rocket hitting the moon one foot away from point q is similar. The audience accepting the approximation believes the one-foot deviation can be accounted for in terms of the projectile's (and the moon's) departure from being perfectly resistanceless point masses.[7]

This view entails that idealizations play more than a merely heuristic role in the scientific enterprise. They also serve to expand the class of explananda covered by adequate explanations for any given audience, by permitting many that are only "approximatively covered" to be included too.

5. Amount of Tolerance

The existence of a tolerance interval is conditioned on the believed presence of idealizations in the explanans. However, the size of that interval is conditioned on the contrasts, made by the audience, with the value mentioned in the explanandum. How tolerant the audience is with approximations to E depends on the members of the contrast class with which m is compared.

Generalizing on the analysis of Section 2.'s examples yields an explicit account of how this works. Tolerance interval d depends on the "spacing" of the contrast class values. Values close together (and hence close to m) indicate that d is small, while values far(ther) apart betray a large(r) tolerance. If h (the difference in values mentioned in E and E') is less than the difference between the audience's contrast values, then the approximative explanation of E appears adequate to them. But if h exceeds the difference between contrast values, the approximation will seem inadequate as an explanation.

Which alternative is the case for any given explanation clearly depends on which items the audience uses for comparison, as well as on the numerical difference between E and E'. This is a "pragmatic" solution to the problem of approximative explanation, because the contrast class members depend in turn on the beliefs and interests of the audience to whom the explanation is addressed (more on this below).

Once again, the pendulum case (from Section 4) illustrates this claim. When requesting an explanation of the pendulum's period, the audience may be making tacit "order of magnitude" comparisons with m = 6.27 seconds (these are often the most useful heuristically). If so, they will want to know why the period is 6.27 seconds "as opposed to" being .627 or 62.7 seconds, for example. In these circumstances, the difference h (= .01 sec) between E and E' is much less than the intervals (of about 6 and 60 seconds, respectively) between m and the implicit contrast values. Since the tolerance width d is so large, Interchange Principle II easily permits the transfer of explanatory adequacy from E' to E.

Under other conditions, the audience might demand to know why the period in question is 6.27 "as opposed to" 6.265 or 6.275 seconds. The difference between m and the contrast class values would then be a time interval of less than .01 second. As a result of adopting this new set of comparisons, tolerance d has shrunk down not only much smaller than before, but also smaller than h. The audience remains willing to honor as adequate approximative explanations of E any non-approximative accounts giving the pendulum's period a value within .005 seconds of 6.27, according to the Interchange Principle, But X is no longer among them: since h exceeds d, this explanation is now approximatively *in*adequate.

Which audience characteristics affect their choice of one contrast class (and hence tolerance interval) rather than another? I think the most important factors are these:

A. Their background information: What the audience already knows (or believes) about the construction of the pendulum, for example, or the overall size of the lunar surface, influences the comparisons they take seriously. Hence, it affects how generous the tolerance interval is that derives from those comparisons.

B. Their goals and interests: Missing a terrestrial target by one foot would thwart most goals associated with shooting bullets (e.g., hunting). Missing a lunar target by the same margin, however, would have no adverse effect on such likely goals as terrain scrutiny or radio transmission. Such a difference in purposes partly inspires the difference in audience contrasts made in each case.

C. Precision of their measurements: Beliefs about the sensitivity (ability to discriminate) and reliability (ability to repeat) of the instruments and techniques used to obtain m can also affect comparison choices. There is no *simple* link, however, between precision and tolerance interval.

Great imprecision in measuring m can sometimes make accepting an approximative explanation harder, not easier. It causes h, the difference in values between E and E', to become very indeterminate. So the audience may have trouble deciding whether or not h exceeds the contrast class "spacing" in size, and also whether or not a departure from ideal conditions could account for h.

Conversely, even when m (and h) are thought to be very precise, the audience may still be willing to accept approximative explanations over a large tolerance interval. This is so if (for reasons A or B above) they employ a contrast class with widely-spaced members, and if they think the difference (h) in values can be explained by deviations from ideal conditions associated with the explanans (as the pendulum case shows).

6. Conclusions

How does solution (3) to the problem of approximative explanation stand now? As expected, we found no single number indicating how close the values in E and E' must be for explanatory adequacy. But, contrary to Barr, we have discovered general criteria by which both the initial plausibility (Section 4) and the range of acceptability (Section 5) of such approximations are judged.

1. Solution (3) is not ad hoc. Fortified by the Tolerance Principle (which replaces E with E_t) or the Interchange Principle (which transfers explanations from E' to E), quite the same story describes the adequacy of approximative event-explanations as can be given for non-approximative ones. The quest for explanatory power plays the same role here, for example, that it plays anywhere.

2. Solution (3) is basically pragmatic. It permits the same approximative explanation of E to be both adequate for one audience and inadequate for another, if their beliefs about which contrast class is appropriate, and hence about how wide the tolerance interval is, differ significantly.

3. Solution (3) is specific and constructive. To the extent that the contrast class members remain unclear or unspecified, to that extent a request for evaluating the adequacy of an explanans qua approximation is also unclear and unanswerable. An explicit answer requires an explicit contrast class. Likewise, the process of making overt which contrast class an audience is tacitly using is the process of making overt the scope of approximations they would accept as adequate explanations.

4. Solution (3) can be generalized. Suppose one seeks to explain not a single event but a putative general law L (such as $P = 2\pi(1/g)^{\frac{1}{2}}$). Perhaps they can at best find explanans entailing only another law L' which approximates L over some range of values.[8] When is an audience willing to consider such a merely approximative explanation for L, and how closely must L and L' agree for success?

I conjecture that the strategy which worked before will work again to answer these questions, if suitably modified because the explanandum is now a law. I suspect that an audience will entertain an approximative account only when they believe that L is itself an idealization, and that the degree to which L' must approach L will depend on what they believe about the comparative roles of each one in the system of accepted theoretical principles.

In other words, the approach outlined above also provides a program for solving the general problem of approximative explanation. But that is a project for another day.

Notes

[1]Adapted from the definition offered in [1] , p. 53.

[2]General explananda, which pose extra problems, will be treated briefly at the end of the paper.

[3]This is only a necessary, not a sufficient condition for the pragmatic adequacy of event-explanations.

[4]The "as opposed to" phrase, but neither this interpretation of it nor its application to approximative explanations, appears in [3] , p. 68. Also cf., [2] , pp. 28-33.

[5]Thus *d* measures the size of the "vicinity of" *m* considered by the audience.

[6]William Barr carefully distinguishes between idealized theories, ideal conditions, and ideal cases. For example, (a) "the distance a body falls divided by the square of the elapsed time equals a constant" is an *idealized theory*, (b) "the friction, buoyancy, and non-gravitational deflection of a body equal zero" is a set of *ideal conditions*, and (c) "if (b), then (a)" is an ideal case. This useful terminology will be adopted below. Barr's technical definitions are found in [1] , pp. 49ff.

[7]Conversely with Newton's derivation of the speed of sound in air, which was within 16% of available experimental values. No one accepted this as an approximative explanation, however, because they believed that condition (2) had never been met. Thomson suggests why in [6] , p. 104. Lindsay has conveniently collected the relevant documents in [4] .

[8]E.g., an *L'* for the pendulum law is given in [5] , pp. 104-107 and 150-152.

References

[1] Barr, William. "A Pragmatic Analysis of Idealizations in Physics." *Philosophy of Science* 41 (1974): 48-64.

[2] Girill, T. R. "Explanatory Pragmatics: A Critical Analysis." *Philosophy Research Archives* 3(1977): 1-50.

[3] Gorovitz, Samuel. "Aspects of the Pragmatics of Explanations." *Nous* 3 (1969): 61-72.

[4] Lindsay, R. Bruce. (ed.). *Acoustics: Historical and Philosophical Development*. Stroudsburg, Pa.: Dowden, Hutchinson and Ross, 1972.

[5] Page, Leigh. *Introduction to Theoretical Physics*. Princeton: D. Van Nostrand Co., 1965.

[6] Thomson, David. "Acoustics." In *Encyclopaedia Britannica* 9th ed., vol. I. New York: Samuel L. Hall, 1878. Pages 100-119.

Part VI

Aspects of Rationality

Four Contributions Values Can Make to the Objectivity of Social Science

Sandra G. Harding[1]

University of Delaware

In his intellectual autobiography, Rudolf Carnap describes a dispute between Otto Neurath and the rest of the members of the Vienna Circle.

> All of us in the Circle were strongly interested in social and political progress. Most of us, myself included, were socialists. But we liked to keep our philosophical work separated from our political aims. In our view, logic, including applied logic, and the theory of knowledge, the analysis of language, and the methodology of science, are, like science itself, neutral with respect to practical aims, whether they are moral aims for the individual, or political aims for a society. Neurath criticized strongly this neutralist attitude, which in his opinion gave aid and comfort to the enemies of social progress. We in turn insisted that the intrusion of practical and especially of political points of view would violate the purity of philosophical methods. ([2], p. 23).

In this paper I will ask whether this still-dominant view of the proper relationship between social values and scientific method could <u>in principle</u> direct inquiries in social science which maximized objectivity. Neurath rejected the value-neutralist thesis on moral/political grounds by claiming it "gave aid and comfort to the enemies of social progress." ([2], p. 23). I shall argue against this thesis on epistemological grounds: the value-neutralist thesis gives aid and comfort to the enemies of <u>scientific</u> progress. If my argument is correct, we must develop a more adequate conception of objectivity.

While this value-neutralist thesis <u>is</u> the dominant view in philosophy of social science, it has not, of course, gone unchallenged. In fact it has been loudly disputed--if not always illuminatingly--from the time of the first attempts to found social inquiry on a scientific basis right down to the present moment. Even the call for a more adequate notion of

---

PSA 1978, Volume 1, pp. 199-209

objectivity is not new. Twenty-five years ago Richard Rudner argued that one consequence of Quine's position in his dispute with Carnap over the nature of "internal" vs. "external" questions in science is that the scientist _qua_ scientist makes value judgments. (This claim has been disputed but not defeated; cf. [7] for an assessment of the current status of the claim.) Rudner thought that recognition of this fact required "a radical reworking of the ideal of scientific objectivity." "...Objectivity for science lies at least in becoming precise about what value judgments are being and might have been made in a given inquiry - and even, to put it in its most challenging form, what value decisions ought to be made;... ." ([14], p. 6). While working social scientists often attempt to specify the value-laden character of particular theoretical approaches alternative to their own, comparatively little has been done on Rudner's other proposal--that is, on identifying what value decisions the scientist _qua_ scientist _ought_ to make. Peter Winch, too, has opposed the value-neutralists' way of thinking about objectivity in social inquiry [15]. However, his account and that of his critic, Ernest Gellner ([10],[11]) are legitimately open to objections, not because they give too large a role to the value judgments of scientists, but because they give too limited a role to such value judgments. Winch identifies one and Gellner two of the ways certain moral/political values ("social values" hence) must play a role in inquiry if the investigator is to provide the most adequate explanation of social phenomena. But the claim here is that there are four such objectivity-increasing roles for values. In all four cases it is not the presence but the suspension or absence of certain social values which contributes to biased accounts of social phenomena: for epistemological reasons the social scientist should be committed to certain social values.

Objectivity has meant different things in different contexts and to different people (cf.,[5]). Here the proposal is that, first, objectivity be thought of as fundamentally a property of inquiries--of the way scientists conduct them--and only derivatively as a property of the descriptions of nature produced by inquiries. This helps distinquish objectivity from truth: objectivity should be thought of not primarily as a property of propositions but, instead, as an attitude toward the conditions under which a hypothesis is to be regarded as true or false. This inquiring attitude is directed by the belief that the truth of the claims of science should not be taken to depend upon who makes the claims, whether "who" is a particular individual or an entire historically-located community. Thus objective inquiries are not compatible with relativism toward knowledge claims. When one conducts an inquiry objectively, one is willing--hopefully, eager--to countenance challenges to the truth of the results of one's inquiry, but objectivity requires also that one be motivated to further inquiry as the result of such challenges.

Secondly, maximal objectivity is a necessary but not a sufficient condition for arriving at the most adequate accounts of nature's regularities and their underlying determinants. It is not sufficient because our attempts to understand nature can be inadequate for us actually to understand nature in spite of our scrupulousness in trying to maximize objectivity. The most carefully conducted inquiries which are possible at any given time often result in incorrect descriptions of nature.

Traditional accounts of objectivity in social science have also divided into defenses of <u>methodological monism</u> and defenses of <u>methodological dualism</u>. Traditional methodological monists think that the natural sciences result in maximally objective accounts of nature. They think that there are no differences in principle between obstacles to achieving objectivity in social inquiry and obstacles in the natural sciences, and so the social sciences should try to duplicate the generally successful methods developed in the natural sciences.[2] Especially important among these are devices found useful in the physical sciences for insuring intersubjective agreement among observers, especially operationalized concepts and emphasis on quantitative measures (cf.,([1], p. 16);[3];([13], pp. 473-502)). For monists, to the extent that the social values of the observer direct social inquiry, the results of inquiry cannot exhibit objectivity. Social values are thought of as "subjective" or as "ideological" and thus as an obstacle to maximizing that intersubjective agreement which is the mark of reliable--that is, objective--inquiry.

In contrast, one-criterion dualists assert that there are key notions in the social sciences which are not compatible either with operationalization or with quantitative measurement insofar as these devices are designed to outrank sincere first-person testimony (cf., e.g.,([12], pp. 54-90)). And they accept the monists' criterion of objectivity. To the extent that social explanations must make use of these concepts, one-criterion dualists think such explanations are not objective at all: the social scientists <u>lack</u> objectivity. The two-criteria dualists think that there are two criteria of objectivity, and that neither the social nor the physical sciences can satisfy both (e.g.,[15]). They think that social inquirers do not need the quantitative methods so successful in physical science in order to achieve an objective understanding of alien ways of social life. It is both necessary and sufficient, they argue, to adopt methods of inquiry which maximize understanding the rules of a society and the institutions structured by these rules within which individual behavior occurs. The meaning of a society's way of life and thus the underlying determinants of social behavior can be grasped by understanding <u>as the natives do</u> the rules structuring the society's institutions.

Let us now examine four ways the social values of the inquirer can contribute to objectivity in social science. These contributions will be characterized as identifying value judgments, agent-critical value judgments, inquirer-critical value judgments, and community-critical value judgments. Where the first two are judgments about the social act which is being explained, the last two are judgments about properties of the inquirer.

Consider the following situation. In an alien culture, a visiting social scientist observes a large number of men living in a camp of some sort. The men are dressed in uniforms and are observed in such activities as marching, exercising, chanting in unison, lining up for food, and so forth. What has the social scientist seen? An army camp? A detention camp for dissidents? A group of workers at leisure? A monastery? A health spa? A training camp for members of the national opera company?

First of all, in order to identify and classify this behavior as human action the observer would have to discover what the men intended to do by their various actions. What marks phenomena as social actions rather than as biological, chemical, or astronomical events is that they strive to reach normative standards. They are explicitly or implicitly not only goal-directed, but directed toward normative goals. Consequently, as Winch, following Collingwood, argued, one cannot even identify _which_ phenomena are social phenomena nor what kind of social phenomena they are without grasping the normative concepts involved for the persons or culture observed. Here identifying value judgments are required.

The social scientist's guide says that this camp is a kind of monastery. Each man has decided that he has acquired too much property, that in this respect he has been avaricious. The guide reports that whenever someone recognizes this particular bad character trait in himself, he arranges his affairs and goes off to the monastery for three months to atone for his socially-destructive behavior and to purify himself of his excessive desire for worldly goods. The sinner also donates to the poor a generous amount of his property when he goes on such a retreat. Later the social scientist gets a chance to spend some time in the monastery, and the natives confirm the guide's account of their behavior. Our observer could not correctly identify this camp as a monastery unless he understood in his home culture what it is to be greedy for worldly goods, to atone for bad behavior, to rid oneself of unwanted desires, to donate to the poor. So the observer must first understand a certain range of social concepts (moral/political concepts, that is) as part of his own value system and, secondly, be willing to judge when others' behavior is structured by these social concepts. For the observer to _refrain_ from employing his home social concepts in identifying and classifying others' acts would result in inaccurate descriptions of what he observed. And if the observer's home society or the one observed _lacks_ a range of social concepts, it is probable that his description of the social regularities in the alien culture will be incomplete and/or biased by the gap in, or superfluous presence of, these values in his home society. (By "home social values" I mean just the inquirer's moral/political values. By "alien social values" or "alien culture" I mean simply values or culture not the inquirer's own. Alien values and culture sometimes exist within one's own political community, of course: the rich and the poor within a community might have different values and culture or might share the same values and culture.)

But an adequate explanation of an act often requires that the inquirer go beyond merely identifying and classifying the act as the actor would. For there are many cases in which an agent does not fully understand the true causes or consequences of his acts. And many acts which seem perfectly rational to the agent, appear irrational to an observer. If the social inquirer asked only the agent for an identification and classification of the act, he would often end up with an incomplete or irrational account. Now the inquirer must critically examine from the perspective of his own social value-system--that is, from the perspective of his own moral/political intuitions--the function that the act and the agent's kind of account of it have in the alien society. Otherwise, suspending such a critical examination, the inquirer biases his account in favor of the agent's value-system and produces just the kind of too-tolerant, particularistic account which scientific inquiry is designed to avoid. Such an examination might require the agent-critical value judgment that some people and societies sometimes are deluded about the nature and reasons for their own behavior, and that the maintenance of such delusions often brings benefits to other members of the society. In the case considered above, the inquirer might discover that the religious campers were in fact all members of a rising mercantile class whose ability to amass property threatened the power of the traditional priestly aristocrats. He might discover that "greed" had been re-defined in the preceding generation so that it was now taken to mean "owning more than your father owned". Citizens were constantly encouraged to undertake self-criticism in this respect, and the occasional mysterious disappearance of merchants who did not retreat to the monastery often enough served to keep the merchant class "voluntarily" eager to identify this bad character trait. The inquirer might conclude that it was very much to the benefit of the priestly class that the merchants want to limit their wealth "voluntarily", and that the "religious" explanation, believed at some level by just about everyone, was in fact useful to the priestly class as a means of controlling the merchants.

A _scientific_ account of social phenomena must invoke the inquirer's values in this second way. Like the natural scientist, the social scientist is skeptical of the mere _apparent_ regularities in nature--in our example, those regularities perceived by the agents. The social scientist drew upon his home values to discover the _real_ regularities and their underlying determinants, regularities which were not apparent to the agents. In this case, the inquirer's home values which contributed to a more objective account were more egalitarian than those of the society he studied. For him, it was not natural that in this inegalitarian society the campers should voluntarily be working against what appeared to be their own interests. He wanted to know how this inequality was maintained with so little overt conflict. Many social phenomena are like this one in that at first they appear to the observer to be irrational, and also in that they are only partially understood by the agents themselves. Often such lack of understanding is itself implicitly or explicitly socially valued.

Identifying and agent-critical value judgments drew on the inquirer's home moral/political intuitions, his home understanding of the difference between voluntarily trying to improve one's character and giving up property under an implied threat of death. Since the inquirer needed to invoke indentifying and agent-critical value judgments in order to identify, classify and (to a first approximation) explain alien social phenomena, these value judgments have themselves become variables influencing the nature of the results of inquiry. Thus some moral/political properties of the inquirer have become the remaining uncharted variables influencing the nature of the results of inquiry. Properties which have been necessary to maximize the objectivity of inquiry may well have in another respect have resulted in value-biased results of inquiry. How can the social scientist identify and critically evaluate the influence these variables have on the results of his inquiry? Two additional kinds of value judgments are here required, but let me just examine more fully why this move is necessary. We all know of notorious cases where the unexamined values of observers led them to mis-identfy and mis-explain alien phenomena, characteristically where there was either great "cultural distance" or a real conflict in interests between observer and observed. Missionary anthropologists misperceived much behavior of non-Christian groups because they uncritically applied their own normative standards to societies with very different normative standards. Social scientists coming from societies with racist, sexist or classist belief systems similarly sometimes have misunderstood the beliefs and behavior of persons in their own societies as well as of people in societies with very different attitudes toward race, sex or class. And, inquirers from scientific cultures have had trouble accurately describing the beliefs and practices of traditional societies. While these sorts of cases lead the value-neutralist to argue that it is crucial that the inquirer suspend his values when he conducts inquiry, that the inquirer's values be banned from inquiry, the argument here is that from a scientific point of view such an approach is not only ineffectual but also incoherent.

Consider that if temperature is a variable which affects the volume of solids and we want to measure the volume of a solid, we do not try to "suspend" temperature variations by ignoring them or not mentioning them, but instead to identify, evaluate the effect of, and control the variations. Refusing to examine the effect of a variable on a phenomenon does not contribute to increasing the reliability of the results of inquiry; it decreases the reliability to the extent that we cannot predict how the variable effects the phenomenon. Similarly, if the values of the investigator help to shape descriptions of social nature, then the values of the investigator should be identified and their effects evaluated.

Undertaking such an inquiry our social scientist might first note that social theories can be classified according to the social values which their conceptual schemes tend to promote. The claim here is that different theories cast "the facts" in such a light that different social policies seem very naturally to follow from the results of research. To take one example of such a classification, the political theorist William Connolly has identified varieties of "consensus" and "conflict" theories in social sciences [4]. The research of consensus theorists in each of the social sciences tends to be dominated by a central problem: "Why, in any particular setting, does such extensive social conflict exist, and how do and can established authorities cope with this conflict while maintaining the ongoing system?"([4], p. 15). Consensus is thus seen as part of the natural, expectable, order of social life and conflict and coercion, when they are detected at all, are seen as problems to be explained. In contrast, conflict theorists expect that in inegalitarian societies one should find conflict, dissension and unrest. For them, an inegalitarian society which has apparent consensus and very little overt conflict is itself a problem to be explained, and they ask: "How are existing inequalities maintained with the minimal use of explicit sanctions? How are the underlying conflicts contained and the most deprived groups suppressed?" ([4), p. 17). For conflict theorists, "established inequalities are viewed critically; the legitimacy of those in authority is seen as marginal; dissident activists are viewed presumptively as creative catalysts of needed social change."([4], p. 17).

Examinations such as Connolly's suggest the need for a third kind of value judgment which contributes to objectivity in social science: inquirer-critical value judgments . Inquirer-critical value judgments increase the objectivity of social science by exposing the alternative value systems prevailing in the home community and influencing inquirers' choices of theories and of conceptual schemes. Our social scientist might note that in his home society, social scientists who had testified on the side of the oil company in a recent anti-trust case were less able or willing to discover that the camps in the alien culture functioned to protect the monopoly interests of the priestly aristocrats. In contrast, the inquirer's own anti-monopoly values in part helped him to "discover" the monopoly-preserving function of the camps.

But is this third kind of value-judgment, in addition to the preceding two, sufficient to maximize objectivity in social science? How can the inquirer expose and examine those of his own social values which <u>he shares with everyone else in his community</u>--the community-wide values? (And a given social scientist might share values with and thus be a member of several only partially-overlapping communities: a scientific community, a university community, a neighborhood political community, a national organization working for certain suffrage reforms, etc.) Values turned up through inquirer-critical value judgments are comparatively easy to detect since these are variations in the values <u>within</u> the culture. But some value systems have been community-wide or close to community-wide in many people's experience--racism, sexism, classism, and speciesism for instance. If everyone I am exposed to simply describes the world in a "factual" manner from a classist perspective, how will I even know that attitudes toward class <u>are</u> values biasing my choice of an appropriate conceptual scheme and thus of the results of my research? How am I to identify the social values I have which, passing through my inquiries, will in fact help to shape my descriptions of others' behavior?

Here is the fourth positive role certain social values can play in increasing the objectivity of social inquiry. What the social scientist must do, it seems, is actively to seek out an alternative perspective on his own moral/political value system in order to decrease the incidence of undetectable, because community-wide, values he holds which shape the results of his inquiries. And, what the society has not provided, he must, for epistemological reasons, seek to create. Were there available to him only "consensus" theories of social behavior, we can see that it would increase the objectivity of his inquiry were he to promote the development of conflict theories. Without such effort, he will not be able to identify and assess the effect of "consensus" values on the results of his research. Thus the social scientist must, for epistemological reasons seek to produce value-systems alternative to the dominant values of the community.

Let me stress the difference between the third and fourth ways in which certain social values of the inquirer can increase the objectivity of social science. In making inquirer-critical value judgments the inquirer is to look around in his home society or in others and expose value systems already existing which are alternative to those shaping the prevailing social theories. He is to use these alternative perspectives to identify and to objectify inquirer-values which have already functioned in the identification and critical explanation of social behaviors. In the case of community-critical value judgments, in contrast, the inquirer must actively work to create a community in which the interests of all of the members of the community are correctly identified (where possible, correctly self-identified) and articulated. (The articulation of interests is in part done through the development of theories of social life.) This fourth case requires the social scientist to work for political egalitarianism; but note that he is required to work for political egalitarianism not for moral or political reasons but for epistemological reasons. In contrast to Neurath, I am concerned not with what makes good societies, but with what makes good science.

The directive to produce community-critical value-judgments is a directive to make the "republic of observers", upon which the objectivity of scientific inquiry rests, more truly representative. Since scientific claims are precisely those the truth of which should not depend upon who makes the claims, both scientific claims and all the evidence sustaining them must be available for everyone to examine. And who is "everyone"? Let me sketch out what we might think of as the "geography" of empirical support.

With respect to a physical object--say, a stick in water--if two people simultaneously perceive it from different locations in space, we can be sure that their perceptions of the stick will be different to the extent that these are actually made from different points of view. However, the many different perceptions of the stick need not--indeed, should not--result in hopelessly incompatible beliefs about the stick.

It is the job of scientific theory to get behind such conflicting appearances, to reconcile them by discovering their underlying determinants.

How similar is the geography of empirical support for claims about social events? What constitutes the "space" within which different "points of view" are located? First of all,even to identify and classify an event as the agents would, as part of a religious ceremony, let us say, require that the observers be "located" within shared/moral/ political "space"--that they share moral/political categories. Additionally, observers of social events are located in social space mapped by the <u>different</u> interests they may have in the event to be explained or in events similar to the one to be explained. An economist employed to help preserve a great industrial fortune may have very different interests in the problem of the camp than would an economist employed by the government to assess how best to redistribute economic benefits through new tax schedules. And not just loyalty to a conceptual scheme dictated by a favored problem, but all the various normative stances taken in everyday life give social scientists various interests in just the sorts of properties of social life for which it is the job of the social scientist to account. Where all the points of view on a physical phenomenon should ultimately be compatible at any given historical moment, a single coherent account of social phenomena is probably neither possible nor desirable. Interests are not just different; they are essentially conflicting in a way perceptions from different spatial locations are not.

Thus the problem is to obtain evidence for or against claims about a social phenomenon from <u>all</u> the significant locations in social space. Note that some physical locations are both significant and difficult to get observation reports from in geology, astronomy, biology. Similarly, some locations in social "interest space" are both significant perspectives and difficult ones from which to get reliable reports about certain aspects of social life. What makes social locations problematic in this way? Well, people in some social locations are less articulate. The very locations are politically remote in that people occupying these interest positions are seldom heard from. For instance, the poor seldom express their "true interests" through theories with which they study how to make the rich happier with less profit; blacks rarely have been in a position to use their values systematically to draw conclusions about the family life of racists; the normative standards of factory workers are rarely used to commission studies of how to improve the performance of factory owners; until recently, housewives had not developed theories to explain the bizarre behavior of psychoanalysts. Our most general views of human behavior are often biased precisely by the lack of such perspectives functioning in the construction of theories and/or the acceptance of hypotheses. Thus for epistemological reasons the social scientist has an interest in increasing the discussion of moral and political issues within his home society and in insuring that those within the society who might have reason to be critical of the dominant hypotheses have an opportunity to present their alternative conceptual schemes and the "facts" these pick out. Creating alternatives to the favored moral/political conceptual schemes and hypotheses in the home society turns out to be a prerequisite to maximizing objectivity of the results of inquiry in the social sciences.

In conclusion, I have suggested four ways in which the maximally objective inquiry necessarily will be one calling upon certain moral and political values. If I am correct, it is incorrect to assign such a negative, minimal, or non-existent epistemological role to the values of the investigator as has been insisted upon by the dominant tradition in Anglo-American social science and philosophy of social science. The tradition directed by the majority views of the Vienna Circle makes it impossible for social scientists to discover those social regularities and their determinants which remain hidden as long as the moral and political intuitions shaping the observations of social scientists remain unexamined. And, it is clear that if value-neutrality and objectivity are not connected in the way traditionally assumed, then we need a new account of what objectivity consists in. Hopefully this paper has provided at least some insights into the shape such an account should take. Finally, if the argument of this paper has been successful, then many interesting questions are open about the appropriate relation between social science and our moral/political life.

Notes

[1]The author wishes to express gratitude to the University of Delaware Research fund for support of part of the work associated with this paper, and to John Connolly for helpful comments.

[2]For examinations of the apparent inability of the physical sciences to force nature to reveal her secrets, cf.,the essays in [8]. For a discussion of the implications of this discovery for the social sciences, cf.,[9]. [6] provides an important critique of the positions referred to here as methodological monism and methodological dualism.

References

[1] Braybrooke, D. "Introduction." Philosophical Problems of the Social Sciences. New York: Macmillan, 1965. Pages 1-18.

[2] Carnap, R. "Intellectual Autobiography." The Philosophy of Rudolf Carnap. Edited by P.A. Schilpp. LaSalle, IL: Open Court, 1963. Pages 3-84.

[3] Chapin, S. "Definitions of Concepts." Social Forces 18 (1939): 153-160.

[4] Connolly, W. "Theoretical Self-Consciousness." Policy 6 (1973): 5-35.

[5] Diesing, P. "Subjectivity and Objectivity in the Social Sciences." Philosophy of the Social Sciences 2 (1972): 147-165.

[6] Fay, B. and Moon, J.D. "What Would an Adequate Philosophy of Social Science Look Like?" Philosophy of Social Science 7 (1977): 209-227.

[7] Gaa, J.C. "Moral Autonomy and the Rationality of Science." Philosophy of Science 44 (1977): 513-541.

[8] Harding, S.G. (ed.) Can Theories be Refuted? Essays on the Duhem-Quine Thesis. Dordrecht: D. Reidel, 1976.

[9] ------------. "Does Objectivity in Social Science Require Value-Neutrality?" Soundings LX (1977): 351-366.

[10] Gellner, E. "Concepts and Society." Transactions of the Fifth World Congress of Sociology 1 (1962): 153-183. (As reprinted in Gellner Cause and Meaning in the Social Sciences. New York: Macmillan, 1965. Pages 18-46.)

[11] ----------. "The New Idealism - Cause and Meaning in the Social Sciences." In Problems in the Philosophy of Science. Edited by I. Lakatos and A. Musgrave. Amsterdam: North-Holland, 1968. Pages 377-406, 426-432.

[12] Malcolm, N. Dreaming. New York: Humanities Press, 1960.

[13] Nagel, E. The Structure of Science. New York: Harcourt, Brace, and World, 1961.

[14] Rudner, R. "The Scientist Qua Scientist Makes Value Judgments." Philosophy of Science 20 (1953): 1-6.

[15] Winch, P. The Idea of a Social Science. New York: Humanities, 1958.

Four Basic Concepts of Medical Science

Caroline Whitbeck[1]

Institute for the Medical Humanities
University of Texas Medical Branch at Galveston

It might be thought that the basic concept in medical science is that of health. However, as it turns out, medicine has little or nothing to say about health. (This is less true of some of the other health sciences and health professions, notably nursing and public health.) Instead it is diseases, injuries, impairments, and, to a lesser extent, symptoms which are the focus of concern in medicine. It is a generally recognized and frequently bemoaned fact that in this country, contemporary medicine and medical education largely ignore, not only the promotion of health, but even the prevention of disease and injury.[2] (Medical educators are among those most acutely aware of this situation.) Instead, medicine is principally concerned with the diagnosis and treatment of medical crisis.

It is small wonder that in explicating the concept of health as it is found in the theories which inform the practice of medicine, Boorse ([1],[2]) arrives at a negative notion, viz., *health as the absence of disease and injury*. While a negative notion does do justice to the concept of health in contemporary medical theory, that tells us more about contemporary medicine than about health. Furthermore, I shall argue that contrary to what Boorse says, the concepts of disease, injury, and impairment found in medical sciences *are* value-laden; I propose an original account of the nature of the normative component in each of these concepts, and one which I believe is more precise than that offered by Margolis and Engelhardt, although in much the same spirit as theirs.

1. Diseases

Diseases, unlike injuries and impairments, are *processes*. The assumption underlying medical theory and practice since Thomas Sydenham has been that these psycho-physiological processes fall into certain natural classes or species, called "disease entities." As I have argued elsewhere, the criteria used for the classification of disease

PSA 1978, Volume 1, pp. 210-222

prominantly includes information regarding prevention and treatment [16]. Now I do not wish to argue, as Engelhardt sometimes has, that to call the processes going on in someone "a case of disease" is to say that it is necessarily, or even *prima facie*, something which it is against the person's interest to have and/or which imposes a therapeutic imperative on others ([5], pp. 102-3).[3] Rather, the centrality of information regarding prevention and treatment in disease classification shows that to call something a case of disease is to say that it is an instance of the *sort* of psycho-physiological process that people *wish to be able* to prevent or terminate. (By "people" I always mean the members of some human society because, like many others, I do not believe that a person can be defined in purely biological terms.) According to this account, *disease* is a value-laden concept, the decision to regard some process taking place in a particular person a case of a disease does not carry with it the presumption that it is against the bearer's interest to have that process. The full definition of human disease which I offer is the following:

A disease is any type of psycho-physiological process such that:

1.1. People wish *to be able* to prevent or terminate the process because it interferes with the bearer's psycho-physiological capacity to do those things that people commonly wish and expect to be able to do;

1.2. Either the process is statistically abnormal in those at risk or people have some other basis for a reasonable hope of finding means to prevent or effectively treat the process; and

1.3. The process is not also *necessary* for doing anything that people commonly want and expect to be able to do.

The first thing to notice about this definition is that it makes the extension of the term "disease" *relative to societal context* in being relative to what people are understood as wanting and expecting to be able to do. However, since the determination of what processes are statistically abnormal is independent of societal context, a process which qualifies as a disease by the first and third criteria, but which is ubiquitous in those at risk in a given locale and thus is not recognized as a disease by cultures within that locale, will still qualify as a disease. Thus the judgment as to what types of processes constitute disease depends on a value judgment of the societal group, *rather than* upon either the judgment of the person afflicted, or simply upon the judgment of the professional whom the society has charged with developing and applying preventive and therapeutic measures (although it is the latter who provide expert testimony about point 1.2 above).

Notice that this definition shows the concept *disease* to be value-laden in a very different way from another class of value-laden concepts, concepts such as that of a misfortune. Diseases, like forest fires and automobile accidents as well as injuries and impairments, are the *types* of things which people wish *to be able* to control in certain ways. With all things of the foregoing types we are particularly interested in being able to prevent, combat, eliminate, terminate, lessen, etc. such things, and it is for this reason that we regard the valuation involved as negative. However, although people do wish *to be able* to prevent, terminate, etc., things of this type, they do not wish to exercise that ability in every case. Although every case of a misfortune is necessarily something which is counter to the interests of its bearer, it may be in the interests of a particular individual to have a case of some disease or to bear a particular injury or impairment. For example consider the good fortune of contracting cowpox when smallpox threatens and there is no safe smallpox vaccine. Other examples will be discussed in the section on impairments. The point I wish to emphasize is that the desirability of *being able* to do something does not imply even a *prima facie* case for desirability of doing. Thus a person's desire *to be able* to drive his or her car long distances does not lead to even a *prima facie* presumption the person wishes to actually make any long trips by car.

You may also notice that the above definition applies only to human disease. I have confined myself to human disease because I agree with Engelhardt that the criteria for human and non-human disease are different. In the case of non-humans (non-humans that we do not anthropomorphize), it is necessary that a psycho-physiologic process threaten the survival of the individual or species *and* be statistically abnormal in order for it to count as a disease [5].

In contrast, when we consider human diseases, we count as diseases some processes which are statistically *normal* in those at risk. Furthermore, medical science has come to classify as diseases, processes which were previously considered normal. Consider especially those universal or near universal changes which in former times were considered to be normal consequences of aging, and which now are regarded as diseases in spite of their universal presence in older people.[4] In some instances the processes were seen to bear some resemblance to others which we already termed "diseases" when they occurred in younger people, in other instances, such as arteriosclerosis, increase in severity led epidemiologists to focus on the causes of this change and hence to raise the possibility of delaying or retarding the progress. In all cases *we came to think of the process as one which we might be able to prevent or treat*, and therefore no longer a necessary feature of the human life cycle. In terms of the foregoing definition of disease we came to regard the conditions outlined in criterion 1.2. as having been met.

Boorse tries to handle the distinction between disease and the consequences of aging with a distinction between processes which are internally caused and those which are externally caused, but this distinction does not help because, except in the case of a few genetic diseases, both internal and external factors are always involved. As I have argued elsewhere, those environmental contributions which are *counted* as causes are the ones that serve our Instrumental interest [16]. My point is illustrated by the fact that a number of the changes which *continue* to be regarded as due to age are due, at least in part, to living in earth's gravitational field over a long period of time, but since we believe that the strength of the gravitational field is not subject to our manipulations, we do not mention it as a cause.

Cases of changed status (from non-disease to disease) contrast with those cases in which medicine simply erred in its construal of the causal links between various signs, symptoms and other processes. Such errors occurred when it was thought normal for a woman to cough blood at the time of menstruation (so that many cases of TB in women went undiagnosed until the advanced stages), or that masturbation caused insanity [6].

I shall return to the subject of the descriptive and normative components in the concept of disease in a moment, but first I wish to define two notions which are similar to, but distinct from, that of disease. These are injury and impairment.

2. Injury

"Injury", "trauma", or "wound" are all synonyms. Any process which is triggered by the infliction of a wound is either a healing process or a disease process (e.g., a staphlycoccal infection) or some combination of the two. In either case the process is *not* conceived as intrinsic to the wound or injury. However, although it differs in its descriptive component the concept of injury is value-laden in exactly the same way as the concept of disease.

I define an injury, trauma, or wound as any type of psycho-physiologic change[5] such that:

2.1. People wish to be able to prevent or correct it because it interferes with the bearer's ability to do those things that people commonly want and expect to be able to do;

2.2. People have some reasonable hope of being able to prevent it or to make its reversal more rapid or complete ; and

2.3. Is not necessary in order to be able to do what people commonly want and expect to be able to do.

3. Impairment

It is useful to employ certain frequently used distinctions among the concept related to those I am discussing. For example, Mervyn Susser offers the following distinctions with which I am in general agreement:

> *Disease* is best thought of as a process that creates a state of physiological and psychological dysfunction...[in] the individual. *Illness* is best thought of as a subjective state, a psychological awareness of dysfunction, also... [in] the individual. *Sickness* is best thought of as a state of social dysfunction, a social role assumed by the individual that is defined by the expectations of society and that, thereby, affects the state of his relations with others. Impairment, disability, and handicap are analogous terms, but they are not synonyms either. They refer to established, stable and persisting states rather than to unstable evolving processes. *Impairment* refers to a persisting physical or psychological defect in the individual. *Disability* refers to persisting physical or psychological dysfunction, also confined to the individual. *Handicap*, like sickness, refers to persisting social dysfunction, a social role assumed by the impaired and disabled individual that is defined by the expectations of society. ([15], pp. 4-5).

One may have a few qualms about the fact that what frequently is impaired in an impairment is an ability, so that to keep the contrast with disability one might wish to substitute the term "defect" with its connotations of attendant structural abnormality. However, to do this would be to ignore the fact that it is in terms of impaired *ability* or *function* that many such impairments or defects are identified. The underlying anatomical or physiological conditions may or may not then be discovered. Sterility is a convenient example. Various anatomical anomalies (including some which are intentionally created) may underlie sterility. Moreover a number of diseases and disorders have sterility as a sign or symptom (e.g., starvation, Simmond's disease, anemic debility, and various uterine, ovarian, and testicular diseases). Given a person who complains of sterility, the problem is to decide if the sterility is a manifestation of a disease process or whether it is due to a stable (though perhaps correctable) anatomical defect (and hence an impairment), or whether it is due to a temporary anatomical change such as inflammation due to injury. (The case for underlying injury is more plausible if we take blindness, lameness, or amnesia, rather than sterility, as our example.) It is the failure to distinguish diseases from impairments and injuries which underlies Boorse's peculiar proposal to count sterility as a disease (along with such items as gunshot wounds and "general crushing") ([1], pp. 550-551). Sterility is first and foremost the name of a sign or symptom.[6] By extension the term is applied to the stable anatomical anomaly, if any, which gives rise to it. Since many, though by no means all, impairments go by the same name as manifestations of disease, any attempt to blur the distinction between disease and impairment threatens to obscure the distinction between diseases and their manifestations (i.e., their clinical and pathological signs and symptoms) - a distinction which medicine has been careful to make at least since the time of Sydenham.[7] Furthermore, a blurring of the distinctions between the terms "disease", "injury" and "impairment" leads one to underestimate the *descriptive* component in the meanings of these terms. I shall return to the sources of Boorse's error presently, but first let us consider the question of values in connection with impairments. I do agree that the normative component is the same in the case of the concepts of disease, injury and impairment. However, unlike Boorse, I think it is not null.

The definition which I offer of impairment is (predictably) the following:

An impairment is a persisting psycho-physiological anomaly such that:

3.1. People wish to be able to prevent or correct the anomaly because it interferes with the ability of the bearer to do what people commonly want and expect to be able to do; and

3.2. Is not necessary for doing what people commonly want and expect to be able to do.[8]

According to this definition, conditions like sterility will count as impairments even though they are states which some people undergo surgery to achieve. I take it as a recommendation for the present explication of the normative component over those of Engelhardt [5] [4] and Margolis [11] that it accounts for and explains why particular cases of disease, impairment, and/or injury need not be undesirable for their bearers. So far I have discussed disease, injury and impairment. Unlike Boorse, I claim that these concepts have normative (as well as descriptive) components even though there are circumstances in which it is to a person's *advantage* to have a given disease, injury, or impairment. I have explicated the normative element in the concepts disease, injury or impairment, to be that these are psycho-physiological processes and states which people (in some societal group) wish *to be able* to prevent, treat, or correct. Therefore, the society develops the means to do this.[9] These means are embodied in the institution of medicine or healing which Fabrega [8] tells us exist within every human culture.

4. Diseases, Injuries and Impairments in Relation to Symptoms

Some cases of disease (and *mutatis mutandis* injuries and impairments) develop without symptoms. This asymptomatic period may or may not be succeeded by a period in which symptoms are manifest. If such an asymptomatic period follows a symptomatic period it is called the "latent stage." If it precedes all symptomatic episodes (episodes of illness or acute episodes), then the disease at that stage is alternatively termed "sub-clinical". Now it is clear that during the sub-clinical phase of a disease there will be no ill (or symptomatic) person to seek treatment.[10] In cases of diseases which are asymptomatic during only part of their course, there is no problem in explaining the desire to *be able* to prevent and treat them, a desire which I have proposed as one essential feature of disease. Let us consider those cases which are subclinical throughout their *entire* course and furthermore leave no significant effects except, perhaps, specific immunity. In these cases it is the fact that the process bears a significant resemblance to other cases (that is, cases of the same disease entity) which *do* have symptomatic stages.

It is a consequence of the definition of disease which I have offered that if an asymptomatic process in a given person were not sufficiently similar to processes which manifest symptoms at *some* stage to be considered a process of the same primary type or species, then we would *not* consider it to be a case of a disease. This has an interesting consequence that, in conformity with medical usage, all responses to vaccination which are typically asymptomatic are classified as non-diseases *in spite of the fact the body's response to the killed or attenuated organism is indistinguishable from its initial response to the fully virulent organism* in cases of the corresponding disease.[11] The response to the *virulent* organism typically is part of a process which is, of course, not entirely asymptomatic and that is the reason why the whole process is classified as a disease. But the fact that the initial physiological response to the fully virulent organism is considered to be the first (and sub-clinical) stage of the disease in question, while the *physiologically indistinguishable* response to vaccination is *not*, shows how far medical usage is from using the term "disease" in a way which is purely descriptive of physiological and anatomical changes. It may be clear that the concept of a disease is intimately related to the concept of an illness, and not definable in purely physiological terms as Boorse ([1], [2]) has held.

5. Diagnostic Categories and Medical Records

Boorse has pointed out that the standard nomenclature manuals include in their listing conditions which do *not* interfere with the ability to do what people commonly want and expect to be able to do. One example is the absence of an appendix. (Since Boorse does not distinguish diseases from injuries and impairments, he calls the absence of an appendix a non-debilitating "disease".) Boorse's mistaken belief that there are non-debilitating impairments or diseases arises from a misunderstanding of the nature and function of nomenclature manuals. He takes the latter to provide an inventory of all and only those conditions which are considered to be diseases (injuries and impairments). However their actual function is to provide a uniform system for keeping records. It is important to keep records of many anomalies even if they are not (as far as we know) impairments. For example, it is important to record the fact that a person has no appendix, if, during an operation, it is observed that this is the case. This will not only explain why no appendectomy was performed, but also rule out the diagnosis of "appendicitis" for any abdominal pain that a person may have in the future. It is of course also possible that we will come to recognize functions of now apparently useless structures in the human body, so one might on principle wish to keep a record of all anomalies to be on the safe side.

The nomenclature manuals, both the 1962 *AMA Standard Nomenclature* which Boorse employs extensively, and the work that has replaced it for the keeping of hospital records in this country, the *Hospital Adaptation of the ICDA (H-ICDA)*[12] do in fact show some appreciation of the distinctions between diseases, injuries, impairments, and simple anomalies. It is true that these manuals do use two crude general heading "operations" on the one hand, and either "diseases" *(AMA Standard Nomenclature)* or "diseases and injuries" (the *H-ICDA)* on the other, but, as the *subheadings* under these general headings clearly indicate, not all that falls under the first is an operation, and not all that falls under the second is a disease (or injury).[13]

6. The Patient

Earlier I said that medicine is not very concerned with health. I was using the term "health" in what I take to be the fundamental sense. However, I recognize that the term is also used to mean merely the absence of disease, injury, or impairment, although this use is common only in the context of a description of the application of medical art, which is a further indication that this use arose by fitting the concept of health to the Procrustian bed of medical concepts. Consider, for example, the expression "restoring the person to health". The preceding account may be regarded as an examination of this so-called negative notion of health. It is, I claim, the only one which is a concern of somatic medicine, and medically-oriented psychiatry. It is this notion of health which is related to that of patient, where "patient" means the host, or *bearer* of a disease, injury, or in certain cases, impairment.[14]

There is another sense of "patient" which contrasts with the one I am considering; namely that of a person who, voluntarily or otherwise, has come under treatment by the healing institutions of the society and thus, to a varying extent, has assumed the sick role. It may be noted that in our society it is common for pregnant women to be patients in the second sense, the sense of interest to the social scientist, without being patients in the first sense, the sense which concerns us at the moment.

In view of the foregoing distinction between the two senses of patient, I wish to conclude by reiterating the role of socially recognized values in determining the extension of the term "patient" where the latter means simply the bearer of a disease or injury. According to the definitions of disease, injury and impairment offered above, the extension of these terms, and hence that of "patient" depends on what people do commonly want and expect to be able to do, the existing basis for a reasonable hope for prevention and treatment. Determination of these factors presupposes a context of assumed knowledge and shared interests (and hence values). Thus the societal context is an important factor in determining the extension of the term "patient",[15] but this is *not* because the social role determines the application of the term in the sense in which I have been discussing it.

Notes

[1] I am indebted to Joseph Margolis for some very useful criticisms of an earlier draft of this paper.

[2] Well-baby care within general pediatrics is something of an exception. However, much of the preventive concern tends to focus on the prevention of specific disease (e.g., via vaccination) rather than the promotion of health which guards against a variety of diseases and injuries. I deal with the distinction between prevention of disease and the promotion of health at length in [18]. In that paper I sketch a theory of health and argue that contrary to what Boorse, Engelhardt and Margolis say, disease (together with injury and impairment) does not form the complement of the concept of health, but that disease and health are not even concepts of the same order, i.e., health does not represent one end of a continuum and disease (together with injury and impairment) the other. The conceptual relationships turn out to be much more complex so that for example health (good health) is compatible with a fair degree of disease, injury and impairment. Therefore I hold that while greater orientation towards prevention, and towards management of chronic disease is possible and desirable in medicine (and in fact exists in some other societies, see Sidel and Sidel, [13]), the primary practice of health promotion is most appropriately conducted by individuals and lay communities with access to the resources to pursue the task.

[3] In discussing the therapeutic imperative, Engelhardt may have been led astray by Parson's classical formulation of the ideal doctor-patient relationship and its attendant view of the sick role. ([12], p. 465). However, the obligation of the practitioner to treat the sick, i.e., those who have taken on the sick *role* is a fact about the social roles. In a similar move, Boorse conflates sickness and illness ([2], p. 61). For the distinction between disease, sickness, and illness, see Section 3. below.

[4] These are among the universal diseases which pose such a problem for Boorse's account. Roughly his approach is to say that disease is a psychophysiologic process which lowers one's functional capacity below the ability of those in one's reference class. His problem is in defining the reference class. He must include age or all babies would be counted as diseased, but then he is at a loss to account for the fact that some processes are counted as diseases in spite of afflicting all older people ([1], pp. 566-568).

[5]Some may find the notion of a psychophysiologic change somewhat vague, and prefer some formulation such as "change in tissue or psychic state", however, such formulations misleadingly suggest that there is some clear dichotomy between the psychic and somatic. The literature on stress is instructive in revealing inseparability of psychic and somatic changes. Elsewhere I have argued that the distinction between somatic diseases and those conditions which are with any justification called "mental diseases" is not very interesting [16]. I maintain that most diseases have both psychic and somatic manifestations and predisposing factors (Tay-Sachs disease is one of the few purely somatic diseases that comes to mind.).

[6]A single phenomenon may be both a sign and a symptom if it is both observable by the practitioner and experienced by the bearer. Symptoms include not only abnormal *sensations* but abnormal functioning or appearance. Symptoms which are apparent *only* to the bearer are termed "subjective symptoms". *Stedman's Medical Dictionary*, ([14], p. 1231).

[7]Elsewhere I have discussed at length the relation between a disease process and its manifestations [16].

[8]The definitions which I have offered of disease, injury and impairment have an obvious application in clarifying what is meant by the charge that some issue or area has been "medicalized". It is that some state or process or persisting condition such as pregnancy, social deviance, or feature of sexual behavior is treated as though it met the defining criteria for being a disease, injury or impairment.

[9]In another paper [10] we have argued that among the kinds of actions one can learn to perform are actions upon the self, although western culture primarily recognizes and supports the mastery of skills in acting upon the non-self. We argue that in particular the state of so-called "selflessness" sought in certain eastern traditions is better understood as the ability to act on oneself in such a way that one learns how to psychically disengage from ones projects (including ones modes of interpretation) without abandoning them as false or useless. (This state of "selflessness" has been argued to be the same as that of Jungian individuation or of Plato's "birth in beauty" [7] or of the end state in psychotherapy [9].) Suppose that this is so, and this state became one which people in western society commonly sought to attain and furthermore that some people seemed congenitally unable to attain it. Given this expanded repertoire of what people would commonly want and expect to be able to do, we would then come to regard such people as impaired, although their impairment might conceivably be of benefit to themselves in specific instances.

[10] Strictly speaking, very minor, *non-diagnostic* symptoms may occur in the course of sub-clinical disease. Thus cases of infantile paralysis which manifest themselves in ways that are indistinguishable from a slight cold are termed "sub-clinical".

[11] I omit from consideration vaccinations which are typically followed by significant symptoms. Depending on the specifics of the example chosen there may be reason to consider some of these to be very mild cases of the disease. Of course it is always possible that some (correctly prepared) vaccine which typically produces no symptoms, will trigger symptoms in a given case. The disease in question may be anything from an allergic reaction, to encephalomyelitis and other neurologic lesions. As their names indicate, all of these diseases are distinguished from the diseases which the vaccination was designed to prevent.

[12] This work is in turn the World Health Organization's 1965 *International Classification of Diseases* and the Public Health Service's Eighth Revision of the *International Classification of Diseases, Adapted for Use in the United States*.

[13] Considering the currently employed H-ICDA (which, in this respect too, is an improvement over the AMA work) we find that of the seventeen sub-headings, twelve of these are headings for what are properly termed diseases, a fact reflected in the presence of the word "disease" in the case of all but one, that sub-heading being "Neoplasms" (i.e., tumors). Two of the sub-headings are mixed diseases and non-diseases presumably grouped together on the basis of the specializations that deal with them. These are "Mental Disorders," and "Delivery and Complications of Pregnancy, Childbirth, and the Puerperium" (which includes such familiar non-diseases as "normal delivery"). The remaining three are, significantly, "Congenital Anomalies" (which includes, of course, those which are debilitating, i.e., impairments, as well as *simple* anomalies like absence of an appendix), "Signs, Symptoms, and Ill-Defined Conditions," and finally, "*Injuries* and Adverse Effects."

[14] In cases in which it is not customary to expect to be able to correct the impairment, its bearer is often spoken of as a "disabled person" rather than a "patient".

[15] The room for actual disagreement between societies about whether a person is a patient is likely to be relatively small when both societies are in possession of the same biological, psychological and sociological information about that person; a society that did not have (or recognize its) interest in the survival, reproduction and cooperation of its members would not survive itself.

References

[1] Boorse, Christopher. "Health as a Theoretical Concept." Philosophy of Science 44(1977): 542-573.

[2] -------------------. "On the Distinction Between Disease and Illness." Philosophy and Public Affairs 5(1975): 49-68.

[3] Commission on Professional and Hospital Activities. Hospital Adaptation of ICDA (H-ICDA), 2nd. ed. Ann Arbor, Michigan: CPHA, 1973.

[4] Engelhardt, H. Tristram, Jr. "The Concepts of Health and Disease." Evaluation and Explanation in the Biomedical Sciences. Edited by Engelhardt and Spicker. Dordrecht: D. Reidel, 1975.

[5] --------------------------. "Is There a Philosophy of Medicine?" In PSA 1976, Volume 2. Edited by F. Suppe and P.D. Asquith. East Lansing: Philosophy of Science Association, 1977. Pages 94-108.

[6] --------------------------. "The Disease of Masturbation: Values and the Concept of Disease." Bulletin of the History of Medicine 48(1974): 234-238.

[7] Erde, Edmund. "Mind-Body and Malady." The Journal of Medicine and Philosophy 2(1977): 177-190.

[8] Fabrega, Horacio. Disease and Social Behavior. Cambridge, MA: MIT Press, 1974.

[9] Fingarette, Herbert. The Self in Transformation: Psychoanalysis, Philosophy, & the Life of the Spirit. New York: Basic Books, 1963.

[10] Garvin, Robert and Whitbeck, Caroline. "Toward a Complete Theory of Agency: Acting on the Self." Unpublished manuscript.

[11] Margolis, Joseph. "The Concept of Disease." Journal of Medicine and Philosophy 1(1976): 238-255.

[12] Parsons, Talcott. The Social System. New York: Free Press, 1951.

[13] Sidel, Ruth and Sidel, Victor W. A Healthy State: An International Perspective on the Crisis in United States Medical Care. New York: Pantheon Books, 1977.

[14] Stedman's Medical Dictionary. 22nd ed. Baltimore: Williams and Wilkins, 1972.

[15] Susser, Mervyn. Causal Thinking in the Health Sciences. New York: Oxford University Press, 1973.

[16] Whitbeck, Caroline. "Causation in Medicine: The Disease Entity Model." Philosophy of Science 4(1977): 619-637.

[17] ------------------. "The Relevance of the Philosophy of Medicine for Philosophy of Science." In PSA 1976, Volume 2. Edited by F. Suppe and P.D. Asquith. East Lansing: Philosophy of Science Association, 1977. Pages 123-135.

[18] ------------------. "Health and Agents." Unpublished manuscript.

Why Astrology Is A Pseudoscience

Paul R. Thagard[1]

University of Michigan-Dearborn

Most philosophers and historians of science agree that astrology is a pseudoscience, but there is little agreement on *why* it is a pseudoscience. Answers range from matters of verifiability and falsifiability, to questions of progress and Kuhnian normal science, to the different sorts of objections raised by a large panel of scientists recently organized by *The Humanist* magazine. Of course there are also Feyerabendian anarchists and others who say that no demarcation of science from pseudoscience is possible. However, I shall propose a complex criterion for distinguishing disciplines as pseudoscientific; this criterion is unlike verificationist and falsificationist attempts in that it introduces social and historical features as well as logical ones.

I begin with a brief description of astrology. It would be most unfair to evaluate astrology by reference to the daily horoscopes found in newspapers and popular magazines. These horoscopes deal only with sun signs, whereas a full horoscope makes reference to the "influences" also of the moon and the planets, while also discussing the ascendant sign and other matters.

Astrology divides the sky into twelve regions, represented by the familiar signs of the Zodiac: Aquarius, Libra and so on. The sun sign represents the part of the sky occupied by the sun at the time of birth. For example, anyone born between September 23 and October 22 is a Libran. The ascendant sign, often assumed to be at least as important as the sun sign, represents the part of the sky rising on the eastern horizon at the time of birth, and therefore changes every two hours. To determine this sign, accurate knowledge of the time and place of birth is essential. The moon and the planets (of which there are five or eight depending on whether Uranus, Neptune and Pluto are taken into account) are also located by means of charts on one of the parts of the Zodiac. Each planet is said to exercise an influence in a special sphere of human activity; for example, Mars governs drive, courage and

---

PSA 1978, Volume 1, pp. 223-234

daring, while Venus governs love and artistic endeavor. The immense number of combinations of sun, ascendant, moon and planetary influences allegedly determines human personality, behavior and fate.

Astrology is an ancient practice, and appears to have its origins in Chaldea, thousands of years B.C. By 700 B.C., the Zodiac was established, and a few centuries later the signs of the Zodiac were very similar to current ones. The conquests of Alexander the Great brought astrology to Greece, and the Romans were exposed in turn. Astrology was very popular during the fall of the Republic, with many notables such as Julius Caesar having their horoscopes cast. However, there was opposition from such men as Lucretius and Cicero.

Astrology underwent a gradual codification culminating in Ptolemy's _Tetrabiblos_[20], written in the second century A.D. This work describes in great detail the powers of the sun, moon and planets, and their significance in people's lives. It is still recognized as a fundamental textbook of astrology. Ptolemy took astrology as seriously as he took his famous work in geography and astronomy; this is evident from the introduction fo the _Tetrabiblos_, where he discusses two available means of making predictions based on the heavens. The first and admittedly more effective of these concerns the relative movements of the sun, moon and planets, which Ptolemy had already treated in his celebrated _Almagest_ [19]. The secondary but still legitimate means of prediction is that in which we use the "natural character" of the aspects of movement of heavenly bodies to "investigate the changes which they bring about in that which they surround." ([20], p. 3). He argues that this method of prediction is possible because of the manifest effects of the sun, moon and planets on the earth, for example on weather and the tides.

The European Renaissance is heralded for the rise of modern science, but occult arts such as astrology and alchemy flourished as well. Arthur Koestler has described Kepler's interest in astrology: not only did astrology provide Kepler with a livelihood, he also pursued it as a serious interest, although he was skeptical of the particular analyses of previous astrologers ([13], pp. 244-248). Astrology was popular both among intellectuals and the general public through the seventeenth century. However, astrology lost most of this popularity in the eighteenth century, when it was attacked by such figures of the Enlightenment as Swift [24] and Voltaire [29]. Only since the 1930's has astrology again gained a huge audience: most people today know at least their sun signs, and a great many believe that the stars and planets exercise an important influence on their lives.

In an attempt to reverse this trend, Bart Bok, Lawrence Jerome and Paul Kurtz drafted in 1975 a statement attacking astrology; the statement was signed by 192 leading scientists, including 19 Nobel prize winners. The statement raises three main issues: astrology originated as part of a magical world view, the planets are too distant for there to be any physical foundation for astrology, and people believe it merely out of longing for comfort ([2], pp. 9f.). None of these objections is ground for condemning astrology as pseudoscience. To show this, I shall briefly discuss articles written by Bok [1] and Jerome [12] in support of the statement.

According to Bok, to work on statistical tests of astrological predictions is a waste of time unless it is demonstrated that astrology has some sort of physical foundation ([1], p. 31). He uses the smallness of gravitational and radiative effects of the stars and planets to suggest that there is no such foundation. He also discusses the psychology of belief in astrology, which is the result of individuals' desperation in seeking solutions to their serious personal problems. Jerome devotes most of his article to the origins of astrology in the magical principle of correspondences. He claims that astrology is a system of magic rather than science, and that it fails "not because of any inherent inaccuracies due to precession or lack of exact knowledge concerning time of birth or conception, but rather because its interpretations and predictions are grounded in the ancients' magical world view."([12], p. 46). He does however discuss some statistical tests of astrology, which I shall return to below.

These objections do not show that astrology is a pseudoscience. First, origins are irrelevant to scientific status. The alchemical origins of chemistry ([11], pp. 10-18) and the occult beginnings of medicine [8] are as magical as those of astrology, and historians have detected mystical influences in the work of many great scientists, including Newton and Einstein. Hence astrology cannot be condemned simply for the magical origins of its principles. Similarly, the psychology of popular belief is also in itself irrelevant to the status of astrology: people often believe even good theories for illegitimate reasons, and even if most people believe astrology for personal, irrational reasons, good reasons may be available.[2] Finally the lack of a physical foundation hardly marks a theory as unscientific ([22], p. 2). Examples: when Wegener [31] proposed continental drift, no mechanism was known, and a link between smoking and cancer has been established statistically [28] though the details of carcinogenesis remain to be discovered. Hence the objections of Bok, Jerome and Kurtz fail to mark astrology as pseudoscience.

Now we must consider the application of the criteria of verifiability and falsifiability to astrology. Roughly, a theory is said to be verifiable if it is possible to deduce observation statements from it. Then in principle, observations can be used to confirm or disconfirm the theory. A theory is scientific only if it is verifiable. The vicissitudes of the verification principle are too well known to recount here ([9], ch. 4). Attempts by A. J. Ayer to articulate the principle failed either by ruling out most of science as unscientific, or by ruling out nothing. Moreover, the theory/observation distinction has increasingly come into question. All that remains is a vague sense that testability somehow is a mark of scientific theories ([9], ch. 4; [10], pp. 30-32).

Well, astrology is vaguely testable. Because of the multitude of influences resting on tendencies rather than laws, astrology is incapable of making precise predictions. Nevertheless, attempts have been made to test the reality of these alleged tendencies, using large scale surveys and statistical evaluation. The pioneer in this area was Michel Gauquelin, who examined the careers and times of birth of 25,000 Frenchmen. Astrology suggests that people born under certain signs or planets are likely to adopt certain occupations: for example, the influence of the warlike planet Mars tends to produce soldiers or athletes, while Venus has an artistic influence. Notably, Gauquelin found <u>no significant correlation</u> between careers and either sun sign, moon sign, or ascendant sign. However, he did find some statistically interesting correlations between certain occupations of people and the position of certain planets at the time of their birth. ([5], ch. 11, [6]). For example, just as astrology would suggest, there is a greater than chance association of athletes and Mars, and a greater than chance association of scientists and Saturn, where the planet is rising or at its zenith at the moment of the individual's birth.

These findings and their interpretation are highly controversial, as are subsequent studies in a similar vein [7]. Even if correct, they hardly verify astrology, especially considering the negative results found for the most important astrological categories. I have mentioned Gauquelin in order to suggest that through the use of statistical techniques astrology is at least <u>verifiable</u>. Hence the verification principle does not mark astrology as pseudoscience.

Because the predictions of astrologers are generally vague, a Popperian would assert that the real problem with astrology is that it is not falsifiable: astrologers can not make predictions which if unfulfilled would lead them to give up their theory. Hence because it is unfalsifiable, astrology is unscientific.

But the doctrine of falsifiability faces serious problems as described by Duhem [4], Quine [21], and Lakatos [15]. Popper himself noticed early that no observation ever guarantees falsification: a theory can always be retained by introducing or modifying auxiliary hypotheses, and even observation statements are not incorrigible ([17], p. 50). Methodological decisions about what can be tampered with are required to block the escape from falsification. However, Lakatos has persuasively argued that making such decision in advance of tests is arbitrary and may often lead to overhasty rejection of a sound theory which <u>ought</u> to be be saved by anti-falsificationist strategems ([15], pp. 112 ff.). Falsification only occurs when a better theory comes along. Then falsifiability is only a matter of replaceability by another theory, and since astrology is in principle replaceable by another theory, falsifiability provides no criterion for rejecting astrology as pseudoscientific. We saw in the discussion of Gauquelin that astrology can be used to make predictions about statistical regularities, but the nonexistence of these regularities does not falsify astrology; but here astrology does not appear worse than the best of scientific theories, which also resist falsification until alternative theories arise.[3]

Astrology can not be condemned as pseudoscientific on the grounds proposed by verificationists, falsificationists, or Bok and Jerome.

But undoubtedly astrology today faces a great many unsolved problems ([32], ch. 5). One is the negative result found by Gauquelin concerning careers and signs. Another is the problem of the precession of the equinoxes, which astrologers generally take into account when heralding the "Age of Aquarius" but totally neglect when figuring their charts. Astrologers do not always agree on the significance of the three planets, Neptune, Uranus and Pluto, that were discovered since Ptolemy. Studies of twins do not show similarities of personality and fate that astrology would suggest. Nor does astrology make sense of mass disasters, where numerous individuals with very different horoscopes come to similar ends.

But problems such as these do not in themselves show that astrology is either false or pseudoscientific. Even the best theories face unsolved problems throughout their history. To get a criterion demarcating astrology from science, we need to consider it in a wider historical and social context.

A demarcation criterion requires a matrix of three elements: [theory, community, historical context]. Under the first heading, "theory", fall familiar matters of structure, prediction, explanation and problem solving. We might also include the issue raised by Bok and Jerome about whether the theory has a physical foundation. Previous demarcationists have concentrated on this theoretical element, evident in the concern of the verification and falsification principles with prediction. But we have seen that this approach is not sufficient for characterizing astrology as pseudoscientific.

We must also consider the <u>community</u> of advocates of the theory, in this case the community of practitioners of astrology. Several questions are important here. First, are the practitioners in agreement on the principles of the theory and on how to go about solving problems which the theory faces? Second, do they care, that is, are they concerned about explaining anomalies and comparing the success of their theory to the record of other theories? Third, are the practitioners actively involved in attempts at confirming and disconfirming their theory?

The question about comparing the success of a theory with that of other theories introduces the third element of the matrix, historical context. The historical work of Kuhn and others has shown that in general a theory is rejected only when (1) it has faced anomalies over a long period of time and (2) it has been challenged by another theory. Hence under the heading of historical context we must consider two factors relevant to demarcation: the record of a theory over time in explaining new facts and dealing with anomalies, and the availability of alternative theories.

We can now propose the following principle of demarcation:

> A theory or discipline which purports to be scientific is <u>pseudoscientific</u> if and only if:

1) it has been less progressive than alternative theories over a long period of time, and faces many unsolved problems; but
2) the community of practitioners makes little attempt to develop the theory towards solutions of the problems, shows no concern for attempts to evaluate the theory in relation to others, and is selective in considering confirmations and disconfirmations.

Progressiveness is a matter of the success of the theory in adding to its set of facts explained and problems solved ([15], p. 118; cf. [26], p. 83).

This principle captures, I believe, what is most importantly unscientific about astrology. First, astrology is dramatically unprogressive, in that it has changed little and has added nothing to its explanatory power since the time of Ptolemy. Second, problems such as the precession of equinoxes are outstanding. Third, there are alternative theories of personality and behavior available: one need not be an uncritical advocate of behaviorist, Freudian, or Gestalt theories to see that since the nineteenth century psychological theories have been expanding to deal with many of the phenomena which astrology explains in terms of heavenly influences. The important point is not that any of these psychological theories is established or true, only that they are growing alternatives to a long-static astrology. Fourth and finally, the community of astrologers is generally unconcerned with advancing astrology to deal with outstanding problems or with evaluating the theory in relation to others.[4] For these reasons, my criterion marks astrology as pseudoscientific.

This demarcation criterion differs from those implicit in Lakatos and Kuhn. Lakatos has said that what makes a series of theories constituting a research program scientific is that it is progressive: each theory in the series has greater corroborated content than its predecessor ([15], p. 118). While I agree with Lakatos that progressiveness is a central notion here, it is not sufficient to distinguish science from pseudoscience. We should not brand a nonprogressive discipline as pseudoscientific unless it is being maintained against more progressive alternatives. Kuhn's discussion of astrology focuses on a different aspect of my criterion. He says that what makes astrology unscientific is the absence of the paradigm-dominated puzzle solving activity characteristic of what he calls normal science ([14], p. 9). But as Watkins has suggested, astrologers are in some respects model normal scientists: they concern themselves with solving puzzles at the level of individual horoscopes, unconcerned with the foundations of their general theory or paradigm ([30], p. 32). Hence that feature of normal science does not distinguish science from pseudoscience. What makes astrology pseudoscientific is not that it lacks periods of Kuhnian normal science, but that its proponents adopt uncritical attitudes of "normal" scientists despite the existence of more progressive alternative theories. (Note that I am not agreeing with Popper [18] that Kuhn's normal scientists are unscientific; they can become unscientific only when an alternative paradigm has been developed.) However, if one looks not at the puzzle solving at the level of particular astrological predictions, but at the level of theoretical problems such as the precession of the equinoxes, there is some agreement between my criterion and Kuhn's; astrologers do not have a paradigm-induced confidence about solving theoretical problems.

Of course, the criterion is intended to have applications beyond astrology. I think that discussion would show that the criterion marks as pseudoscientific such practices as witchcraft and pyramidology, while leaving contemporary physics, chemistry and biology unthreatened. The current fad of biorhythms, implausibly based like astrology on date of birth, can not be branded as pseudoscientific because we lack alternative theories giving more detailed accounts of cyclical variations in human beings, although much research is in progress.[5]

One interesting consequence of the above criterion is that a theory can be scientific at one time but pseudoscientific at another. In the time of Ptolemy or even Kepler, astrology had few alternatives in the explanation of human personality and behavior. Existing alternatives were scarcely more sophisticated or corroborated than astrology. Hence astrology should be judged as not pseudoscientific in classical or Renaissance times, even though it is pseudoscientific today. Astrology was not simply a perverse sideline of Ptolemy and Kepler, but part of their scientific activity, even if a physicist involved with astrology today should be looked at askance. Only when the historical and social aspects of science are neglected does it become plausible that pseudoscience is an unchanging category. Rationality is not a property of ideas eternally: ideas, like actions, can be rational at time but irrational at others. Hence relativizing the science/pseudoscience distinction to historical periods is a desirable result.

But there remains a challenging historical problem. According to my criterion, astrology only became pseudoscientific with the rise of modern psychology in the nineteenth century. But astrology was already virtually excised from scientific circles by the beginning of the eighteenth. How could this be? The simple answer is that a theory can take on the appearance of an unpromising project well before it deserves the label of pseudoscience. The Copernican revolution and the mechanism of Newton, Descartes and Hobbes undermined the plausibility of astrology.[6] Lynn Thorndike [27] has described how the Newtonian theory pushed aside what had been accepted as a universal natural law, that inferiors such as inhabitants of earth are ruled and governed by superiors such as the stars and the planets. William Stahlman [23] has described how the immense growth of science in the seventeenth century contrasted with stagnation of astrology. These developments provided good reasons for discarding astrology as a promising pursuit, but they were not yet enough to brand it as pseudoscientific, or even to refute it.

Because of its social aspect, my criterion might suggest a kind of cultural relativism. Suppose there is an isolated group of astrologers in the jungles of South America, practicing their art with no awareness of alternatives. Are we to say that astrology is <u>for them</u> scientific?

Or, going in the other direction, should we count as alternative theories ones which are available to extraterrestrial beings, or which someday will be conceived? This wide construal of "alternative" would have the result that our best current theories are probably pseudoscientific. These two questions employ, respectively, a too narrow and a too broad view of alternatives. By an alternative theory I mean one generally available in the world. This assumes first that there is some kind of communication network to which a community has, or should have, access. Second, it assumes that the onus is on individuals and communities to find out about alternatives. I would argue (perhaps against Kuhn) that this second assumption is a general feature of rationality; it is at least sufficient to preclude ostrichism as a defense against being judged pseudoscientific.

In conclusion, I would like to say why I think the question of what constitutes a pseudoscience is important. Unlike the logical positivists, I am not grinding an anti-metaphysical ax, and unlike Popper, I am not grinding an anti-Freudian or anti-Marxian one.[7] My concern is social: society faces the twin problems of lack of public concern with the advancement of science, and lack of public concern with the important ethical issues now arising in science and technology, for example around the topic of genetic engineering. One reason for this dual lack of concern is the wide popularity of pseudoscience and the occult among the general public. Elucidation of how science differs from pseudoscience is the philosophical side of an attempt to overcome public neglect of genuine science.

Notes

[1]I am grateful to Dan Hausman and Elias Baumgarten for comments.

[2]However, astrology would doubtlessly have many fewer supporters if horoscopes tended less toward compliments and pleasant predictions and more toward the kind of analysis included in the following satirical horoscope from the December, 1977, issue of Mother Jones: VIRGO (Aug. 23-Sept. 22). You are the logical type and hate disorder. This nitpicking is sickening to your friends. You are cold and unemotional and sometimes fall asleep while making love. Virgos make good bus drivers.

[3]For an account of the comparative evaluation of theories, see [26].

[4]There appear to be a few exceptions; see [32].

[5]The fad of biorhythms, now assuming a place beside astrology in the popular press, must be distinguished from the very interesting work of Frank Brown and others on biological rhythms. For a survey, see [5].

[6]Plausibility is in part a matter of a hypothesis being of an appropriate kind, and is relevant even to the acceptance of a theory. See [26], p. 90, and [25].

[7]On psychoanalysis see [3]. I would argue that Cioffi neglects the question of alternatives to psychoanalysis and the question of its progressiveness.

References

[1] Bok, Bart J. "A Critical Look at Astrology." In [2]. Pages 21-33.

[2] -----------, Jerome, Lawrence E., and Kurtz, Paul. Objections to Astrology. Buffalo: Prometheus Books, 1975.

[3] Cioffi, Frank. "Freud and the Idea of a Pseudoscience." In Explanation in the Behavioral Sciences. Edited by R. Borger and F. Cioffi. Cambridge: Cambridge University Press, 1970. Pages 471-499.

[4] Duhem, P. The Aim and Structure of Physical Theory. (trans.) P. Wiener. New York: Atheneum, 1954. (Translated from 2nd edition of La Théorie Physique: Son Object Sa Structure. Paris: Marcel Rivière & Cie, 1914.)

[5] Gauquelin, Michel. The Cosmic Clocks. Chicago: Henry Regnery, 1967.

[6] -----------------. The Scientific Basis of Astrology. New York: Stein and Day, 1969.

[7] -----------------. "The Zelen Test of the Mars Effect." The Humanist 37(1977): 30-35.

[8] Haggard, Howard W. Mystery, Magic, and Medicine. Garden City: Doubleday, Doran & Company, 1933.

[9] Hempel, Carl. Aspects of Scientific Explanation. New York: The Free Press, 1965.

[10] ------------. Philosophy of Natural Science. Englewood Cliffs: Prentice-Hall, 1966.

[11] Ihde, Aaron J. The Development of Modern Chemistry. New York: Harper and Row, 1964.

[12] Jerome, Lawrence E. "Astrology: Magic or Science?" In [2]. Pages 37-62.

[13] Koestler, Arthur. The Sleepwalkers. Harmondsworth: Penguin, 1964.

[14] Kuhn, T.S. "Logic of Discovery or Psychology of Research." In [16]. Pages 1-23.

[15] Lakatos, Imre. "Falsification and the Methodology of Scientific Research Programmes." In [16]. Pages 91-195.

[16] ------------- and Musgrave, Alan.(eds.). Criticism and the Growth

Knowledge. Cambridge: Cambridge University Press, 1970.

[17] Popper, Karl. The Logic of Scientific Discovery. London: Hutchinson, 1959. (Originally published as Logik der Forschung. Vienna: J. Springer, 1935.)

[18] ------------. "Normal Science and its Dangers." In [16]. Pages 51-58.

[19] Ptolemy. The Almagest (The Mathematical Composition). (As printed in Hutchins, Robert Maynard (ed.). Great Books of the Western World, Volume 16. Chicago: Encyclopedia Britannica, Inc., 1952. Pages 1-478.)

[20] -------. Tetrabiblos. Edited and translated by F.E. Robbins. Cambridge: Harvard University Press, 1940.

[21] Quine, W.V.O. "Two Dogmas of Empiricism." In From a Logical Point of View. New York: Harper & Row, 1963. Pages 20-46. (Originally published in The Philosophical Review 60(1951): 20-43.)

[22] Sagan, Carl. "Letter." The Humanist 36(1976): 2.

[23] Stahlman, William D. "Astrology in Colonial America: An Extended Query." William and Mary Quarterly 13(1956): 551-563.

[24] Swift, Jonathan. "The Partridge Papers." In The Prose Works of Jonathan Swift, Volume 2. Oxford: Basil Blackwell, 1940-1968. Pages 139-170.

[25] Thagard, Paul R. "The Autonomy of a Logic of Discovery." Forthcoming in the Festschrift for T.A. Goudge.

[26] ---------------. "The Best Explanation: Criteria for Theory Choice." Journal of Philosophy 75(1978): 76-92.

[27] Thorndike, Lynn. "The True Place of Astrology in the History of Science." Isis 46(1955): 273-278.

[28] U.S. Department of Health, Education and Welfare. Smoking and Health: Report of the Advisory Committee to the Surgeon General of the Public Health Service. Washington, D.C.: U.S. Government Printing Office, 1964.

[29] Voltaire. "Astrologie" and "Astronomie". Dictionnaire Philosophique. In Oeuvres Complètes de Voltaire, Volume XVII. Paris: Garnier Frères, 1878-1885. Pages 446-453.

[30] Watkins, J.W.N. "Against 'Normal Science'." In [16]. Pages 25-37.

[31] Wegener, Alfred. "Die Entstehung der Kontinente." *Petermanns Geographische Mitteilung* 58(1912): 185-195, 253-256, 305-309.

[32] West, J.A. and Toonder, J.G. *The Case for Astrology*. Harmondsworth: Penguin, 1973.

Rhetoric and Scientific Rationality

Maurice A. Finocchiaro [1]

University of Nevada- Las Vegas

The problem of scientific rationality and Feyerabend's views on the subject need no special introduction in this context. What does need emphasis is the desirability of resisting the widespread tendency either to dismiss him as a Dadaist or to take his rhetoric seriously and despair about the rationality of science. Both tendencies betray a failure to appreciate his peculiar contributions, which are a (relatively) novel theory of rationality, and a (relatively) novel problem. The theory is that scientific rationality is essentially rhetorical, in the sense that scientific method ultimately reduces to a matter of techniques of persuasion (as distinct from, say, a matter of "method"); and the problem, which remains even for those who may reject the theory, is that of the role of rhetoric in science.

Since Feyerabend's latest book [1] [2] is full of the rhetoric of irrationalism, it is best to begin by purifying the air surrounding such rhetoric. The first point to be made here is that even if he were otherwise right, his "irrationalism" would be wrong. That is, even if it were true that "reason" was frequently overruled in the development of science (pp. 145, 155), this would show that it is sometimes reasonable to overrule "reason", which can therefore be only what was thought to be, but is not really, reason. Thus, in this type of argument, far from it being the case that Feyerabend can't lose, as one critic has stated in another context [7], he can't win.

Feyerabend's considerations against "reason" can be divided into three arguments, which we may call: the insufficiency of reason argument; the argument from incommensurability; and the historical argument. In the first Feyerabend argues as follows. Since the teaching of small children is not exclusively a matter of argument, and since there are processes like the mastery of a language which look like the result of reason but are due partly to indoctrination and partly to natural processes of growth, it follows that nonrational growth is possible both in adults and in institutions. Therefore "even the most puritanical

PSA 1978, Volume 1, pp. 235-246

rationalist will then be forced to stop reasoning and use propoganda and coercion" (p. 25) whenever conditions are such that "forms of argumentation turn out to be too weak" (p. 25) to accomplish his ends. Feyerabend's argument may be accepted with the following qualifications. There is nothing irrational in sometimes stopping reasoning; there would be only if one assumed that reasoning is the only proper activity human beings can engage in, an assumption no rationalist needs to make. Moreover, there is no need to resort to propaganda and coercion, for propaganda is a perversion of rhetoric, while coercion is a misuse of noncognitive causes, and such perversions and misuses may or may not be used by the rationalist; the only thing that follows is that rhetoric and noncognitive causes may have to be used; but such things are nonrational rather than irrational, and they are susceptible of being handled properly as well as improperly. So the conclusion to be reached is that reasons and arguments are not the only causes that affect human actions and thought; there are also rhetoric and noncognitive causes that operate.

Now at this point, Feyerabend would like to add that therefore it would be very strange if arguments were the only thing operating in scientific rationality, so strange that even if that were the case, then one should rebel against such unnatural restrictions and help to create a more human science in which the other factors are not excised. To this we may reply by asking why should every human activity be a microcosm of the entire human life. Why can't there be some activity, e.g., science, where the only rules of the game involve argumentation? As long as we remember that there are other things in life, why should this activity of reasoning which we have labeled "science" be mixed with the others? I believe Feyerabend's only plausible answer could be, because as a matter of (historical) fact what we label "science" does contain a mixture of elements. Here we would leave Feyerabend's general considerations and go to his historical argument, which therefore is more crucial to his position than he makes it seem.

Before examining Feyerabend's historical argument, however, let us look at the one from incommensurability, which he thinks "creates problems for all theories of rationality" (p. 214). Feyerabend calls two theories incommensurable (pp. 223,228-9, 269) when the subject matter to which they refer differs so radically that (1) the existence of the entities and processes presupposed by one theory implies the nonexistence of the entities and processes presupposed by the other theory, and (2) it makes no sense to say that one theory is a generalization of the other or that they overlap. The main step in the argument consists in trying to establish that there are scientific changes where the theories involved are incommensurable, e.g., Aristotelian and Galilean physics, classical and quantum physics, and Newtonian and relativisitic mechanics (pp. 224-5,271,276-7); given this incommensurability, Feyerabend concludes that one cannot say that the new theory which replaces the old is epistemologically better. However, if we distinguish progress from rationality (cf., [2]), then we realize that Feyerabend's conclusion does not imply that the agents involved in the transition did not act rationally. In fact, while answering a number of philosophical objections against his incommensurability claims (pp. 277-85), Feyerabend himself argues cogently that, in spite of incommensurability, it is still possible to behave rationally; for example, a self inconsistent observation report would disconfirm the corresponding theory (p. 278); one could learn the meaning of the new theory the way anthropologists learn the language of newly-discovered tribes (pp. 278-82); and crucial experiments are still possible (pp. 282-3). So Feyerabend has not shown that incommensurability creates problems for theories of rationality, but rather that it creates problems for theories of progress. Moreover, since, besides being distinct, rationality and progress are related, the possibility of rationality opens up a minimal possibility of progress, namely that changes involving incommensurable theories are changes for the better in the sense that they result from the behavior of rational agents.

But how shall we test for the actuality of these possibilities still left open after Feyerabend's arguments? The answer, whose suggestion can be found in Feyerabend himself, is that the test is to be conducted by anthropological field work: "Let us commence field work in this domain also, and let us study the language of new theories not in the definition-factories of the double-language model, but in the company of those metaphysicians, physicists, playwrights, courtesans, who have constructed new world views!" (p. 282). When the event under consideration is relatively far into the past, like the emergence of modern science in the 17th century, the only kind of field work possible is the analysis of historical records.

This brings us to his historical argument against reason. Here he tries to show that "<u>essential ingredients of modern science survived only because reason was frequently overruled in the past</u>" (p. 145). When we look at the details of the argument (e.g., pp. 145-61, 179-80, 196-200) we discover that the reason which was overruled was usually what empiricists, inductivists, Popperian critical rationalists, or Imre Lakatos <u>think</u> is reason. In other cases, there was no <u>overriding</u> of reason but rather a combination of it with nonrational factors; nor does Feyerabend show that such nonrational factors were improper from their own point of view. All of the strengths and all of the limitations of Feyerabend's position are present in his analysis of Galileo, and so it will be valuable to examine this in detail.

The account consists of four elements: an interpretation of Galileo's work in terms of a definite procedure which he allegedly follows; a description of this procedure in terms of such concepts as theory, observation, propaganda, appeals to emotion and prejudices, etc; a historical explanation of the fact of Galileo's success as resulting from the procedures he used; and a favorable evaluation of Galileo insofar as the procedure he follows is "<u>perfectly reasonable</u>" (p. 143).

One need not have mentioned the first item in this list were it not for the fact that all of Feyerabend's talk against method may easily be interpreted as showing that the concept of method is a useless one in the philosophy of science. It is clear, however, from Feyerabend's "actual procedure" (his method, as it were) that he is only arguing against "the idea of a method that contains firm, unchanging, and absolutely binding principles for conducting the business of science" (p.23) i.e., "the idea that science can, and should be run according to fixed and universal rules" (p. 295). In a few places (e.g., pp. 145 and 163) he even uses the forbidden word "method" to refer to Galileo's procedure So Feyerabend's point is really that no general theory of method is defensible, which says that all scientists in all situations use a certain definite method; instead different scientists use different methods on different occasions. It is important to note, however, that Feyerabend's account presupposes that the concept of method is a very useful one in understanding what a scientist does; a method becomes the effective cause (in a nondeterministic sense) in the historical explanation of a given scientific success. I do not think that such method-explanations of success are objectionable on general grounds; one may argue that such accounts provide historical understanding and can even be used to ground philosophical theses ([4] , pp. 225-28).

Let us now see how Galileo's method is described. Feyerabend deserves great credit for enlarging Galileo's method beyond the epistemological factors usually considered and for including rhetorical and artistic components. The epistemological factors Feyerabend considers are the following. (1) Galileo introduces and elaborates hypothese (e.g., motion of the earth, relativity of motion, celestial reliability of telescopes) which are inconsistent with well-established theories (e.g., geocentrism, operative character of motion, optics) and with well-established facts (e.g., vertical fall, apparent size of Venus and Mars); this procedure is labeled "counterinduction" (pp. 29,77,99-101). (2) Galileo lets inadequate views mutually support each other, e.g., (a) "an inadequate view, the Copernican theory, is supported by another inadequate view, the idea of the nonoperative character of shared motion" (p. 89), (b) he lets one refuted view--Copernicanism--support another refuted view--the idea that telescopic phenomena are faithful images of the sky (pp.141-2), and (c) "Galileo changed his view about the 'neutral motions--he made them permanent and 'natural'--in order to make them compatible with the rotation of the earth" (p. 96); Feyerabend gives no special name to this aspect of Galileo's procedure, which may be labeled "dialectical", to adapt Amos Funkenstein's terminology [5] . (3) Ad hocness: Galileo uncritically accepts any and all ideas and observations which support Copernicanism (e.g., pp. 90,93-8). (4) Galileo drasticall reduced the <u>content</u> of dynamics by replacing the Aristotelian comprehensive theory of change, including locomotion, qualitative change, and generation and corruption by a theory dealing merely with the locomotion of matter (pp. 99-100,160-1); this is what Feyerabend labels the "step-back " (pp. 176, 153, 113).

The rhetorical aspects of Galileo's method are allegedly the following: deceptive tactics (pp. 70,81,87,160); utterances which are arguments in appearances only (p. 81); propaganda (pp. 81,90,157,160); psychological tricks (pp. 81,88,154); persuasion by confusion (p. 84); clever techniques of persuasion (pp. 141,143); misleading insinuations (p. 160); distortion (p. 160); appeal to emotion and prejudice (p. 154); jokes (p. 154); non-sequiturs (p. 154).

The artistic, aesthetic, and literary factors include "style, elegance of expression, simplicity of expression, tension of plot and narrative" (p. 157), "a sense of humor, an elasticity...and awareness of the valuable weaknesses of human thinking" (p. 161).

The epistemological factors emphasized by Feyerabend are ones that go against almost every principle held by orthodox philosophers of science; the rhetorical factors seem to violate the ideas of the basic honesty and decency of science, widely held by scientists and laymen alike; the aesthetic factors contradict the alleged gap between the arts and the sciences. Indeed Feyerabend delights in being contrary. His contrariness reaches its highest pitch when he argues that such Galilean procedure is "perfectly reasonable" in itself (pp. 143,145-61) and fruitful in other fields (pp. 163-4). In other words, Feyerabend's _evaluation_ of Galileo's method is the orthodox one, it is his descriptive interpretation of the features of that method that is unorthodox; and since his evaluation is partly grounded on his interpretation, that means that _his reasons_ for his evaluation are also unorthodox. Feyerabend does not always _effectively_ distinguish his description and his evaluation, though he often seems to or tries (pp. 143,156). Thus we often find him arguing that because it can be argued (by Feyerabend himself) that a certain view was refuted or inadequate, and because therefore Galileo _should_ have seen that it was inadequate, therefore it follows that Galileo regarded it as inadequate, and hence that in espousing it he was acting counterinductively; or that because it can be argued (by Feyerabend himself) that a certain argument is incorrect, and because therefore Galileo _should_ have known that it was incorrect, therefore it follows that he knew it to be incorrect, and hence that his using it amounted to deceptive trickery. We might say that Feyerabend's alleged descriptive interpretations are really evaluations, and hence his whole case is historically unfounded; and since, as we argued earlier, his two other arguments (from the insufficiency of reason, and from incommensurability) depended on the historical argument, we might conclude that Feyerabend's views do not merit any further consideration.

This, however, would be a mistake. We need not deny our inclination to think that Feyerabend is doubly in the wrong, namely that his historical interpretations are descriptively wrong, and that _if_ they were right, then that would show that Galileo was wrong (irrational). The fact is, or rather my suspicion is, that Feyerabend turns out to be _right in spite of himself_. That is, his account of Galileo is not really irrationalistic, but pseudo-irrationalistic, and in reality Galileo proceeds rationally for Feyerabend; however, our idea of scientific rationality must be expanded first to allow aesthetic and rhetorical factors, and second epistemological practices proscribed by orthodox philosophies of science; moreover the aesthetic and rhetorical factors are by themselves merely _alogical_, and they must be judged by their own criteria; finally, whatever unorthodox epistemological practices Galileo may engage in, the real test of their rationality or propriety is their correspondence to basic and elementary forms of reasoning and argumentation, rather than to philosophically articulated _theories_ of scientific rationality, most of which presuppose the theory/

observation distinction, for example. Keeping this in mind, I now proceed to test my suspicion. I shall do so by examining Feyerabend's account of the tower argument.

The tower argument was one of the classical objections to the earth's rotation. Galileo states and criticizes it in a passage ([6], pp. 138-40) which may be analyzed as follows. The argument, which according to Galileo's spokesman, Salviati, begs the question, may be reconstructed as follows:

(A1) Bodies fall vertically.
(A2) If the earth rotated, bodies could not fall vertically,
because
(A3) If the earth rotated, then while a body was falling the place on the earth directly below it would be carried along toward the east and the body would land to the west of where it was originally dropped.
Therefore,
(A4) The earth does not rotate.

Galileo has no objection to the (formal) validity of the last step in this argument (the one from (A1) and (A2) to (A4)); however he questions its soundness in terms of whether premise (A1) is justified. The Aristotelian spokesman, Simplicio, gives the following justification:

(A1) Bodies fall vertically
because
(A5) Bodies _are_ _seen_ to fall vertically.

Now, with respect to this argument Galileo accepts the premise but questions its (formal) validity. This validity depends on whether or not apparent vertical fall implies actual vertical fall. The implication needs justification because it would _not_ hold on a rotating earth. In fact, if the earth _were_ rotating and bodies were _seen_ to fall vertically, then in actuality they would be following a path slanted to the earth's surface; that is, on a rotating earth, apparent vertical fall would _not_ imply actual vertical fall. Now, how could one possibly justify the implication? The only relevant argument would seem to be the following one:

(A6) If the earth does not rotate, then apparent vertical fall implies actual vertical fall.
(A7) The earth does not rotate.
Therefore,
(A8) Apparent vertical fall implies actual vertical fall.

Premise (A6) is indeed true, and the argument is (formally) valid. Unfortunately premise (A7) is identical with conclusion (A4) of the original argument, which is being examined. In short, the vertical fall argument, from (A1) and (A2) to (A4), is such that one of its premises (A1) is being justified on the basis of the same proposition (A7) [=A4] it has for conclusion.

The important issues can be put into focus in terms of the following argument, which combines the three separate segments just discussed, and where some obvious symbolic abbreviations have been made and the proposition numbers correspond:

(A6) If not-R, then S implies V.
(A7) Not-R.
(A8) ∴ S implies V.
(A5) S.
(A1) ∴ V.
(A2) If R, then not-V
(A4) ∴ Not-R.

Logically speaking, this is a very interesting deduction; in particular its three steps are valid. Rhetorically speaking, however, the identity of propositions (A7) and (A4) makes it worthless: someone who does not adhere to (A4), will not be any more favorably inclined toward it after this argument, which requires such adherence already at the very beginning.

In Feyerabend's account four main theses are involved. First (pp. 70-75), the Aristotelian tower argument presupposes naive realism with respect to motion (p. 75); and it is important to notice that this Aristotelian presupposition takes the form of what Feyerabend calls a "natural interpretation"; that is, "one does not first distinguish the apparent motion from the real motion and then connect the two by a correspondence rule. One rather describes, perceives, acts towards motion as if it were already the real thing" (p. 75). Second (pp. 75-8), Galileo discovered this Aristotelian natural interpretation counterinductively. Third (pp. 78-81), Galileo tests and examines the Aristotelian identification of real and apparent motion by introducing a different interpretation which "restores the senses to their position as instruments of exploration, <u>but only with respect to the reality of relative motion</u>. Motion 'among things which share it in common' is 'non-operative', that is, 'it remains insensible, imperceptible, and without any effect whatever' " (p. 78). Fourth (pp. 81-92), Galileo uses, and needs to use, propaganda, psychological tricks, and deception in elaborating this new concept of the relativity of motion.

It is a great insight for Feyerabend to have arrived to the first of these conclusions, though it must be qualified to restrict it to vertical fall, so as to say that the tower argument presupposes, as a "natural interpretation," the identification of real and apparent <u>vertical fall</u>. Feyerabend himself goes through the motions of making the qualification (p. 75), but he then forgets its importance in the rest of his discussion. At any rate, it is possible to give a textual and logical proof of the qualified conclusion (cf., [3]). However, that same analysis also shows that Feyerabend is completely wrong in his second thesis; Galileo is simply not aware of this Aristotelian presupposition of the tower argument, so he neither discovers it counterinductively, nor does he discover it at all. <u>We</u> can discover it by logical manipulations, but these are merely suggested by the text, nowhere contained in it. Feyerabend formulates his second thesis because he approaches the philosophy of science with the spectacles of the theory/observation distinction; so he wants to reconstruct some theory which the Aristotelians held, and the one that Galileo introduced. If Feyerabend had approached the tower argument as an argument, which is the way Galileo does, then he would have been inclined to interpret Galileo's discovery as one about an argument, namely that the tower argument is a *petitio principii*. Though Galileo could have proceeded to inquire about general theories and natural interpretations, the fact is that he does not do so; and he does not do so because he seems to be interested in arguments.

Third, Galileo introduces his concept of the relativity of motion not to test a presupposition of the tower argument, but to answer the objection to the earth's motion from the deception of the senses. Feyerabend thinks as he does partly because he fails to distinguish two different arguments against the earth's motion; namely the one from vertical fall and the one from the deception of the senses. The logic of the two arguments is different. Galileo correctly criticizes the first as being circular, the second as being groundless ([6] , pp. 167-71, and 233-47) in the sense that the earth's motion would not involve any deception of the senses; and there would be no deception of the senses because of the relativity of motion; in other words, it's not a case where one does not see something he ought to see (the real slanted path on a rotating earth); there is no reason why one ought to see such a slanted path because our eyes can only detect motion relative to them; that our eyes are so built can be proved by experiences on moving systems such as ships. That is, Galileo refutes the objection by arguing that there would be no deception of the senses if we experienced apparent vertical fall on a rotating earth, because only motion relative to the observer is perceivable. In this critique Galileo is certainly subsuming vertical fall under what Feyerabend calls "Paradigm II" (p. 87), namely the paradigm of motion in a moving system, like a boat. But this is not a "tricky" or "deceptive" introduction of a different paradigm of motion because this is the same paradigm used in the statement of the objection, which is that on a rotating earth our senses would be deceived in experiencing the appearance of vertical fall while in actuality the fall was over a slanted path. In other words, this different argument against the motion of the earth cannot even be made without using Paradigm II. Hence, in answering it, it is quite proper to use the same paradigm.

On the other hand, in criticizing the tower argument, the relativity of motion is not needed or used by Galileo in his charge of *petitio*; hence, here too, trickery and deception would be beside the point. Feyerabend thinks otherwise probably on account of his mistaken belief that Galileo's critique of the tower argument is meant to constitute positive evidence for the earth's motion, rather than a mere refutation. The situation here is rather complicated but may be summarized as follows. Feyerabend is right that Galileo's critique implies that naive realism is presupposed by the Aristotelian argument. The presupposition could be stated in the form ([3] , pp. 147-48):

not-R if and only if (S if and only if V),

using the earlier abbreviations. This, in turn implies that

R if and only if not-(S if and only if V).

Then Feyerabend would have Galileo support the right side of this second biconditional with the following argument: apparent and actual vertical fall are distinct on moving systems like a ship, since only relative motion is visible in such systems; hence apparent and actual fall _with respect to the earth_ are distinct. This would yield an argument concluding with "R." Feyerabend correctly points out that this amounts to subsuming vertical motion under "Paradigm II: Motion of objects in boats, coaches and other moving systems" (p. 87) instead of under "Paradigm I: Motion of compact objects in stable surroundings of great spatial extention--deer observed by the hunter" (p. 87). He is also right in saying that the gap between these two paradigms is unbridgeable, in the sense that there is no noncircular way of choosing the former for the purpose of applying it to vertical fall. He is wrong, however, in attributing such an argument to Galileo, who wisely avoided it.

And this brings us to the sound element of Feyerabend's fourth thesis. He is certainly right that Galileo uses rhetorical alogical considerations which are interesting to examine and which play an important function. However, their propriety is a very complicated affair. In order to begin to understand this matter, it is essential to refer to the passage ([6], pp. 140-43) which immediately follows the one analyzed above and which discusses the argument from _apparent_ vertical fall.

This argument may be reconstructed as follows:

(B1) If the earth rotated, then bodies could not appear (to an observer on its surface) to fall vertically, since (B2) if the earth rotated and bodies appeared to fall vertically then bodies would have mixed natural motions simultaneously around and toward the center, (B3) which is impossible; since (B4) bodies do appear to fall vertically, we may conclude that (B5) the earth does not rotate.

It was certainly a clever move for Galileo to begin his critique of the tower argument (in the earlier passage) by interpreting it to refer to _actual_ vertical fall. Feyerabend is right in saying that the Aristotelian argument simply interchanges apparent and actual fall without distinguishing them; so Galileo could have begun with either version of the argument. The fault of the argument from _apparent_ vertical fall is that it depends on a premise (B3) which is as much in need of proof as the conclusion at issue. This fault is less serious than the circularity of the argument from actual vertical fall. It is indeed more effective to start with a more serious criticism, and so Galileo's first rhetorical move is a judicious one.

Having focused on the actual vertical fall version, Galileo has in front of him a formally valid instance of denying the consequent (_modus tollens_). Of the two premises of this argument, propositions (A1) and (A2) above, Galileo could have questioned either one. If he had questioned the conditional proposition, the circularity could not have been exhibited as easily since the justification of this conditional premise is circular only insofar as it depends ultimately on the nonconditional premise of the original modus tollens, namely the proposition (A1) that bodies fall vertically ([3], pp. 144-45). It is certainly rhetorically effective, but certainly not improper, to exhibit the failure of an argument in the easiest possible way.

Third, when Galileo comes around to discuss the apparent vertical fall version of the argument, this is made to look as a revised version of the original argument, and hence to some extent as ad hoc. This seems to be a merely rhetorical fault, rather than a logical one. However, the argument is also shown to be dependent on a premise (B3) which is as much in need of proof as its conclusion (B5). This feature, though not purely logical, is not purely rhetorical either. For though questions of the comparative knowability of propositions are context-dependent, the context is an epistemological one, rather than one dependent merely on who is speaking to whom.

Fourth, the way that the discussion changes from apparent vertical fall to the ship experiment involves rhetorical considerations. Someone might say, following Feyerabend's evaluation of Galileo's rhetoric, that besides introducing the argument from apparent vertical fall in a prejudicial manner, he drops the topic in such a way as to make it look as if Simplicio was giving up on the argument and relying on a new one ([6], p. 141). This rhetorical evaluation would depend on interpreting Galileo's criticism of the new topic, the earth/ship analogy ([6], pp. 141-43), as part of his criticism of the ship experiment argument against the earth's motion ([6], pp. 143-49). This interpretation in turn would involve attributing to Galileo another deception, namely that he would be trying to show that a rock dropped from the top of the mast of a moving ship would land at the foot of the mast, and similarly on a rotating earth, while he does not believe in the analogy. However, it is itself a prejudicial interpretation to regard the passage where Galileo criticizes the ship/earth analogy as a logical part of his discussion of the ship experiment argument. No sound reason can be given for this interpretation, whereas evidence can be given to show that Galileo wants his criticism of the analogy to be part of his criticism of the apparent vertical fall argument. The evidence consists, first, of the logical fact that such a connection exists; for the alleged results of the ship experiment can be cited as evidence that bodies refuse to move simultaneously toward and around the center (proposition B3), and hence as support for the corresponding premise in the apparent vertical fall argument, namely the proposition (B1) that if the earth rotated, bodies could not be seen to fall vertically. Moreover, there is textual evidence that this is what Galileo has in mind. For example, the first time that the ship experiment is mentioned ([6], p. 126) is in the context of an argument where the impossibility of mixed motion is given as the explanation for the alleged deflection on the moving ship. Finally, the way the switch is made from the criticism of the analogy to the criticism of the alleged experimental results ([6], p. 143) suggests the same thing: Salviati agrees that "up to this point" the ship experiment as an argument purely from analogy has not been considered. In other words, in this passage Galileo is saying that though the ship experiment has just been criticized insofar as it is the basis of an explanatory argument against mixed motion, it is still standing as a pure analogy argument against the earth's motion. I believe Galileo is right here, and that a pure analogy argument has some force even though the mechanism of the analogy is unknown, i.e., even though we do not know why the analogy holds, as long as we are prepared to claim that it holds. At any rate, the conclusion is inescapable that interpretations of the rhetorical situation must be grounded on interpretations of the logical situation. This also means that Feyerabend's switch from method to rhetoric suggests a greater role for logical analysis. Thus Feyerabend's approach represents good news not only for rhetorical scholars (understanding rhetoric to refer to the theory and practice of verbal techniques of persuasion) but also for logicians (if logic is understood as the theory and practice of reasoning). Above all, of course, the news is good for all those who seek a deeper understanding of science.

Notes

[1]The author gratefully acknowledges the support received from the National Science Foundation on Grant No. SOC76-10220 and from the University of Nevada, Las Vegas for a sabbatical leave in 1976-77.

[2]References to this book are so numerous in what follows that the notation "[1]" will be dropped hereafter and only the page numbers will be given.

References

[1] Feyerabend, Paul K. Against Method. Atlantic Highlands, N.J.: Humanities Press, Inc., 1975.

[2] Finocchiaro, Maurice A. "Essay-review of Criticism and the Growth of Knowledge." Studies in History and Philosophy of Science 3 (1972-73): 357-372.

[3] ______________________. "Galileo as a Logician." Physis 16 (1974): 129-148.

[4] ______________________. History of Science as Explanation. Detroit, Michigan: Wayne State University Press, 1973.

[5] Funkenstein, Amos. "The Dialectical Preparation for Scientific Revolutions." In The Copernican Achievement. (UCLA Center for Medieval and Renaissance Studies Contributions, Volume VII.) Edited by Robert S. Westman. Berkeley: University of California Press, 1975. Pages 165-203.

[6] Galilei, Galileo. Dialogue Concerning the Two Chief World Systems. Translated by S. Drake. Berkeley: University of California Press, 1953, 1962 and 1967.

[7] Gellner, Ernest. "Beyond Truth and Falsehood." British Journal for the Philosophy of Science 26(1975): 331-342.

Part VII

Quantum Logic

The Metalogic of Quantum Logic

Peter Mittelstaedt

Institut für Theoretische Physik
der Universität zu Köln, Germany

1. The Object Quantum Logic

In order to establish the object language of quantum physics, we begin with a quantum mechanical system S (atom, nucleus, elementary particle) and consider propositions A,B,... about this system S which can be proved or disproved by measuring processes. These propositions will be called elementary propositions. We will assume here that elementary propositions are value definite, i.e.,there always exists an experimental testing procedure which decides between truth and falsity of the respective proposition.

The logical connectives $A \wedge B$ (A and B), $A \vee B$ (A or B), $A \rightarrow B$ (A then B) and $\neg A$ (not A) will be defined by means of dialogs. A dialog is a formalized kind of discussion between two participants, the proponent P who asserts a certain proposition and the opponent O who attempts to refute it [3]. The concept of dialog can be defined by general frame-rules ([4],[7]) which are briefly summarized as follows: After assertion of the initial argument by P the dialog consists of a series of arguments which are stated in turn by P and O and which are either attacks on or defences of previous arguments. Attacks are rights which may be taken at any time in the dialog; defences are commitments which must be carried out in the reverse order of the corresponding attacks and at the latest, when there is no longer a right for an attack. If one of the participants has no argument to continue, he looses and the other one wins the dialog. In addition to the frame-rules the logical connectives are defined by the possibilities of attack and defence in a dialog:

connective	possibilities of attack	possibilities of defence
conjunction $A \wedge B$	1? , 2?	A , B
disjunction $A \vee B$	?	A , B
material implication $A \rightarrow B$	A	B
negation $\neg A$	A	

PSA 1978, Volume 1, pp. 249-256

The logical connectives can be used in order to extend the set of elementary propositions $\mathcal{S}_e$ by incorporating arbitrary iterations of the logical connectives. The resulting set of propositions will be denoted by $\mathcal{S}$. Formally the logical connectives $\wedge$, $\vee$, $\rightarrow$ and $\neg$ are two and one-place operations respectively on the set $\mathcal{S}$.

The length of the dialogs about the connectives are so far unrestricted since attacks and defences can be repeated an arbitrary number of times. It could be limited to a finite length if, e.g., in the dialog about $A \wedge B$ - there would be any guarantee that after a successful proof of A and B the result of the A-proof is still available and thus a repetition of the A-proof would lead to the same result. In order to test whether the proof result of a proposition A is still available after a proof of B, we introduce availability propositions k(A,B) (commensurability) and $\bar{k}$(A,B) (incommensurability) which state whether A is available after a B-proof or not. If these availability propositions are incorporated as additional possibilities into the attack and defence scheme of the logical connectives, the dialogs about the connectives can be limited to a finite length ([4],[7]).

The possibilities of attacking and defending elementary propositions, logical connectives and avilability propositions are summarized in the rules of the material dialog-game D_m. This game is called "material" since the respective dialogs contain the (non-dialogical) "material" proofs of elementary and availability propositions. If for a compound proposition A there exists a strategy of success within the dialog-game D_m we write $\vdash_{D_m} A$. It is obvious that the success of a dialog about A depends on the elementary propositions and the availability propositions contained in it ([4],[6]).

There are, however, compound propositions C which can successfully be defended in a material dialog irrespective of the elementary propositions contained in it and irrespective of the availability propositions k(A,B) and $\bar{k}$(A,B) of arbitrary subpropositions A and B of C. These propositions C are called formally true propositions. The proof procedures of these propositions can be summarized in the formal dialog-game D_f such that a proposition C can successfully be defended in a formal dialog if and only if C is formally true. In this case we write $\vdash_{D_f} C$. The formally true propositions C form the effective quantum logic and they can be derived from the calculus Q_{eff} of effective quantum logic. This will be denoted by $\vdash_{Q_{eff}} V \leq A$ [3]. If the value definiteness of the elementary propositions is taken into account, the calculus Q_{eff} can be extended by incorporating the principle of excluded middle for all propositions. Thus one arrives at the calculus Q of full quantum logic [6].

2. The Meta-Logic of Formal Meta-Propositions

Statements about the truth or the falsity of an object proposition with respect to a given dialog-game will be called meta-propositions. According to the dialog-games D_m and D_f respectively we have two kinds of meta-propositions

$\vdash_{D_m}$ A: "A is materially true"

and

$\vdash_{D_f}$ A: "A is formally true"

The falsity of a proposition A can always be expressed by the corresponding truth of the negation $\neg A$. These statements, i.e., $\vdash_{D_m}$ A and $\vdash_{D_f}$ A, will be called <u>elementary meta-propositions</u>, the proof of which has to be carried out within the framework of the object-dialog-games D_m and D_f.

In order to further extend our meta-language, we introduce the <u>meta-dialog-game</u> $\bar{D}$ which is again governed by the frame-rules mentioned above but which deals with meta-propositions instead of object-propositions $A \in \mathcal{S}$. Meta-propositions will be denoted by $\bar{A},\bar{B},\bar{C},\ldots$. In the next step we then define the <u>meta-logical</u> connectives by means of the possibilities of attack and defence in the meta-dialog-game D. The dialogic definition of the meta-connectives, which will be denoted by $\barwedge$, $\veebar$, $\Rightarrow$ and $\bar{\neg}$, is in complete analogy to the definition of the connectives in the object language, namely by the following attack-and-defence scheme:

Meta-connective	attack	defence
$\bar{A} \barwedge \bar{B}$	1? , 2?	$\bar{A}$, $\bar{B}$
$\bar{A} \veebar \bar{B}$	1?	$\bar{A}$, $\bar{B}$
$\bar{A} \Rightarrow \bar{B}$	$\bar{A}$	$\bar{B}$
$\bar{\neg}\bar{A}$	$\bar{A}$	

In a similar way as in the object language the meta-connectives $\barwedge$, $\veebar$, $\Rightarrow$ and $\bar{\neg}$ can be used in order to extend the set $\bar{\mathcal{S}}_e$ of elementary meta-propositions by incorporating arbitrary iterations of the meta-connectives.

In this section we consider <u>formal</u> elementary meta-propositions $\bar{A} = \vdash_{D_f} A$ and compound meta-propositions which contain elementary meta-propositions of this kind and an arbitrary number of iterations of the meta-connectives. Formal elementary meta-propositions $\bar{A}$ can either be expressed in terms of dialogs by $\vdash_{D_f}$ A or using the calculus Q_{eff} by $\vdash_{Q_{eff}} V \leq A$. In both formulations it becomes evident that all elementary meta-propositions are mutually commensurable, i.e., $k(\bar{A},\bar{B})$ is valid for all meta-propositions of this kind. In fact, the first formulation $\bar{A}$ asserts the existence of a strategy of success for the proposition A in D_f whereas in the second formulation $\bar{A}$ states that A can be deduced within the calculus Q_{eff}.

It is a consequence of this commensurability property that in contrast to the object-language dialogs the respective meta-dialogs for the connectives can always be restricted to a finite number of steps. Consequently, all compound meta-propositions containing one connective are commensurable as well. Hence, compound meta-propositions which contain more than one meta-connective can again be defined by meta-dialogs with a finite number of steps. In this way one arrives at the important result that all finitely compounded meta-propositions are mutually commensurable and thus unrestrictedly available in a meta-dialog.

Analogous to the situation in the object-language, one can define formally true meta-propositions. The totality of all formally true meta-propositions will then constitute the formal meta-logic of quantum logic. Since meta-propositions are mutually commensurable, one thus obtains the logic of unrestrictedly available propositions, i.e., the ordinary effective logic. The principle of excluded middle can generally be proved here no more than in the well known dialogic justification of ordinary logic (cf., [3],[4]). It is an open question whether formal elementary meta-propositions are value definite and thus the principle of excluded middle can be justified for all finitely compounded meta-propositions.

Hence we find that the formal meta-logic of quantum logic agrees precisely with the ordinary effective logic. The principle of excluded middle cannot be justified by our methods; it will, however, be assumed to be valid in the following considerations. Consequently, classical propositional logic will be the logic of formal meta-propositions. Quantum logic is not the meta-logic of quantum logic. It should be emphasized that this result is obtained from the same dialog-game D_f which was used in order to establish the calculus Q of the object quantum logic.

3. The Quantum Logical Modalities

In addition to the formal meta-propositions $\vdash_{D_f} A$ (A is formally true) and $\vdash_{D_f} \neg A$ (A is formally false), we now consider material meta-propositions $\vdash_{D_m} A$ (A is materially true). A certain class of material meta-propositions defined later will turn out to have a strong similarity to the logical modalities "necessary" and "possible". Hence, we will call these meta-propositions quantum logical modalities. Starting from these quantum logical modalities, a logical calculus can be established which will be called the modal logic M(Q) of quantum logic, and which is found to be quite different from the modal calculi of ordinary logic [1].

A situation in which modalities can meaningfully be applied is given, if with respect to a certain object-system S, some knowledge W is available which, however, is not sufficient in order to determine all possible properties of this system. In particular, this is the case if a quantum mechanical system S is known to possess a property $\mathcal{P}_W$ which corresponds to a proposition W. In fact even if W is given by a one-dimensional subspace ϕ of the Hilbert space $\mathcal{H}(S)$ of the system S (which is a maximal information about S), it is not possible to determine the values of all measurable properties of S from this proposition ϕ. This is only the case for those propositions A, the projection operators P_A of which commute with the projection operator P_ϕ of ϕ, i.e., for which the relation $[P_\phi, P_A]_- = 0$ hold. Propositions A of this kind are also called objective with respect to ϕ [3].

A system S which has the property $\mathcal{P}_W$ will be denoted by S(W). If for a system S(W) the truth of a proposition A can be determined with certainty, this proposition A will be called necessary with respect to W, which will be denoted by $\Delta_W A$ [2]. Since the most general proof procedure of a proposition A is given by a material dialog, a proposition A is necessary with respect to W if and only if there is a strategy of success for the material implication $W \to A$ within the material dialog-game. Hence we define the modality of necessity:

A is necessary with respect to W : $\Delta_W A$ iff $\vdash_{D_m} W \to A$

According to this definition the modality $\Delta_W A$ is in fact a material meta-proposition. If once the modality $\Delta_W A$ is introduced in this way, one can obtain further modalities by means of $\Delta_W A$ and the meta-logical connectives $\curlywedge$, $\curlyvee$, $\Rightarrow$ and $\neg$. Here we define three further modalities, the "possibility" $\nabla_W A$ and the "contingency" $\Xi_W A$, which are known from the traditional modal logic, and the "objectivity" $\Box_W A$ which is of special importance for quantum mechanical propositions [2].

A is possible with respect to W : $\nabla_W A$ iff $\neg \Delta_W \neg A$

A is contingent with respect to W : $\Xi_W A$ iff $\neg \Delta_W A \curlywedge \neg \Delta_W \neg A$

A is objective with respect to W : $\Box_W A$ iff $\Delta_W A \curlyvee \Delta_W \neg A$

If classical logic is applied to meta-propositions, it follows that $\Xi_W A \Leftrightarrow \neg \Box_W A$, i.e., the contingency of a proposition A is equivalent to its "non-objectivity" with respect to W.

It is interesting to express these logical definitions of the modalities in terms of quantum theory. Since, in this theory, the propositions W and A correspond to projection operators P_W and P_A, we find for the most interesting case of a maximal information $W = \phi$ (using well known theorems on projection operators) [5]:

a) $\Delta_\phi A$ iff $P_\phi = P_\phi P_A$

b) ∇A iff $P_\phi P_A \neq 0$

c) $\Xi_\phi A$ iff $[P_A, P_\phi]_- \neq 0$

d) $\Box_\phi A$ iff $[P_A, P_\phi]_- = 0$

These equivalences show that the denotations "objective" and "non-objective" for $\Box_\phi A$ and $\Xi_\phi A$ respectively are in fact in accordance with the terminology used in quantum mechanics [3].

Since the modalities defined here can be reduced to $\Delta_W A$ we will consider only Δ-modalities in the following investigations. Starting from "elementary modalities" $\Delta_W A$ one can then obtain "compound modalities" by means of the meta-logical connectives. The proof of these compound modalities has to be carried out within the meta-dialog-game $\bar{D}$, whereas the elementary modalities have to be proved within the object-dialog-game D_m. As an example, we consider the compound modality $\Delta_W A \Rightarrow \Delta_W A \wedge B$, the proof of which consists of a meta-dialog about the material meta-implication "$\Rightarrow$" and the object-dialogic proofs of the elementary modalities. It is obvious that the truth of this compound modality depends on the proposition W.

4. The Modal Logic of Quantum Logic

There are, however, some compound modalities which can successfully be defended within a meta-dialog irrespective of the proposition W, to which the elementary modalities are related. As an example we consider the meta-dialog for $\Delta_W A \,\bar{\wedge}\, \Delta_W (A \rightarrow B) \Rightarrow \Delta_W B$. After the successful proof of $\Delta_W A$ and $\Delta_W (A \rightarrow B)$ by the opponent, the proponent has to prove $\Delta_W B$. This is, however, always possible irrespective of the proposition W. In fact $\vdash_{D_m} W \rightarrow A$ and $\vdash_{D_m} W \rightarrow (A \rightarrow B)$ implies $\vdash_{D_m} W \rightarrow (A \wedge (A \rightarrow B))$ and hence $\vdash_{D_m} W \rightarrow B$. In this derivation we have made use of the <u>modus-ponens-law</u> for object-propositions $(A \wedge (A \rightarrow B)) \rightarrow B$ which can be derived within the calculus Q of object quantum logic [3].

In the <u>modal logic of quantum logic</u> such compound modalities are investigated which can be successfully defended within the meta-dialog-game D irrespective of the object proposition W contained in it. The example mentioned illustrates the method which can generally be used in order to prove "formally true modalities" of this kind: From the premises $\Delta_W A$ and $\Delta_W (A \rightarrow B)$ one can obtain the conclusion $\Delta_W B$ if and only if the corresponding object-premises A and $A \rightarrow B$ (without Δ) imply the respective object-conclusion B within the framework of the calculus Q of object quantum logic. Hence we arrive at the following result.

<u>Theorem</u>: If in a position of a meta-dialog the opponent O had proved the modalities $\Delta_W A_1, \Delta_W A_2, \ldots \Delta_W A_n$ and the proponent P is challenged to prove $\Delta_W B$ - then P can win the dialog for all W if and only if in the formal object-dialog-game D_f the proponent can win a dialog about B in which $A_1, A_2 \ldots A_n$ are available as propositions of O.

(For details on the proof see [5].) Due to this theorem a general proof procedure for formally true modalities is given, which reduces the proofs of the meta-proposition to the proofs of corresponding object propositions. Since the object propositions must be defended within the formal quantum dialog-game D_f, the provable object propositions are just the formally true propositions of the object-quantum logic. Consequently the structural properties of the quantum logical calculi Q_{eff} and Q - in particular the well known restrictions due to the possible incommensurability of object propositions - will be reproduced in the modal logic of quantum logic. This modal logic of quantum logic can be represented most conveniently by a calculus $M(Q_{eff})$ or $M(Q)$ respectively, which allows for the derivation of all formally true modalities. It is obvious from the considerations mentioned above that the calculi $M(Q_{eff})$ and

M(Q) are essentially different from the modal calculi of intuitionistic and classical logic which are known from the literature ([1],[2]).

In order to illustrate the difference between $M(Q_{eff})$ and the modal calculi of ordinary logic, we mention some examples of compound modalities which are not valid in $M(Q_{eff})$ but which are usually assumed to be true in ordinary modal logic (since we are dealing here with *formally true modalities*, we omit the index W at Δ_W [5]:

1) $(\Delta(A \vee B) \wedge \Delta(\neg B)) \Rightarrow \Delta A$
2) $(\Delta(A \rightarrow B) \wedge \Delta(\neg B)) \Rightarrow \Delta \neg A$
3) $\Delta A \Rightarrow \Delta(B \rightarrow A)$

These counter-examples show that the modal logic of quantum logic is in fact quite different from the well known modal calculi [1] of ordinary logic (e.g., M, S4). Thus we arrive at the interesting result that the *formal meta-logic* of quantum logic is equivalent to the meta-logic of ordinary logic - but that the *modal-logic* of quantum logic is essentially different from the corresponding classical structure.

Finally it should be mentioned that the modality $\Delta_W A \veebar \Delta_W \neg A$, i.e., the "*objectivity*" mentioned above is not a formally true modality. Therefore, the law $\Delta A \veebar \Delta \neg A$, sometimes - and misleadingly - called "principle of excluded middle", does not hold in M(Q) in spite of the deducibility of $A \vee \neg A$ in Q. In terms of quantum theory this means (for $W = \phi$) that $[P_\phi, P_A] = 0$ is not valid for arbitrary ϕ, i.e., the proposition A is not "*objective*" with respect to any $S(\phi)$. Hence the lack of $\Delta A \veebar \Delta \neg A$ in M(Q) seems to be specific for the situation in quantum modal logic. This is, however, not the case since $\Delta A \veebar \Delta \neg A$ does not hold even in classical modal logic [1]. The very difference between the quantum and the classical case is rather as follows:
Whereas for propositions A about a classical system there always exists at least one proposition W_o such that every A is "*objective*" with respect to W_o, for a quantum system $S(\phi)$ there never exists a proposition W_o such that $\Delta_{W_o} A \veebar \Delta_{W_o} \neg A$ holds for every A [5].

References

[1] Hughes, G.E. and Cresswell, M.J. An Introduction of Modal Logic. London: Methuen and Co., Ltd., 1968.

[2] Mittelstaedt, P. "Quantenlogik." Fortschritte der Physik 9(1961) 106-147.

[3] ---------------. Philosophical Problems of Modern Physics. (trans.) W. Riemer and revised R. Cohen. (Boston Studies in the Philosophy of Science, Volume XVIII). Dordrecht: D. Reidel, 1976. (Translated from the 4th edition of Philosophische Probleme Der Modernen Physik.)

[4] ---------------. "Quantum Logic." In Problems in the Foundations of Physics. (Proceedings of the International School of Physics "Enrico Fermi", Varenna (1977) Course LXVII). Forthcoming.

[5] ---------------. "The Modal Logic of Quantum Logic." Journal of Philosophical Logic (forthcoming).

[6] --------------- and Stachow, E.W. "The Principle of Excluded Middle in Quantum Logic." Journal of Philosophical Logic 7(1978): 181-208.

[7] Stachow, E.W. "The Completeness of Quantum Logic." Journal of Philosophical Logic 5(1976): 237-279.

A Role for Categories in the Foundations of Quantum Theory

David Geoffrey Holdsworth[1]

Institut für Theoretische Physik, Universität zu Köln
Department of Philosophy, The University of Western Ontario

In this paper I shall outline a possibility for incorporating category theory in an essential way into the foundations of quantum theory. The role for category theory which I shall sketch here is strictly foundational in the sense that it offers no new perspectives for a physicist working within the quantum formalism, but rather provides a way of interpreting the algebraic structure by giving an alternative axiomatization of the underlying logic of the theory. The axiomatization is given without recourse to any classical set theory by invoking methods closely connected with the work of F. W. Lawvere ([5],[6]). More precisely, then, a possible approach to the foundations of physics which axiomatizes its algebraic structure in a way which is independent of set theory and hence points, as it were, to a structuralist foundation of physical theories, will be proposed. The present paper consists of (1) a brief remark on the historical background to such a possibility; (2) a summary of the detailed axiomatization of the logical structure of classical theories; and (3) an explication within the framework thus obtained of van Fraassen's modal interpretation of quantum logical propositions.

1. Historical Prelude

Category theory was first introduced by Eilenberg and MacLane [1] in 1946, partly in response to certain very general problems in the theory of homological algebras. One of the most important mathematical problems which the theory was able to solve concerns the explication of "natural" in the notion of a natural isomorphism. Physicists and philosophers of science are most familiar with this problem in the context of the theory of vector spaces, where, as is well-known, a vector space and its first dual are isomorphic, but not naturally isomorphic. A vector space and

---

PSA 1978, Volume 1, pp. 257-267

its second dual are, on the other hand, naturally isomorphic. Now this notion can be explicated within vector space theory as follows: To say that a vector space V and its second dual V^{**} are naturally isomorphic is to say that the isomorphism is independent of the choice of basis in which the isomorphism is expressed. Eilenberg and MacLane sought to explicate this notion more generally in terms of criteria which do not depend on concepts from the particular mathematical theory (e.g., bases), since equivalent concepts arise in different mathematical contexts. Using the language of category theory, we get the following generalization: A vector space V and its second dual V^{**} are naturally isomorphic precisely because there exists a natural transformation from the second dual functor on N-LIN-SP to the identity functor on N-LIN-SP which is an isomorphism.[2] (See Herrlich-Strecker [2] or my [3].)

Since the original introduction of category theory a great deal of work has been done towards both (1) providing an adequate set-theoretic foundation for the theory, and (2) construing category theory itself as an independent foundation for mathematics. Although this latter possibility is not yet on all-fours, substantial progress has been made by the work of F. W. Lawvere ([5],[6]), and it is within this framework that I shall work in this paper.

For the details of Lawvere's programme the reader is referred to the work cited above. We consider here only the essential features, since, for one thing, the detailed way in which a foundation for mathematics is provided will not concern us. Lawvere's work proceeds in two stages. First, an elementary axiomatization for the category of sets is given in which classical elementhood plays no role [5]. This is the work which will concern us principally below, and we shall return to it shortly. Second, a similar axiomatization of the category of categories is given [6]. The original theory of the category of sets is then regained by embedding it into the category of categories. But the latter theory is stronger. For example, the addition of certain non-elementary axioms yields objects in the category of categories which correspond to the category of sets, the category of groups, etc. Moreover, the category is sufficiently large to yield such objects as arbitrary functor categories (by virtue of the exponentiation axiom) and other objects which have proven useful in contemporary mathematical practice.

In this paper I shall not be concerned with the category of categories. Rather, I shall operate strictly at the level of elementary axiomatizations and seek such an axiomatization independently for each kind of mathematical object required by physical theories. For example, classical mechanics, whose proposition system has the structure of a Boolean algebra, calls for the use of mathematical objects with precisely this structure. Thus we shall consider below - as an example - the detailed axiomatization for the category of complete atomic Boolean algebras. But let us first consider a few elementary details from Lawvere's original theory, if only to establish the notation and terminology to be used below.

As already pointed out, Lawvere seeks to axiomatize the category of sets without any recourse to the classical notion of elementhood. Indeed, the new primitive notions with which we shall work are <u>mapping</u>, <u>domain</u>, <u>codomain</u> and <u>composition</u>. The term "mapping" (map or morphism) merely refers to 'elements' of the universe of our theory.[3] The other three terms derive their meaning from certain elementary axioms for a category [1], which we now state:

A1 $\Delta_i(\Delta_j(x)) = \Delta_j(x) \quad i,j=0,1$;

A2 $\Gamma(x,y:u)$ and $\Gamma(x,y:u') \Rightarrow u=u'$;

A3 $\exists u[\Gamma(x,y:u)] \Leftrightarrow \Delta_1(x)=\Delta_0(y)$;

A4 $\Gamma(x,y:u) \Rightarrow \Delta_0(u)=\Delta_0(x)$ and $\Delta_1(u)=\Delta_1(y)$;

A5 (Identity) $\Gamma(\Delta_0(x),x:x)$, $\Gamma(x,\Delta_1(x):x)$;

A6 (Associativity) $\Gamma(x,y:u)$ and $\Gamma(y,z:w)$ and $\Gamma(x,w:f)$ and $\Gamma(u,z:g) \Rightarrow f=g$.

The notation here is Lawvere's [6]. We read $\Delta_0(x)$ as "the domain of x", $\Delta_1(x)$ as "the codomain of x", $\Gamma(x,y:u)$ as "u is the composition of x and y". We work with a standard first-order language whose formulae are defined recursively and in which the variables x,y, ... range over our domain of mappings. It is important to note that we have only mappings in our categories. Objects are effectively identified with identity maps via the definition:

$$\mathrm{Obj}(x) \text{ iff } \Delta_i(x)=x,\ i=0,1.$$

Thus an <u>object</u> is any mapping which is its own domain and codomain. Objects will be denoted by upper-case letters. The above notation is useful only for expressing these elementary axioms. Hence, we introduce:

(a) $A \xrightarrow{f} B$ iff $\Delta_0(f)=A$ and $\Delta_1(f)=B$;

(b) $fg=h$ iff $\Gamma(f,g:h)$;

(c) $\begin{array}{ccc} A & \xrightarrow{f} & B \\ & {\scriptstyle h}\searrow & \downarrow{\scriptstyle g} \\ & & C \end{array}$ iff $\begin{cases} \Delta_0(f)=\Delta_0(h)=A \text{ and} \\ \Delta_1(f)=\Delta_0(g)=B \text{ and} \\ \Delta_1(g)=\Delta_1(h)=C \text{ and } \Gamma(f,g:h). \end{cases}$

Now Lawvere proceeds to add to these elementary axioms for a category further axioms which characterize the category of sets. I shall not state them here since my axioms below for Boolean algebras parallel them in a certain sense to be made clear shortly. It is important here only that the existence of a terminal object is guaranteed by the axiom of finite roots; we denote it 1 and introduce the definition:

$$x \in A \text{ iff } 1 \xrightarrow{x} A.$$

Note that for classical point sets the terminal objects are the one-element sets, and also that in the present axiomatization we seek only the skeleton of the category; hence, we shall talk about <u>the</u> terminal object. It is clear that for classical point-sets there is a one-one correspondence between elements of a point-set A and functions from a one-element set to A.[4] This definition motivates in a perspicuous way the use of the algebraic notation fg=h, as opposed to the analytic notation gof=h. For example, suppose x∈A and $A \xrightarrow{f} B$. Then $1 \xrightarrow{x} A \xrightarrow{f} B$. That is, $1 \xrightarrow{xf} B$. We thus see that xf∈B. Indeed, we shall use the notation xf, where, in analysis, one writes f(x); but here, of course, x is itself a mapping and the notation represents composition. The most fortuitous aspect of this notation is that we shall not be plagued with the unintuitive direction-reversal in the analytic notation, although I use the analytical notation below whenever giving proofs for point-set objects.

2. Category Theory and the Foundations of Physics

Space limitations prohibit the inclusion of further details of Lawvere's theory *per se*. In this section I shall explain why I think Lawvere's theory is significant for physics, and along the way I shall sketch in greater detail an axiomatization which is of the same kind.

In extant physical theories (classical and quantum mechanics) one works with a variety of mathematical objects. In classical physics we work with sets (phase-space) function algebras (observables) and logical structures (Boolean algebras). In quantum physics we work with normed linear spaces (both the Hilbert space and the observables) and again logical structures (orthocomplemented, quasimodular[5] lattices or partial Boolean algebras). In [3] I have described how these collections of mathematical objects "hang together" within the framework of the language of Eilenberg-MacLane category theory. There I introduced the notion of the precategory of a physical theory. By this I mean simply the category whose objects are those which reflect the fundamental algebraic structure of the theory - SET for classical mechanics and N-LIN-SP for quantum mechanics. All other objects arising in the respective theories are obtained under functors from the precategory, as described in [3] for kinematical structures.

The programme I am presently engaged in seeks to axiomatize each such category after the fashion of Lawvere. In this way the following point of view is made more precise: A physical theory is, in part, a collection of structures which, insofar as the theory is true (on the right track), reflect real structural features of the world. Having axiomatized a particular category, the point-set objects (or at least the totality of them which satisfy the point-set axiomatizations) are models for the theory obtained. In actual scientific practice, one works, of course, with the classical representations (function space representation of the Hilbert space, etc.), but these are human artifacts, the tools of the trade, necessary for actual calculation, but themselves only models for the relevant structures. In keeping with recent trends in philosophy of science (theory of theories) I wish to say that the physical theory itself is something covering both the axiomatizations involved above, and much more. But this theme must be developed elsewhere; here I wish to exhibit in some detail the formal progress made so far.

To this end I have a complete axiomatization of the category of complete atomic Boolean algebras and a partial axiomatization of the category of groups. I describe below the former in some detail and make some brief remarks about the latter.

The following theory is worked out in [4], where theorems and other details are given with complete proofs. The strategy in working out the axiomatization is based on the following result from orthodox category theory: The opposite category to the category of sets[6] and the category of complete atomic Boolean algebras are equivalent. Thus certain of Lawvere's axioms for the category of sets should be such that their dual is true in the category of complete atomic Boolean algebras, namely those which assert the existence of objects with universal mapping properties which are preserved by equivalences and certain adjoint operations. Thus I take over the dual of precisely those axioms as axioms for my theory and add to these additional axioms to complete the theory. I enumerate now the axioms, writing "(Lawvere)" whenever they are the dual of Lawvere's axioms and making appropriate comments:

<u>Axiom 1</u> (Lawvere): All finite roots exist. That is, there exists a terminal object $\underline{1}$ and an initial object $\underline{2}$. Given any two objects A and B, the product $(A \times B, P_A, P_B)$ and the coproduct $(I_A, I_B, A+B)$ exist. Given any pair of mappings $A \overset{f}{\underset{g}{\rightrightarrows}} B$, the equalizer and the coequalizer exist.

<u>Axiom 2</u> (Lawvere): The coexponential ${}^{A}B$ of any pair of objects A and B exists.

Axioms 1 and 2 are straight-forward. In the presence of an axiom of infinity, which quarantees the existence of a special object with countably infinite many atoms, axiom 2 quarantees even larger Boolean algebras, for the coexponential has exactly as many atoms as there are Boolean homomorphisms from A to B. The detailed construction of the coexponential for classical point-set Boolean algebras is carried out in [4, Appendix 2].

<u>Axiom 3</u>: Axiom of infinity.

<u>Axiom 4</u> (Lawvere): $\underline{2}$ is a cogenerator,

$$\text{i.e., } A \overset{f}{\underset{g}{\rightrightarrows}} B \Rightarrow (f \neq g \Rightarrow \exists v (B \xrightarrow{v} \underline{2} \text{ and } fv \neq gv)).$$

<u>Axiom 5</u> (Characterization of $\underline{4}$): The product $\underline{2} \times \underline{2}$, which exists by axiom 1, has the following properties:

(1) The projections $\underline{4} \overset{\partial_0}{\underset{\partial_1}{\rightrightarrows}} \underline{2}$ are distinct and there are no other mappings from $\underline{4}$ to $\underline{2}$;

(2) There are distinct mappings $\underline{4} \overset{d_0}{\underset{d_1}{\rightrightarrows}} \underline{4}$ satisfying

$$\begin{array}{ccc} \underline{4} & \xrightarrow{\partial_i} & \underline{2} \\ & {\scriptstyle d_i}\searrow & \downarrow{\scriptstyle u} \\ & & \underline{4} \end{array}, \quad i=0,1;$$

(3) In addition to $\underline{4}$ and d_i there is exactly one mapping $\underline{4} \xrightarrow{\Delta} \underline{4}$ from $\underline{4}$ to $\underline{4}$.

It is required that the mappings satisfy:

(i) $d_i \partial_i = \partial_i$ $(i=0,1)$;

(ii) $\Delta\Delta = \underline{4}$;

(iii) $\Delta \partial_i = \partial_{i+1}$ (addition mod 2);

(iv) $d_i \Delta = d_i$ $(i=0,1)$.

Heuristically, 4 is the 4-element Boolean algebra. The ∂_i are clear. The d_i send a to O and a' to O respectively. Δ interchanges a and a'. However, this point-set Boolean algebra is to be thought of as a model for our axiom and the heuristic dropped.

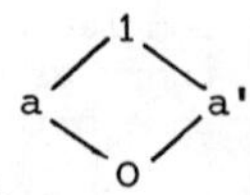

Definition 1: $a \in A$ iff $4 \xrightarrow{a} A$ (Read "a is an element of A")

Definition 2: For any object A, O_A and 1_A are the unique elements of A such that

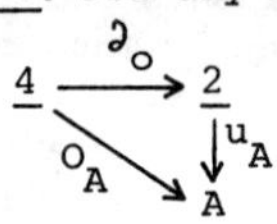

and

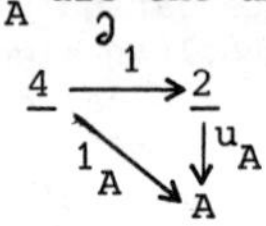

commute.

Here u_A denotes the unique mapping from 2 to A. We have written u above for u_4.

Definition 3: By a valuation we shall mean any morphism with codomain 2.

Definition 4: By a proposition we shall mean any morphism with domain 4.

Definition 5: A proposition a is true under a valuation v iff $av=\partial_1$.

Definition 6: If $a \in A$, then $a' \in A$ is the unique element of A, Δa.

Axiom 6: For every $a,b \in A$ there exists an $h \in A$ such that

$$(\forall A \xrightarrow{v} 2)(hv=\partial_1 \text{ iff } av=\partial_1 \text{ and } bv=\partial_1).$$

In [4, lemma 1] I prove h is unique. Thus we can write $h=a \wedge b$.

Definition 7: $a \vee b =_{df} (a' \wedge b')'$.

Definition 8: a is an atom of A iff $a \in A$ and $(\exists! v)(A \xrightarrow{v} 2$ and $av=\partial_1)$.

Axiom 7 (Atomicity): For every object A

$$(\forall A \xrightarrow{v} 2)(\exists! a)(a \text{ is an atom and } av=\partial_1).$$

For completeness I include axiom 8. However, definitions from [4] are necessary to understand it.

Axiom 8: Every valuation on a coproduct is a valuation over both the injections.

Axiom 9 (Lawvere): If $A \neq 1$ then there are valuations on A.

In [4] many propositions are proven. The more general ones are identified as theorems and are given below without proof:

Theorem 1: $1 \neq 2$.

Theorem 2: Every $A \xrightarrow{f} B$ preserves Boolean operations.

Theorem 3: 4 is a generator.

Theorem 4: If A and B have the same number of atoms they are isomorphic.

Corollary: For every pair of objects A and B there is a C such that

$$A+B \cong A \times B \times C$$

The provision of this axiomatization is only a first step in the programme described above. Next I wish to generalize the foregoing results to lattices and partial Boolean algebras to obtain the logical structures for quantum theory. At present, however, I am working on the category of groups, which is the first step towards establishing normed linear spaces and Hilbert spaces. For groups it turns out that the generator is a countably infinite object, namely the group of integers, and I now have an axiom which characterizes it completely in terms of its mapping properties. Note that GRP and N-LIN-SP have zero objects; this plays an important role in the axiomatization. The important thing for us is that we obtain a special vector space, namely the one-dimentional vector space over the 2-element field, which I shall denote $\vec{1}$. $\vec{1}$ is a vector space with exactly two elements, and exactly two subspaces, namely $\vec{0}$ and $\vec{1}$ itself. Within a theory of the category of normed linear spaces we shall have the definition:

$$a \in A \text{ iff } \vec{1} \xrightarrow{\vec{a}} A \text{ (Read "a is an element of A").}$$

This special vector space will play a role in the discussion below.

3. Direct Applications to Quantum Logic

In this final section I shall exploit the foregoing notions to give an explication of "truth in a (classical) physical theory" in terms of a commuting diagramme. I then generalize this definition to quantum mechanics and establish connections with van Fraassen's modal interpretation.

States in classical mechanics are elements of the set of states. In the framework being used here we can thus represent them by mappings from the terminal object of SET, 1, to an object of SET, S, which we shall call the <u>state-set</u>. Observables are mappings $S \xrightarrow{f} R$, from the state-set to an object R, the <u>result-set</u>. R is usually identified with the set of real numbers, R; in this framework the set of real numbers represents R. We suppose further that we obtain the logic of classical mechanics under a functor, namely the contravariant power-set functor from SET to SET^{OP}. (See my [3].) In the diagramme below I assume that the algebra of subsets of R is appropriately restricted to the borel σ-algebra without making the restriction precise. Borel sets thus get represented by mappings from $\underline{4}$ to B(R). Moreover, the initial object $\underline{2}$ of BOO-ALG is identified with P(1).[7]

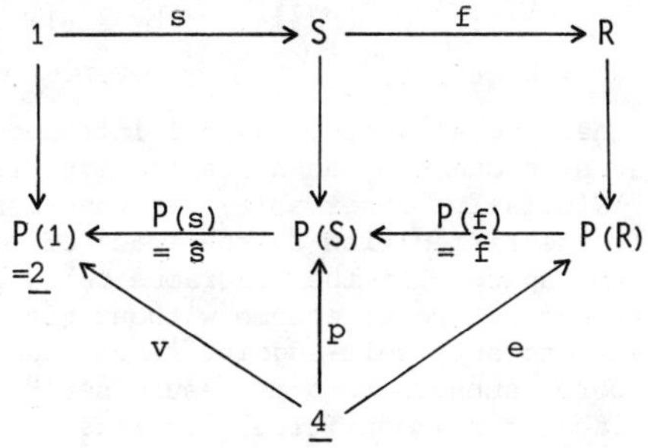

We now define "truth in a (classical) physical theory":

<u>Definition</u>: The proposition p is <u>true in the state s</u> iff $p\hat{s}=\partial_1$;
<u>false in the state s</u> otherwise.

Now let [f,e] be the proposition which is read "observable f takes a value in the borel set e". Then clearly [f,e] is precisely the proposition $e\hat{f}$, and

$$[f,e] \text{ is true in the state } s \text{ iff } e\hat{f}\hat{s}=\partial_1 .$$

We now seek to generalize the foregoing definition to quantum mechanics. To this end we introduce the following functor: let HIL-SP be the category of separable Hilbert spaces and LAT the category of lattices. Then if A,B,f $\in$ HIL-SP with $A \xrightarrow{f} B$, we define:

(1) L(A) as the lattice of subspaces of A ; and

(2) L(f), with $L(B) \xrightarrow{L(f)} L(A)$, as the lattice homomorphism such that

$$L(f)(C) = \bigvee\{f^{-1}[C]\} .$$

Thus L is a contravariant functor. Three conventions have been introduced here: (1) Whenever a vector space appears within square brackets the underlying set of vectors is to be understood ; (2) Whenever a linear transformation is written with the inverse symbol the inverse of the underlying function is to be understood ; (3) Whenever a set of vectors appears within curly brackets an arbitrary denumerable subbasis of that set is to be understood. The symbol "$\bigvee$" stands for the vector space span.

We note that since C above is a vector space and f a linear transformation, it can be shown that

$$[\bigvee\{g^{-1}[C]\}] = g^{-1}[C] . \tag{3-1}$$

We can now show that L is a functor.

(1) L preserves identities: Suppose A=B=f. Then

$$L(A)(C) = \bigvee\{A^{-1}[C]\} = \bigvee\{[C]\} = C .$$

(2) L preserves compositions contravariantly: Suppose $A \xrightarrow{f} B \xrightarrow{g} C$. Then

$$L(g \circ f)(D) = \bigvee\{(g \circ f)^{-1}[D]\} = \bigvee\{f^{-1}[g^{-1}[D]]\} .$$

But
$$\begin{aligned}(L(f) \circ L(g))(D) &= \bigvee\{f^{-1}[\bigvee\{g^{-1}[D]\}]\} \\ &= \bigvee\{f^{-1}[g^{-1}[D]]\} && \text{by (3-1)} \\ &= L(g \circ f)(D) && \text{Q.E.D.}\end{aligned}$$

We now invoke the special vector space $\vec{1}$ introduced in section 2 and say that pure states in quantum mechanics are maps from $\vec{1}$ to a separable Hilbert space H. Similarly, observables are functionals from H to C, where C is usually identified with ¢, the space of complex numbers considered as a Hilbert space. In the diagramme below <u>2</u> and <u>4</u> are the 2- and 4-element lattices. Here we assume without giving any details that we restrict as usual to only self-adjoint functionals and hence obtain a restriction to borel subsets of some result-set R. Also, maps from <u>4</u> to <u>2</u> have been labeled m (modalities) for reasons to be made clear next

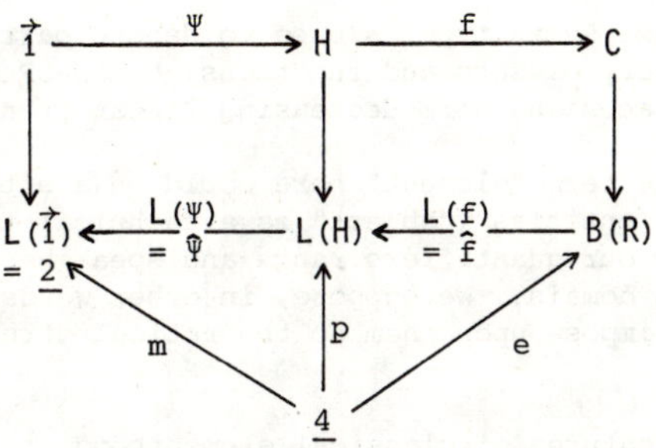

<u>Definition</u>: A proposition p is <u>necessary in the state Ψ</u> iff $p\hat{\Psi}=\partial_1$; <u>not necessary in the state Ψ</u> otherwise.

As before, $[f,e]=e\hat{f}$ is necessary in the state Ψ exactly if $e\hat{f}\hat{\Psi}=\partial_1$.

Van Fraassen [7] introduces the propositions $[m,E]=\{\alpha\in H:P^{\alpha}_{m}(E)=1\}$. They are read "observable m necessarily has a value lying in the Borel set E." This holds in a state α iff $\alpha\in[m,E]$. The following example shows that my definition above captures exactly van Fraassen's notion: Let $H=H_3$.

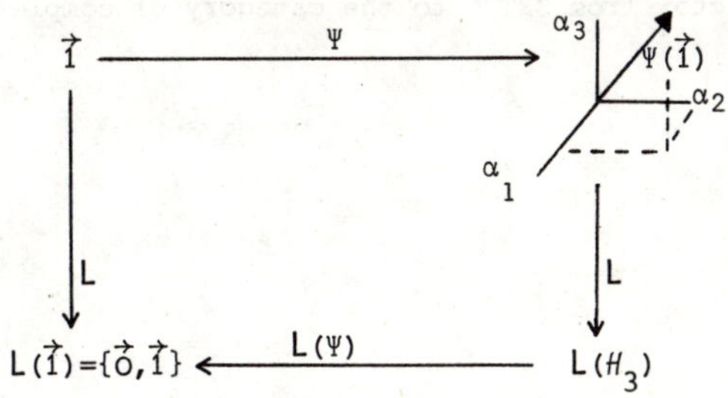

Let Φ be a subspace of H_3 (i.e., a proposition). Then

$$L(\Psi)(\Phi)\begin{cases}\vec{1} \text{ if } \Psi(\vec{1})\in\Phi \\ \vec{0} \text{ if } \Psi(\vec{1})\notin\Phi\end{cases}$$

It should be noted that the functor L is such that L(f) does not preserve orthocomplements in general. For example, if $\Phi=\alpha_1\vee\alpha_2$ and $\Psi(\vec{1})$ is as illustrated above, neither Φ nor $\Phi^{\perp}$ are necessary. Note that the van Fraassen notion of necessity, if applied to classical mechanics, is equivalent to truth. Insofar as a classical state makes a proposition true it does so necessarily. What the analysis given here shows is that when we try to formally generalize the notion to quantum mechanics, we get necessity, not truth. Truth in quantum mechanics, as van Fraassen's modal interpretation makes clear, is something else.

Notes

[1]Work done while supported by the Heinrich Hertz Stiftung, W. Germany.

[2] Here upper-case Roman type is used to denote particular categories. SET is the category of sets and functions; N-LIN-SP is the category of normed linear spaces and norm-decreasing linear transformations.

[3] The use of the term "element" here could be a bit misleading. We suppose a domain of entities ("object" is a technical term which I avoid here) over which our quantifiers range and speak here of the mappings as elements of this domain. We suppose, in other words, the existence of aggregates, but impose upon them no theoretical structure of the classical variety.

[4] Many formal features of classical elementhood are not shared by the notion of elementhood introduced by Lawvere. But this will not limit the generality of our theory.

[5] Mittelstaedt [7] emphasizes quasi- as opposed to ortho-modularity, since quasimodularity does not imply modularity.

[6] "Equivalent" means here that they have isomorphic skeletons.

[7] Actually, this identification depends on the existence of a certain algebraic functor from $\mathrm{SET}^{\mathrm{OP}}$ to the category of complete atomic Boolean algebras.

References

[1] Eilenberg, S. and MacLane, S. "General Theory of Natural Equivalences." Transactions of the American Mathematical Society. 58(1945): 231-294.

[2] Herrlich, H. and Strecker, G.E. Category Theory. Boston: Allyn and Bacon Inc., 1973.

[3] Holdsworth, D.G. "Category Theory and Quantum Mechanics." Journal of Philosophical Logic. 6(1977): 441-453.

[4] ----------------. "An Elementary Theory of the Category of Complete Atomic Boolean Algebras." Unpublished manuscript.

[5] Lawvere, F.W. "An Elementary Theory of the Category of Sets." Proceedings of the National Academy of Sciences. 52(1964): 1506-1511. (A more extensive treatment is to be found in a paper by the same title, mimeographed notes, University of Chicago, 1965.)

[6] -------------. "The Category of Categories as a Foundation for Mathematics." Proceedings of the Conference on Categorical Algebra, La Jolla. Heidelberg/New York: Springer-Verlag, 1966.

[7] Middlestaedt, P. Quantum Logic. Dordrecht: D. Reidel, forthcoming.

[8] van Fraassen, B.C. "Semantic Analysis of Quantum Logic." Contemporary Research in the Foundations and Philosophy of Quantum Theory. (The University of Western Ontario Series in Philosophy of Science, Volume 2). Edited by C.A. Hooker. Dordrecht: D. Reidel, 1973. Pages 80-113.

Only If Quanta Had Logic

James H. McGrath[1]

Indiana University - South Bend

Perhaps no claim in the philosophy of logic has attracted as much attention in the last decade as one defended by Hilary Putnam: some of the so-called 'necessary truths' of logic may turn out to be false for empirical reasons and consequently "logic is, in a certain sense, a natural science." ([15], p. 216). Putnam and others have attempted to reinforce this position by citing findings from quantum mechanics, findings which are intended to provide empirical reasons leading to the conclusion that particular aspects of classical logic have in fact turned out to be false. In this paper I will challenge one of the major procedures leading to the conclusion that quantum mechanics provides ample reason to revise logic. Quantum mechanics, I will argue, has shown neither that logic is empirical nor that some aspect of classical logic is false.

The procedure I am about to challenge is fundamental to quantum logic and is used by its main advocates: Putnam, Patrick Heelan ([9], [10], [11]), David Finkelstein ([3], [4], [5]) and E. G. Beltrametti and G. Cassinelli [1]. It is called the procedure of reading off quantum logic from quantum mechanics, or simply the read off procedure. In rough outline it can be divided into three steps. Step One. The structure of the class of some important quantum entities is determined. Typical examples of such classes are those of actual and possible measurements, systems, events, event languages or closed subspaces of Hilbert space.[2] The structure of these classes has been determined by quantum logicians to be a set-theoretic lattice of some particular kind. Because the class 'carries' the structure which is determined, I will call the class the carrier set and the structure simply the set-theoretic structure. Once the set-theoretic structure of the carrier set has been determined, quantum logicians proceed to Step Two. Members of the carrier set are mapped one to one onto members of some set of propositions in a structure preserving way. In other words, each carrier set element, an entity in the quantum world, is associated with a linguistic counterpart in a manner which preserves the set-theoretic

PSA 1978, Volume 1, pp. 268-276

structure discovered in Step One. Finally in Step Three the set-theoretic structure attributed to the propositional set is claimed to be the quantum logic of the propositional set.

I hope this process is clear. If so, forgive the following analogy. Step One: The structure of the class of American highways is determined. Step Two: Each highway is associated in a structure preserving way with a line drawn on paper. Step Three: The paper having the structure of the highways is claimed to be a road map. So the analogous aspect of the quantum logic procedure and road map procedure is that a structure discovered in one class is transferred to another class.

Before treating the details of the quantum logic reading off process, let us pause for a note intended to forestall confusions. Some researchers, especially mathematicians, use 'quantum logic' as a technical mathematical term, one which denotes what I called the set-theoretic or lattice structure of step one. So used the term has nothing to do with propositions and does not refer to a logic at all as philosophers most usually use the term 'logic'. On the other hand, philosophers, most notably those mentioned above, intend by 'quantum logic' a propositional logic differing from classical logic only in detail and not in kind. In this usage quantum logic is a logic as usually understood by philosophers. Endeavoring to stress this distinction between the two uses of 'quantum logic' I will make the following stipulations. I will abandon the mathematical use of 'quantum logic' altogether, speaking instead of set-theoretic structure. And I will speak of quantum propositional logic to reinforce the philosophers' use of the term.

I intend to discredit the read off process in a most general way and to show that the process is fundamentally misconceived. Accordingly for my present purposes, only a little attention needs to be directed to specific details of reading off; and, here we can gloss over some details which would be crucial in other treatments. So warned, let us see specifically how quantum logic is claimed to have been read off quantum mechanics by considering a particular use of the process.

Finkelstein, ([3], [4], [5]), for example, claims to have examined the carrier set of actual and possible physical systems of quantum mechanics with the following results. Any two systems are ordered by a relation called inclusion. System A is said to be included in system B in case all preparations of A pass tests by B. Since inclusion is a transitive, reflexive, anti-symmetric relation, it is said, in set-theoretic terms, to be a partial ordering and the set of quantum mechanical systems which admits of that relation is said to be a partially ordered set, or for short, a POset. Finkelstein also argues that for any two members of the POset of systems there exists a meet or intersection and join or union also in the POset. Accordingly he claims the set of systems to be a lattice. Each lattice member has a complement. And actual examples of quantum mechanical systems show that the lattice is non-distributive. So Finkelstein (and other quantum logicians as well) has concluded that various carrier sets of important quantum mechanical entities have the set-theoretic structure of a complemented non-distributive lattice. (The current dispute among quantum logicians concerning the extent of the degree of the weakening of the distributive law is not relevant to the purposes of this paper.) This finding concludes what I have called step one.

Here I must interject the following. Quantum logicians, in my opinion, have already at this point in their procedure made a serious mistake because the important carrier sets, despite almost universal opinions to the contrary, are not lattices at all. The objection can be traced to George Mackey ([13], p. 62) or even to Birkhoff and von Neumann [2]: what physical significance can one attach to the intersection of two systems which are not simultaneously testable? If none, there is no physical justification for regarding the POset of systems as a lattice.

Having noted this protest, let us continue, allowing the assumption that quantum logicians correctly determine important carrier sets to be complemented non-distributive lattices. In step two Finkelstein maps the lattice structure one to one onto some set of propositions in a structure preserving way. Assuming the correctness of step one, this step is unobjectionable. In step three, the set of propositions is said to have the same lattice structure as the carrier set of step one. This is also unobjectionable because the mapping of step two was a structure preserving one. But if it is unobjectionable to claim that some set of propositions could have a particular lattice structure, it is also neither novel nor interesting to make such a claim. However the read off procedure is not completed until an additional assertion is made, one which is both novel and interesting. The quantum logician completes the read off procedure by claiming that the structure of the propositions is quantum propositional logic. I find this last claim objectionable.

The claim I am disputing is the one which completes the read off process. Quantum logicians defend and I deny the claim that the propositional set onto which the lattice structure of the carrier set was mapped has a structure rightfully regarded to be logic. Of course one way this claim could be true is if when using the term 'quantum logic' the philosophers were adopting the mathematical usage, thereby intending only some set-theoretic structure. But we must be clear that this is not the case. They intend quantum propositional logic and they intend the read off procedure to provide empirical reasons for revising classical propositional logic in accord with the findings of quantum mechanics. Specifically, they intend to redefine the classical operations of conjunction, disjunction or negation in a way which reflects how operations of intersection, union and complementation function in a complemented non-distributive lattice. Having done this, different logical formulas become tautologies or contradictions and different logical theorems are provable. The logic is different: it is a revisio of classical logic. So make no mistake that quantum logic, for the philosophers we are considering, is quantum propositional logic.

The reasons why I think the read off procedure is fundamentally misconceived and the claim to have arrived at quantum propositional logic is false are implicit in the following fable about a friend who is a neophyte acrobat and also an amateur mathematician. (See Goldstein's [6] text especially pages 52-55.) She has perfected only two active acrobatic operations. She can flip herself standing to a headstand (and back again) and she can rotate herself clockwise (and back counter clockwise) through 90° while facing front (imagine clock hands advancing in three hour increments). In addition, in counterpoint to these vigorous operations, she sometimes simply performs the operation of standing still. Reflecting in a mathematical mood she discovers the following. Calling operations such as flip twice, rotate four times and stand still as equivalent because they have equivalent end results, she discovers that she has only eight operations in her repertoire. These she calls the carrier set of acrobatic operations. Any operations can be combined according to an associative composition function. In other words, flip * (rotate * stand) is equivalent to (flip * rotate) * stand. Standing is an identity operation and any operation has an inverse operation. Having made these discoveries, she concludes that the carrier set of acrobatic operations is a group under composition. She then discovers that the group is non-commutative since (flip * rotate) is not equivalent to (rotate * flip).[3] Adopting a term she has heard her mathematical friends use, she calls the structure a non-abelian group.

(Begin step two.) Consider associating each acrobatic operation with a proposition by a one to one onto mapping. Then associate the composition function between two operations with a propositional function called 'acrobatic and' such that 'p acrobatic and q' is interpreted in a way which reflects the structure of a non-abelian group of propositions. (Begin step three.) Call the structure of the set of propositions acrobatic logic. And notice that in this revision of classical logic some classical theorems are provable; yet others do not hold and are not provable. For example the classical theorem 'not (p and q) is equivalent to not (q and p)' does not hold in acrobatic logic.

Returning from the fanciful, one ought to say one of three things: (i) One accepts both acrobatic logic and quantum logic. Both were the consequence of a justifiable read off procedure; (ii) One accepts neither, the procedure is unjustified; or (iii) One accepts only one (most likely quantum logic) because there is a <u>fundamental and conceptually important</u> difference between reading off quantum logic and reading off acrobatic logic. Neither defenders nor critics of quantum logic would take the first option: to accept both. I invoked my acrobatic example to support option two and quantum logicians most likely would opt for three; they would endeavor to find an important difference between my fable and their actual procedure. So if I am to maintain my case against reading off quantum logic, I am obliged, at this point, to refute the arguments which aim to show that my example is not faithful to quantum logic procedures.

The defender of quantum logic might invoke any or all of the following three reasons intending to show that reading off quantum logic is fundamentally different. (a) Both the carrier set and propositional set in acrobatic logic are simply *ad hoc* and fanciful examples while those of quantum logic are actual sets universally recognizable by physicists. (b) Acrobatic logic is without motivation. Quantum logic is read off with the purpose of developing a propositional logic which allows resolution of the paradoxes of quantum mechanics. Or (c) Quantum logic is read off quantum physics in a way strictly analogous to the way classical logic can be read off classical physics. In both cases the operative set-theoretic structure is a well known lattice. But acrobatic logic is read off a non-abelian group (whatever that is!) which has no classical analogy nor relevance. Each of these possible rejoinders by the defender of quantum logic is off target, as I will now show.

Of course the carrier set of acrobatic operations is *ad hoc* and fanciful. But are the carrier sets of quantum mechanics any *different*? More importantly, what sanctions the "set of *some* propositions" to which the carrier set is associated? Logic aims to structure all propositions, or at least all of a certain category. But here we see the proposal that quantum logic structures *some* arbitrary and limited set of propositions. This point can be restated in terms developed by Professor Susan Haack [7] in her study of non-standard logics, *Deviant Logic*. A reform of logic is either local or global, depending on the scope of the reform. I have tried to make clear that quantum logic aspires only for local reform because the scope is limited to a set of only *some* propositions. So the acrobat logician, faced with the *ad hoc* and fanciful objection of the quantum logician, can reply *tu quoque*. For quantum logic faces the same objection. The correct position is the following. Quantum logicians have discovered an interesting set-theoretic structure, one of course which has as its model or scope, the important aspects of the quantum world. But there is no reason to use these results to justify the claim to have discovered quantum propositional logic.

The second possible rejoinder of the quantum logician fares no better. Admittedly acrobatic logic is without motivation (except the present polemic one), while quantum logic is motivated by the hope of resolving paradoxes. However it is simply a *non sequitur* to argue that the read off process is justified *because* the resulting logic avoids paradoxes. If quantum logic resolves paradoxes, then it is justified because it does, not because it was read off. One could discover quantum logic in a dream. It would be justified because it solved paradoxes, if it did, not because it was dreamed. Admittedly reading off may be justified as a fruitful heuristic process, one which suggests various solutions (as may be dreaming). And if reading off is viewed as no more than this, it is less objectionable. But in that case quantum logic is only suggested. In addition I think the situation is worse still because quantum logic cannot adequately resolve the paradoxes of quantum mechanics. A formal proof is offered to support this claim in the author's [14].

The final possible rejoinder of the quantum logician is based on what is probably the most prevalent and fundamental misconception of the reading off program. Admittedly, *if* classical logic is read off a lattice discovered in classical physics and if quantum physics leads to the discovery of a different type of lattice, then it might be reasonable to expect a different type of logic to be read off quantum physics. But in what sense is classical logic read off classical physics? Obviously it is not a historical fact that classical logic was discovered by reading it off classical physics. So when quantum logicians speak of discovering quantum logic by reading it off they cannot support their procedure with the classical analogy. Nor is it true that classical logic is justified because it can be read off classical physics. For whatever reason we have adopted the logic we have it is not because it is read off physics. Isn't it more reasonable to say that classical logic and classical physics share a common set-theoretic structure? And that is all. There is no warrant to deduce the conclusion that classical logic has structure B from the premise that classical physics has structure B. So I would block the analogy proposed by the defender of quantum logic by simply pointing out that there is no useful sense in which classical logic is read off classical physics.

Let us tie all this together. The thesis that logic is empirical has been thought by some in recent years to be supported by the claim that quantum mechanics has given evidence that some aspect of classical logic is false. This evidence is thought to be provided by the three step read off procedure I have described. I have used the fable of the acrobat to show that the read off procedure leads to absurd misuses. And I have attempted to discredit counter arguments of the defender of quantum logic which would show the acrobat process inappropriate. In meeting these counter arguments I have tried to sketch a case against local reform of logic, to argue that alleged paradox-resolving ability cannot prop up an otherwise unjustified read off procedure, and to disavow a classical motivation for reading off quantum logic. In addition (I have claimed without argument) reading off rests on an important technical mistake and quantum logic, however arrived at, cannot adequately resolve the paradoxes of quantum mechanics. Accordingly an unsound read off procedure has led philosophers to quantum propositional logic which is unmotivated and useless.

Of course by discrediting quantum propositional logic I have said nothing about the quantum logic of the mathematician. That program of investigation is quite different from the one I have been criticizing and it promises fruitful results. See, for example, the anthology edited by C. A. Hooker [12] where notable investigations of this type are collected. And obviously to discredit quantum logic is to say nothing against the logic-is-empirical thesis. Had we found reason to accept a quantum logical revision of classical logic we would have had reason to think logic is empirical. However not finding that particular reason gives no cause to deny that logic is empirical. Logic might be empirical, but not because of current arguments derived from quantum mechanics.

Notes

[1] I gratefully acknowledge encouragement and helpful comments from Roger T. Simonds and Bas C. van Fraassen and financial support from the Borden P. Bowne Foundation.

[2]Finkelstein has used both classes of qualities of physical systems and closed subspaces of Hilbert space. Heelan has as carrier sets both quantum events and event languages as well as closed Hilbert space subspaces. Beltrametti and Cassinelli examine possible yes/no experiments performable in quantum systems. The choice of carrier sets is irrelevant for the arguments of the present paper.

[3]Flip is about a horizontal axis while rotate is within one plane (say the plane of this page) such that

F:  and R: 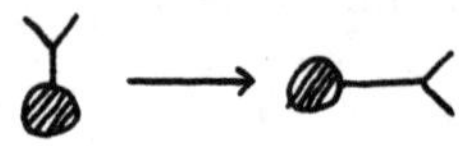

While

R: 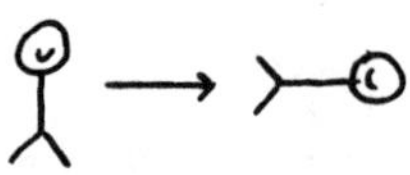 and F: 

References

[1] Beltrametti, E. G. and Cassinelli, G. "Logical and Mathematical Structures of Quantum Mechanics." Forthcoming in Il Nuovo Cimento.

[2] Birkhoff, G. and von Neumann, J. "The Logic of Quantum Mechanics." Annals of Mathematics 37 (1936): 823-843. (As reprinted in [12]. Pages 1-26).

[3] Finkelstein, D. "The Logic of Quantum Physics." Transactions of the New York Academy of Sciences 25 (1962): 621-637.

[4] -----------. "Matter, Space and Logic." In Boston Studies in the Philosophy of Sciences, Vol. 5. Edited by R. S. Cohen and M. Wartofsky. Dordrecht: Reidel, 1970. Pages 199-215.

[5] -----------. "The Physics of Logic." In Paradigms and Paradoxes. (University of Pittsburgh Series in the Philosophy of Science, Volume 5.) Edited by R. Colodny. Pittsburgh: University of Pittsburgh Press, 1972. Pages 47-66.

[6] Goldstein, Larry. Abstract Algebra. New Jersey: Prentice-Hall, 1973.

[7] Haack, Susan. Deviant Logic. Cambridge: Cambridge University Press, 1974.

[8] Halmos, Paul. Lectures on Boolean Algebras. London: Van Nostrand Reinhold, 1972.

[9] Heelan, P. "Complementarity, Context Dependence, And Quantum Logic." Foundations of Physics 1 (1970): 95-110.

[10] ------. "Quantum and Classical Logic: Their Respective Roles." Synthese 21 (1970): 2-33.

[11] ------. "The Role of Subjectivity in Natural Sciences." The Proceedings of the American Catholic Philosophical Association 43 (1969): 187-194.

[12] Hooker, C. A. (ed.). The Logico-Algebraic Approach to Quantum Mechanics. Dordrecht: Reidel, 1975.

[13] Mackey, G. The Mathematical Foundations of Quantum Mechanics. New York: Benjamin, 1963.

[14] McGrath, J. Quantum Mechanical Paradoxes and Non-Standard Logics. Unpublished Ph.D. Dissertation, The American University, 1977.

[15] Putnam, H. "Is Logic Empirical?" In Boston Studies in the Philosophy of Science, Vol. 5. Edited by R. S. Cohen and M. Wartofsky. Dordrecht: Reidel, 1969, Pages 216-241.

Part VIII

Applications of Statistical Ideas

On the Likelihood Principle and a Supposed Antinomy

Barry Loewer, Robert Laddaga, and Roger Rosenkrantz

It is natural to regard a given datum, x, as favoring that one of two rival hypotheses, H or K, which affords x higher probability, so that H is favored iff the likelihood ratio: $P(x|H):P(x|K)$ exceeds one. More generally, the relative support x accords the different members of a partition of hypotheses, $H_1,\ldots,H_n$, is assessed by comparing their likelihoods, $P(x|H_i)$. Considered as a function of the H_i, $L(H_i) = P(x|H_i)$ is called the likelihood function. If we view the likelihood function associated with an observed experimental outcome, x, as conveying the entire evidential import of the experiment, then it follows that the two outcomes, x and y, whether of the same or different experiments, are evidentially equivalent just in case they give rise to the same relative likelihoods, $P(x|H_i):P(x|H_j) = P(y|H_i):P(y|H_j)$ for all i,j, or , in other words, iff $P(x|H_i) \propto P(y|H_i)$ for all i. Writing Ev(E,x) for the evidential import of outcome x of experiment E, we may formulate this as the LIKELIHOOD PRINCIPLE: Ev(E′,x′) = Ev(E",x") iff x′ and x" give rise to proportional likelihood functions.

The Likelihood Principle is an immediate consequence of Bayes′ Rule for updating the probabilities of hypotheses. For that rule tells us that the posterior probability of H_i is proportional to the prior probability, $P(H_i)$, and the likelihood, $P(x|H_i)$, hence, if two experimenters assign the same prior probabilities to the H_i, and the outcomes they observe give rise to proportional likelihoods, they will obtain proportional posterior probabilities, $P(H_i|x) = kP(H_i|y)$, for all i. Since the posterior probabilities must continue to sum to one, k = 1, and the posterior probabilties of the two experimenters are the same.

Orthodox (non-Bayesian) statistical procedures are easily seen to violate the Likelihood Principle. Consider the following example. Two orthodox statisticians, Moe and Joe, are interested in the lot proportion defective in a batch of manufactured articles. Moe elects to sample nine items and finds two defectives. Joe decides to sample until he finds two defectives, but, by coincidence, finds his second defective

PSA 1978, Volume 1, pp. 279-286

on the ninth trial (where sampling is with replacement). Their likelihood functions are, respectively, binomial and negative binomial, viz., $36p^2(1-p)^7$ for Moe, and $8p^2(1-p)^7$ for Joe. And since these are proportional, the Likelihood Principle qualifies their results as equivalent. (Had they started out with the same prior distribution of p, they would be led to the same posterior distribution.) But now suppose that they are interested in testing the hypothesis H that $p = 0.5$ against the alternative K that $p = 0.1$, using a test whose probability of erroneously rejecting H (the so-called 'type I error probability') is $< .05$. As you can easily check, Moe's test will reject H when x, the observed number of defectives, is less than two, while Joe's test will reject H when the second defective occurs on or after the ninth trial. Hence, Moe's test accepts H while Joe's test rejects H. Their different acceptance decisions are traceable to the fact that there are more ways, viz., 36, of obtaining two defectives in nine Bernoulli trials than there are ways, viz., eight, of obtaining the second defective on the ninth trial. But that consideration seems irrelevant, and all the more so if Moe and Joe happen to observe the same sequence of trial outcomes, say one in which the two defectives occur on the first and ninth trials. Only their <u>intentions</u> when to stop sampling would then be different, and, as Edwards, Lindman and Savage insist, "the intentions of the experimenter are irrelevant to the interpretation of the data once collected, though of course they are crucial to the design of experiments."([7], p. 565).

Another sort of violation might occur in our example. Suppose that Moe wants a test whose type I error probability, α is exactly .05 (a so-called <u>exact</u> 5% test). If he rejects H when $x < 2$, $\alpha = .0195$, while $\alpha = .0898$ if he rejects H when $x < 3$. If, however, he rejects H when $x < 2$ and with probability .4339 when $x = 2$, then $\alpha = P(x<2|H) + .4339P(x=2|H) = .05$. This trick is called <u>randomizing at the boundary</u>, and it clearly violates the Likelihood Principle, for, in this example, it makes the import of $x = 2$ depend on the outcome of an extraneous random experiment whose two outcomes have probabilities .4339 and $1 - .4339 = .5661$. Even many orthodox statisticians have balked at allowing the import of the observed outcome to depend on the outcome of an unrelated random device, like a coin flip. But there are stronger criticisms of this than the bald claim that it is counterintuitive.

Moe's willingness to randomize connotes his preference for the resulting error probabilities over those attainable without randomizing. He considers that $\alpha = .05$ provides just the right amount of protection against erroneous rejections of H, while $\alpha = .0898$ provides too little. Again, the test which rejects H when $x < 2$ provides even better protection against erroneous rejections (viz., .0195), but it purchases this greater security only by running a higher risk of erroneous acceptances of H (the so-called 'type II error'), and Moe considers this higher type II error probability too high a price to pay for the reduction of type I error probability to an unnecessarily low level.

In effect, statistical tests or rejection rules are being compared like commodity bundles, where the commodities in each bundle are the type I and type II error probabilities. They can be considered as points in a Cartesian plane, where the coordinates of a point are its two error probabilities, usually denoted α and β. The ordinary rationality assumptions should govern preferences among these commodity bundles. But suppose that Moe prefers an exact test, Q, to tests P′ and P′′, for reasons of the sort sketched above. Then the usual <u>mixture assumption</u> states that P′ should be preferred (respectively dispreferred) to any mixture, kP′+(1-k)P′′, of P′ and P′′ if P′ is preferred (respectively dispreferred) to P′′. (A `mixture′, kP′+(1-k)P′′, refers to a rejection rule which applies P′ or P′′ with probabilities k and 1-k, 0<k<1.) Either way, by transitivity of preference (a normative assumption!), Q should be preferred to any mixture of P′ and P′′. But it may happen that Q lies above the line joining P′ to P′′ (figure 1). In that case, the line from the origin to Q will intersect P′P′′ in a point P whose coordinates are both smaller than the corresponding coordinate of Q. P then has smaller type I error probability than Q and smaller type II error probability than Q, and so, the Principle of Admissibility requires that P be preferred to Q, contradicting what we found above.

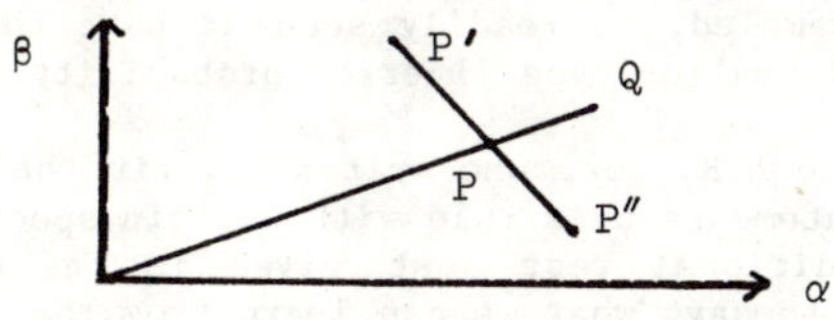

<u>FIGURE 1.</u>

Allan Birnbaum [2] proposes to meet this objection by rejecting the decision theoretic interpretation of a test in favor of an evidential interpretation. That is, he views the proper output of a test as a statement of the form, (Reject H for K, α , β), where the indicated error probabilities provide an informal index of the strength of the evidence against H and in favor of K. Birnbaum then argues that mixing assumptions fail when tests are evidentially interpreted. In testing whether a gene lies on chromosome 1 or chromosome 2, for example, I am interested in decisive results without regard to which hypothesis is favored. In that case, I might strongly prefer a test with error probabilities (.05,.05) to a test with error probabilities (.1,0), and, given the symmetry of the situation, the latter might be indifferent to a test with error probabilities (0,.1). "In such situations", Birnbaum writes, "I particularly value the guarantee, which is provided by use of (.05,.05), that strong evidence will be obtained." ([2], pp. 37 ff.). This pattern of preferences is clearly incompatible with the mixing assumption. For if the tests (.1,0) and (0,.1) are indifferent, either should be indifferent to a 50-50 mixture of the two, and hence, the latter should be dispreferred to (.05,.05). Yet, the error probabilities associated with a 50-50 mixture of (.1,0) and (0,.1) are (.05,.05), and consequently, equating a test with its error probabilities would lead to the absurdity that (.05,.05) is dispreferred to itself. On the evidential interpretation, then, both the mixing assumption and the identification of a test with its error probabilities fail. Two tests with the same error probabilities, even when they lead to the same evidence statements of the form, (Reject H for K, α , β), may be indicative of quite disparate degrees of evidential support for an accepted hypothesis. This point is further accentuated by examples of the following sort.

We are given two normal populations with the same (unknown) mean θ , but quite different standard deviations, $\sigma_1 >> \sigma_2$. A fair coin is flipped, and we sample one population or the other according as it lands heads or tails. We wish to test the null hypothesis $\theta = 0$ against the simple alternative $\theta = \theta' \doteq \sigma_1$, at, say, the 5% level of significance. It is assumed that we know which of the two populations it is we are sampling. Now a conditional test bases calculations of the error probabilities on the particular population known to have been sampled, and thus rejects the null hypothesis if $x > 1.64\ \sigma_1$ when the first population is sampled, and rejects it if $x > 1.64\ \sigma_2$ when the second population is sampled. However, the test which rejects when $x > 1.28\ \sigma_1$, if the first population is sampled, $x > 5\ \sigma_2$, if the second population is sampled, is readily seen to have the same type I error probability and a smaller type II error probability.

The example is due to D.R. Cox, who writes: "...if the object of the analysis is to make statements by a rule with certain specified long-run properties, the unconditional test just given is in order... . If, however, our object is to say 'what we can learn from the data we have', the unconditional test is surely no good. Suppose we have an observation from (θ , σ_1). The unconditional test says that we can assign this a higher level of significance than we ordinarily do, because if we were to repeat the experiment, we might sample some quite different distribution. But this fact seems irrelevant to the interpretation of an observation which we know came from a distribution with variance σ_1^2." ([5], p.360-361).

Cox's intuition about his example can be formulated as the ANCILLARITY PRINCIPLE: $Ev(E,(E',x')) = Ev(E',x')$ and $Ev(E,(E'',x'')) = Ev(E'',x'')$, when E is a mixture of E′ and E″. Here we write (E,(E′,x′)) when the mixed experiment leads to performance of E′ and the latter issues in outcome x′. The Ancillarity Principle was introduced by R.A. Fisher, an outspoken and often acerbic critic of the decision theoretic interpretation of a test, and drives a wedge between that interpretation and the evidential interpretation which Fisher [8] favored. Birnbaum has shown [3] that, together with another almost universally accepted principle (sufficiency), the Ancillarity Principle implies the Likelihood Principle. Given that the Ancillarity Principle is virtually what distinguishes the evidential from the decision theoretic interpretation of an orthodox test, Birnbaum's derivation can be taken (and certainly was taken) to have shown that orthodox statisticians who espouse the evidential interpretation are, in essence, committed to the Likelihood Principle. And espousal of that principle is nearly an espousal of the Bayesian approach. But not quite. One can regard the likelihood function as conveying the entire import of the evidence and still eschew the use of prior probabilities. That position on the board was occupied by R.A. Fisher [8], later by George Barnard [1], and more recently by Ian Hacking [9], A.W.F. Edwards [6] and others.

Given his derivation of the Likelihood Principle and his espousal of the evidential interpretation, Allan Birnbaum became identified as one of the leading spokesmen of this 'likelihood approach'. But, suprisingly, Birnbaum did <u>not</u> favor this approach and soon said so in print [4]. Evidently, he took the Likelihood Principle to be incompatible with a requirement he regarded as even more fundamental, namely, the requirement that an adequate concept of statistical evidence interpret given data as being strongly opposed to a true hypothesis (and favorable to a false hypothesis) with low probability. He regarded this clash of fundamental principles as a genuine antinomy, and indicative of a real crisis in the foundations of statistics. The main burden of this paper is to argue that Birnbaum's alleged counterexample to the Likelihood Principle is unconvincing. We begin with an example that is rather transparently not a genuine counterexample. It will then be seen that our example is like Birnbaum's in all essentials.

A deck of cards is either normal (H) or anomalous (K), where all 52 cards of an anomalous deck are identical. A card is drawn at random and proves to be an ace of spades. This outcome has probabilities 1/52 on H and 1 on the hypothesis that all 52 cards in the deck are aces of spades, indicating strong evidence in favor of the latter hypothesis over all others, and (paralleling Birnbaum's analysis) this carries over to evidence that the deck is anaomalous rather than normal. But strong evidence of anomaly would be obtained, on a likelihood criterion, whatever card we drew from the deck, even if the deck were in fact normal. This leads Birnbaum to conclude that "the likelihood concept cannot be construed so as to allow useful appraisal and thereby possible control, of probabilities of erroneous interpretations." ([4], p. 128).

Birnbaum's very similar example involves testing the hypothesis H′ that a random variable has mean μ and zero variance against the hypothesis K′ that it has mean μ and a much larger variance, where the mean is unknown. Our initial reaction to his example was to doubt whether anything could be learned about the variance on the basis of a single measurement when the mean is unknown, and to distinguish misleading (or uninformative) data from misleading interpretations of non-misleading data. However, a better answer can be given by noting that the analysis of the card example is defective from a Bayesian point of view. For K is a composite hypothesis, one with 52 simple constituents. To adequately assess its evidential support, we must average the likelihoods of its 52 special cases (viz., all cards are aces of spades, all cards are aces of clubs, etc.), and the result, assuming all 52 cases to be equiprobable is, of course, 1/52. For 'all cards are aces of spades' has likelihood one, while the other 51 simple constituents of K have likelihood zero. Hence, both H (a simple hypothesis) and K have the same average likelihood, and we see that, far from providing strong evidence against H in favor of K, the drawing of a single card is completely inconclusive. This both accords with and explains our intuition that no light can be shed upon the question whether the deck is normal or anomalous by drawing just a single card. If two cards were drawn (without replacement) and both proved to be the same (say, aces of hearts), the average likelihood of K would still be 1/52, but the likelihood of H would decrease to $(1/52)^2$, providing fairly strong evidence in favor of K.

The use of average likelihoods has an obvious Bayesian rationale. If H is a composite hypothesis with simple constituents, h_i, $i = 1,\ldots,n$, then the prior (respectively posterior) probability of H is the sum of the prior (respectively posterior) probabilities of the h_i. To obtain the posterior probability of H, therefore, we sum those of the h_i, and the latter are proportional to $P(x|h_i)P(h_i)$. But if all h_i have equal prior probabilities, $P(H|x)$ is proportional to $P(h_i) \quad \Sigma\, P(x|h_i) = P(H) \quad \Sigma\, P(x|h_i)/n$ so that $P(H|x)$ is proportional to the product of $P(H)$ by the average likelihood, $\Sigma\, P(x|h_i)/n$. The same justification applies to a composite hypothesis with continuum many simple constituents (i.e., to probability models or distributions with one or more continuous parameters), the summations giving way to integrals. This case is needed for Birnbaum's example ([4], pp. 127-128).

The mean of a random variable, X, is known to lie between 10^{-10} and 10^{10}. The standard deviation, σ, of X, is either zero or quite large. Specifically, let H assert that the distribution of X has mean μ and zero variance, and let K assert that X has a wide triangular distribution centered at μ, where μ is unknown. (The triangle must, of course, have unit area for this to be a probability distribution.) Given a single measurement x of X, the parameter values $\mu = x$, $\sigma = 0$ have likelihood one, while the likelihoods of the simple constituents of K vary from 0 to the height of the triangle, which is quite small (e.g., it is .01 when the base of the triangle has length 200). Consequently, Birnbaum contends, x provides strong evidence in favor of $\mu = x$ and $\sigma = 0$, and he says: "This includes, in particular, that the value $\sigma = 0$ seems to be supported against the alternative $\sigma = 100$. Evidently the numerical values of the likelihood ratios referred to, all exceeding 100, represent strong evidence in some sense." ([4], p. 128).

The example is parallel to our card example and lends itself to a similar analysis in terms of average likelihood. Here, both H and K are composite hypotheses (with parameter μ). Let us compute and compare their average likelihoods.

We obtain the special cases of K by sliding the unit triangle in question along the x-axis between 10^{-10} and 10^{10}, μ being the midpoint of the base. The likelihood of each member of this family of triangles is 0 if x lies outside the triangle and is otherwise equal to the height of the triangle at the abscissa x. Thus, the likelihood function over the constituents of K is zero everywhere in the interval, except along the base of a triangle centered at x, where it is given by the two sides of the triangle. The average height of this curve (i.e., the average likelihood of K) is equal to the area under the curve divided by the length of the interval, or $1/(2 \times 10^{10})$ since the triangle has unit area.

To compute the average likelihood of H, consider first a less extreme (more realistic) hypothesis, H′, which assigns X a very narrow triangular distribution centered at its unknown mean, μ. However wide or narrow, this triangle must have unit area to qualify as a distribution, and so the above analysis shows its average likelihood to be the same as that of K. H itself is best thought of as a limiting triangular distribution, letting the base shrink to zero and the height climb to one. Its average likelihood must then be the limit of the average likelihoods of these ever narrower triangular distributions, hence, it, too, must be the same as that of K. (This limit argument is the only plausible way of assigning an average likelihood to H.) Again we conclude that, far from providing strong evidence for H against K, a single measurement is completely equivocal, and, in particular, it tells us nothing about the variance. (As in the card example, two or more observations may, of course, tell us a good deal.)

We conclude with one or two brief remarks about the evidential interpretation. As Lindley [10] and Pratt [11] point out in their discussions of Birnbaum′s recent paper, evidence statements of the form Birnbaum favored can be quite unrevealing regarding the strength of the evidence. For example, if the observation lies midway between two normal means, μ_1 and μ_2, where both populations have unit variances, σ_1 and σ_2, the usual test, which rejects $\mu = \mu_1$ when $X > \mu_1 + 1.64$, would lead to the evidence statement, (Reject H for K, .05,.05), and yet, the observation is quite equivocal ([11], pp. 63-4). Similarly, let H assert that X is uniformly distributed in the interval [0,1], and let K assert that X is uniform in [0.9,1.9]. Let X = 0.97 be observed, again, a very equivocal outcome, since it falls in the overlap of the two intervals. But, again, it leads to the (presumably) strong evidence statement, (Reject H for K,.05,.05), using the test which rejects H iff X > 0.95. Moreover, the very same evidence statement is forthcoming if X = 1.01 is observed, but this outcome is entirely decisive. In short, use of such evidence statements or of error probabilities per se, fails to distinguish evidence of very different strengths. This criticism, together with others we could lodge, suggests that the evidential interpretation requires a good deal of unpacking before it can serve as a very firm foundation for statistical practice. Lacking any strong intuitions about how to go about this task, we are more inclined to believe that the Likelihood Principle captures what is sound in the evidential interpretation, and that Birnbaum′s supposed antinomy disappears upon closer inspection.

References

[1] Barnard, G. A., Jenkins, G. M., and Winsten, C. B. "Likelihood inference and time series." *Journal of the Royal Statistical Society* 125(1962):321-327.

[2] Birnbaum, A. "The Neyman-Pearson theory as decision theory, and as inference theory; with a criticism of the Lindley-Savage argument for Bayesian theory." *Synthese* 36(1977): 19-50.

[3] ------------. "On the foundations of statistical inference." *Journal of the American Statistical Association* 57(1962): 269-306 (with discussion).

[4] ------------. "Concepts of statistical evidence." In *Philosophy, Science and Method*. Edited by S. Morgenbesser, P. Suppes, and M. White. New York: St. Martin's Press, 1969. Pages 112-143.

[5] Cox, D.R. "Some problems connected with statistical inference." *Annals of Mathematcal Statistics* 29(1958): 357-372.

[6] Edwards, A. W. F. *Likelihood*. Cambridge: Cambridge University Press, 1972.

[7] Edwards, W., Lindman, H., and Savage, L.J. "Bayesian statistical inference for psychological research." *Psychological Review* 70(1963): 193-242.

[8] Fisher, R. A. *Statistical Methods and Scientific Inference*. Edinburgh: Oliver and Boyd, 1956.

[9] Hacking, I. *The Logic of Statistical Inference*. Cambridge: Cambridge University Press, 1965.

[10] Lindley, D.V. "The distinction between inference and decision." *Synthese* 36(1977): 51-58.

[11] Pratt, J.W. "'Decisions' as statistical evidence and Birnbaum's 'confidence concept'." *Synthese* 36(1977): 59-70.

Belief and the Incremental Confirmation of One Hypothesis Relative to Another

Andre Mirabelli[1]

Boston University

I will present a theory of confirmation (support or corroboration) which can be considered both an extension of the work of Ian Hacking on support [3] and of Karl Popper on corroboration ([5], [6], [7]). However, it will be clear that my theory is not a "mere conjunction of these two theories" ([7], p. 202) but a formally simple and intuitive synthesis that explains their strengths and corrects their weaknesses.

1. Popper's Approach

Popper's explication of corroboration relies on the expression $P(e,h) - P(e)$[2] normalized in different ways ([5], p. 410). He goes on to state the following additional criteria for the corroboration of h by e: (a) we can only "rationally justify a preference for one theory out of a set of competing theories ... so far proposed" ([7], p. 82); (b) e must be the result of a severe test ([5], pp. 414, 418); (c) more-falsifiable hypotheses must be preferred ([5], p. 41); (d) if h is a probabilistic hypothesis then e can only be a statistically interpreted finding, and thus e is often not allowed to be the total available evidence ([5], p. 413). Popper never completely incorporates these criteria formally into his criterion for the

PSA 1978, Volume 1, pp. 287-301

explication of confirmation, and in fact often denies that they can be formalized ([5], pp. 402, 418).

2. Hacking's Approach

The postulate that Hacking most often uses in concrete examples is the law of likelihood: if $P(e,h_1)/P(e,h_2) > 1$, then e supports h_1 better than h_2 ([3], p. 59). However, he adds an additional criterion for the consideration of complex hypotheses (e.g. '$h_1 * h_2$')[2] ([3], p. 90). He also embeds this postulate in a theory of a two-place predicate for "the degree to which e supports h" which, among other restrictions, satisfies Kolmogoroff's axioms for the probability calculus ([3], p. 134). (It should be noted that if we define $C(e,h_1,h_2) = P(e,h_1)/P(e,h_2)$ and if we take Popper's definition of explanatory power ([5], pp. 400, 401) and relativize it to a second hypothesis in the following way, $E(e,h_1,h_2) = [P(e,h_1) - P(e,h_2)] \div [P(e,h_1) + P(e,h_2)]$, then we find that $E(e,h_1,h_2)$ is a strictly monotone increasing function of $C(e,h_1,h_2)$. This shows them to be essentially the same measure.)

3. The Criteria

Knowing that these and many similar ideas have existed in various combinations for a long time, I will put aside the question of origins. I will formulate a formally simple and complete theory of confirmation and (rational) belief that resembles much of what has gone before but dramatically differs from every past theory of confirmation in a way that avoids all the pitfalls of which I am aware. After presenting the theory I will show how it formally incorporates Popper's criteria (a), (b) and (c), and disproves (d), none of which, he claims, can be formalized. I will also show how Hacking's additional criteria are unnecessary and produce inconsistencies. Finally I will exhibit a few of the other interesting results of the theory.

I formally explicate two independent concepts:

Criterion I: The degree to which e incrementally confirms h_1 relative to h_2, given the background information b and the total past evidence E, is

$$C(e,h_1,h_2) = P(e,h_1\&E\&b)/P(e,h_2\&E\&b) = P(e,h_1)/P(e,h_2)^2.$$

Iff $P(e,h_1)/P(e,h_2) > 1$, we simply say e incrementally confirms h_1 relative to h_2 or that e has increased the credibility of h_1 relative to h_2.

Criterion II: The degree of total support of h_1 relative to h_2 is $P(E,h_1\&b)/P(E,h_2\&b)$, where E is the total past evidence. If $P(E,h_1\&b)/P(E,h_j\&b) >> 1$ for all h_j in the non-empty set $CSPh_1$ of all seriously-proposed hypotheses that compete with h_1, then h_1 is added to the background b, and thus to our belief system. Also, all h_j belonging to $CSPh_1$ that were part of our belief system are deleted from it. (Two hypotheses, h_1 and h_2, compete iff $P(h_1,h_2) = P(h_2,h_1) = 0$.)

Appendix i shows that Criterion I,

$$P(e,h_1\&E\&b)/P(e,h_2\&E\&b) > 1,$$

is formally implied by

$$P(e\&E,h_1\&b)/P(e\&E,h_2\&b) > P(E,h_1\&b)/P(E,h_2\&b),$$

and is also implied by

$$P(h_1,e\&E\&b)/P(h_2,e\&E\&b) > P(h_1,E\&b)/P(h_2,E\&b).$$

Since all three of these criteria are intuitively persuasive explications of the incremental confirmation of h_1 relative to h_2 by e, it is a very good sign that the latter two imply Criterion I. The last inequality requires evaluation of posterior probabilities $P(h_j,e\&E\&b)$ and prior probabilities $P(h_j,E\&b)$. I believe these are almost never obtainable, and it is a major strength of my explication that it never requires evaluating them.

I will show that if e incrementally confirms h_1 relative to h_2, and h_1 logically implies h_3, it does not follow that e incrementally confirms h_3 relative to h_2. However, I stipulate that if h_1 is part of a (rational) belief system, then anything that is implied by h_1 is also part of that belief system. Since any hypothesis implies any tautology, then any non-empty belief system must contain all tautologies and thus all of formal logic. For a logical contradiction, $\bar{h}_t$, it is never the case that $P(h,\bar{h}_t) = 0$ since $\bar{h}_t$ implies every hypothesis. It follows that no hypothesis competes with $\bar{h}_t$ and thus $\bar{h}_t$ can never fulfill Criterion II and be added to a belief system. Further, it is clear from Criterion II that two competing hypotheses could never both belong to a single belief system and thus no hypothesis in a belief system could ever imply $\bar{h}_t$.

A number of important properties of Criteria I and II should be pointed out before applying them to particular problems. There is no way in which a piece of evidence can confirm a hypothesis in isolation; a hypothesis is only incrementally confirmed relative to another hypothesis or relative to a set of other hypotheses. This three-place predicate for the degree of incremental confirmation, $C(e,h_1,h_2)$, does not satisfy the Kolomogoroff axioms for the probability calculus and could not satisfy axioms for any two-place predicate $C(e,h)$. This is the case even though $C(e,h_1,h_2)$ is defined to be $P(e,h_1)/P(e,h_2)$, and the two-place predicate $P(e,h)$ satisfies any standard set of axioms for conditional probabilities. Some have taken it to be a criticism of such approaches that they are limited to comparisons between hypotheses ([1], p. 129 and [3], p. 134). However, it has never been shown that a two-place predicate for confirmation is necessary. I believe Criteria I and II show that a two-place predicate is neither necessary nor desirable since a three-place predicate avoids all paradoxes and fulfills all the roles required of it by science. Although Criteria I and II can be formulated exclusively in terms of likelihood ratios $P(e,h_1)/P(e,h_2)$, it does not follow that statistical concepts such as 'confidence' [2] cannot be formulated solely in terms of absolute likelihoods $P(e,h)$.

Criteria I and II do not depend at all on the prior or posterior probabilities of hypotheses; they are defined solely in terms of likelihoods. However, a hypothesis h stating the prior probabilities of other hypotheses, h_i, can become part of a belief system by explicit use of Criterion II. Only then can these priors, $P(h_i,B)$, be used to evaluate the posterior probabilities $P(h_i,e\&B)$. Therefore, hypotheses h_i that do not satisfy $P(E,h_i)/P(E,h_j) >> 1$ for all h_j in $CSPh_i$ might still satisfy the derivative criterion $P(h_i,E)/P(h_j,E) >> 1$ and thus be added to the belief system. The statements e used to describe the evidence must be unambiguous enough for use in the relevant likelihood functions; that is, $P(e,h)$ must be meaningful for the relevant hypotheses. Appendix ii states the definitions of many of the hypotheses and evidence statements to be used in the remainder of this paper and gives some important examples.

4. Complex Hypotheses

I will now consider the case of complex hypotheses. Let e be the statement of some evidence, and let h_1 and h_2 be two hypotheses such that e confirms h_1 relative to h_2, and h_2 does not imply h_1. I show in appendix iii that if h_1 and h_2 compete or if e does not confirm h_2 relative to '$h_1\&h_2$', then e confirms h_1 relative to 'h_1 or h_2.' In my notation this theorem reads that if $P(e,h_1)/P(e,h_2) > 1$, $P(h_1,h_2) \neq 1$ and either $P(h_1,h_2) = P(h_2,h_1) = 0$ or $P(e,h_1\&h_2)/P(e,h_2) \geq 1$, then $P(e,h_1)/P(e,h_1*h_2) > 1$. In stark contrast to this implication of the law of likelihood itself, Hacking's approach to complex hypotheses implies that $P(e,h_1*h_2)/P(e,h_1)$ is always greater than or equal to one ([3], p. 90).

An example will help make this clear. Using the notation of appendix ii we see that since $P(h_a,h_c) = P(h_c,h_a) = 0$ and $P(e_1,h_a)/P(e_1,h_c) > 1$ then $P(e_1,h_a)/P(e_1,h_a*h_c) > 1$. This means that since the hypotheses 'All ravens are black' and 'All ravens are red' compete, and since the evidence consisting of a randomly selected raven which is observed to be black confirms the first hypothesis relative to the second, then it follows that the evidence confirms the intuitively more falsifiable hypothesis 'All ravens are black' relative to the complex hypothesis 'All ravens are black or all ravens are red'. Hacking's additional criterion for complex hypotheses implies 'All ravens are black' can never be incrementally confirmed relative to 'All ravens are black or all ravens are red'.

Further, $P(e_1,h_a) = 1$ and $P(e_1,h_a{*}h_c) = n < 1$ allow us to pick a hypothesis h_z with $z = (1 + n)/2$ so that

$P(e_1,h_a)/P(e_1,h_z) = 1/(1 + n)/2 = 2/(1 + n) > 1$ and

$P(e_1,h_a{*}h_c)/P(e_1,h_z) = n/(1 + n)/2 = 2n/(1 + n) < 1.$

However, h_a implies '$h_a{*}h_c$.' Thus we see that, contrary to Hempel ([4], p. 103), Hacking and Koopman ([3], p. 33) and many others, if h_a implies h_d ($h_d = h_a{*}h_c$ in this case) and e confirms h_a relative to h_z, it is not necessarily the case that e must also confirm h_d relative to h_z. This is both a counter-example to the consequence condition of Hempel, Hacking and Koopman, and an example of how Criterion I already formally incorporates the preference for more-falsifiable hypotheses which Popper held could not be formalized. While it is impossible to <u>prove</u> that Criterion I captures the full content of an unformalized falsifiability criterion, anyone can check that the intuitions regarding his or her favorite case agree with the results obtained from Criterion I.

5. Unique and Existential Hypotheses

Consider the hypothesis h_u 'The next raven observed will be black'. We see that $P(e_1,h_u) = P(e_1)$, thus not allowing previous evidence e_1 to confirm h_u relative to any hypothesis that increases the probability of e_1 above its probability on the background alone. How then do we come to believe h_u? It is simply the case that h_a implies h_u. So that if E has added h_a to our belief system, then by the stipulated consequence condition for our belief system (not for incremental confirmation) we get that the unconfirmed h_u is believed. A similar argument also holds for the unfalsifiable existential hypothesis h_e 'There exists a black raven'. If e_1 belongs to E, then E implies h_e ('A raven is found to be black' implies 'There exists a black raven'), putting h_e in our belief system. This does not prevent $P(e_1,h_a)/P(e_1,h_e)$ from being greater than one ('A raven is found to be black' can incrementally confirm 'All ravens are black' relative to 'There exists a black raven'). Thus, while e_1 implies h_e and does not imply h_a, it is still the case that e_1 incrementally confirms the more falsifiable h_a relative to the unfalsifiable h_e.

6. Severe Tests, etc.

I have shown how Criteria I and II reflect our intuitive preference for the more falsifiable hypotheses. In appendix iv I give an example showing that they also reflect our intuition that more severe tests confirm to a greater degree. By keeping in mind that evidence can only confirm a hypothesis relative to other hypotheses, it is also easy to show that ten ravens observed to be black confirm h_1 relative to h_z to a larger degree than one raven observed to be black. Similar arguments can show that evidence of greater variety is preferable, as we would again intuit, and that Popper's attack ([5], pp. 413) on the use of the total information gained in an observation instead of a statistical summary is incorrect even in the case of probabilistic hypotheses.

7. The Tram Example

An important example for any explication of confirmation in terms of likelihoods is the tank or tram problem. Say you arrive in a city that has just installed a new tram system. You know that the total number of tram cars in the system is less than or equal to some number M. You also know that the trams are numbered consecutively starting from one. You then entertain the following set of competing hypotheses: $\{ h_x \mid x$ is a positive integer less than M & h_w = 'The total number of tram cars is w' $\}$. Upon seeing your first tram car you notice that its number is N. Call this evidence e. Thus, if $x < N$ then $P(e,h_x) = 0$, and if $N \leq x \leq M$ then $P(e,h_x) = 1/x$, showing that h_N is the hypothesis with the greatest likelihood. Since h_N is therefore incrementally confirmed relative to all other competing hypotheses in the set, are we to add h_N to our belief system? Our intuitions strongly urge us against such foolishness.

For $N \leq x \leq M$, Criterion II asserts that the degree of total support of h_N relative to h_x is $P(e,h_N)/P(e,h_x) = 1/N/1/x = x/N$. x/N is much greater than one iff x is much greater than N. Surely any number less than or equal to two is not much greater than one. (This is consistent with intuitions such as the one that does not allow e_1 alone to add h_a to our belief system if h_z, $z = 1/2$, is the only competing seriously-proposed hypothesis.) Thus Criterion II shows that the total support of h_N relative to the specified set of competing hypotheses is insufficient for us to add it to our belief system.

However, consider an alternative set of seriously-proposed hypotheses: $\{ h_x^N \mid$ x is a positive integer & h_x^N = 'the total number of tram cars is greater than or equal to $N(xy - y + 1)$ but less than $N(xy + 1)$ and, if $N(xy - y + 1) \leq i < j < N(xy + 1)$, then $P(h_i,h_x^N)/P(h_j,h_x^N)$ is neither very much less than or very much greater than one' $\}$ where y is a number much greater than one. Then $P(e,h_1^N)/P(e,h_x^N) \gg 1$ for all $x > 1$, and h_1^N is added to our belief system, in perfect agreement with our intuitions.

8. Scientific Progress

Another consequence of Criterion II should be stated here. Progress in theoretical science may be defined as the change from there not being to there being in our belief system a seriously-proposed hypothesis which competes with or implies some hypothesis h, when the set of seriously-proposed hypotheses or the range of available data is increased. This definition seems to capture the intuitive meaning of the concept without relying on the existence of absolute truth or knowledge.

9. What Remains

Using the perspective embodied in Criteria I and II we now have necessary and sufficient conditions, first, for the incremental confirmation of hypothesis h_1 relative to h_j or the set of $\{h_j\}$ by evidence statement e:

$$P(e,h_1)/P(e,h_j) > 1$$

and second, for adding h_1 to our rational belief system:

For all h_j in $CSPh_1$

$$P(E,h_1)/P(E,h_j) \gg 1$$

or

$$P(h_1,E\&e)/P(h_j,E\&e) \gg 1$$

(when $P(h_i,E)$ are already in B)

or

h implies h_1

(when h is already in B).

This explication of incremental confirmation and belief appears to avoid all past paradoxes surrounding confirmation and introduce no new ones. The questions which remain are: 1) how to formally characterize the content of our belief system, which is connected to the degree of clarity of our statement of the evidence and hypotheses with which we are concerned; 2) how to delineate the set of seriously-proposed hypotheses, which depends on a solution to Goodman's new problem of induction; 3) how to relate the confidence concept [2] to incremental confirmation and belief. While I do not believe the first to be a serious question, I do believe the second and third still remain pressing problems.

Appendix i: A Proof

From any set of axioms for conditional probability it follows that:

$$P(e\&E,h\&b) = P(e,h\&E\&b)\ P(E,h\&b)$$

Therefore,

$$\frac{P(e\&E,h_1\&b)}{P(e\&E,h_2\&b)} = \frac{P(e,h_1\&E\&b)}{P(e,h_2\&E\&b)} \frac{P(E,h_1\&b)}{P(E,h_2\&b)}.$$

Bayes' Theorem states:

$$P(h,e\&E\&b) = P(e,h\&E\&b)\ P(h,E\&b)/P(e,E\&b).$$

Therefore,

$$\frac{P(h_1,e\&E\&b)}{P(h_2,e\&E\&b)} = \frac{P(e,h_1\&E\&b)}{P(e,h_2\&E\&b)} \frac{P(h_1,E\&b)}{P(h_2,E\&b)}.$$

Thus,

$$\frac{P(e,h_1\&E\&b)}{P(e,h_2\&E\&b)} > 1$$

is implied by each of the other two criteria given in the text.

Appendix ii: Examples

The likelihood P(e,h) is an important ingredient of this paper and in the theories of Fisher, Hacking, Popper and many others. Thus it is crucial that P(e,h) be unambiguous. Consider the following evidence statements and hypotheses:

e_x = x is a black raven.

e_1 = An object randomly picked from the class of ravens is observed to be black.

e_2 = The largest raven in the Hartford Zoo is observed to be black.

e_3 = An object randomly picked from the class of black objects is observed to be a raven.

e_4 = An object randomly picked from the class of non-black objects is observed to be a non-raven.

e_5 = An object picked from the class of shoes on my feet is observed to be white.

h_a = All ravens are black.

h_b = All ravens observed before 2000 A.D. are black; the rest are white.

h_c = All ravens are red.

h_z = z100% of all ravens are black; the rest are red, $0 < z < 1$.

All occurrences of the above evidence and hypothesis symbols that appear in this paper refer to the definitions given here.

It is important to notice that $P(e_x,h_a)$ is meaningless. Only when we decide to refer to something such as $P(e_1,h_a)$ or $P(e_3,h_a)$ can we make sense of $P(e_x,h_a)$. However $P(e_1,h_a) = 1$ and $P(e_3,h_a) << 1$. Hence the ambiguity of e_x absolutely prohibits a term such as $P(e_x,h_a)$.

We can go on to notice regarding Criterion I that
$P(e_1,h_a)/P(e_1,h_c) = 1/0 >> 1$ and $P(e_4,h_a)/P(e_4,h_c) = 1 + q$, where q is the ratio of the number of ravens to the number of non-black non-ravens. Our background knowledge tells us that q is so small that $1 + q$ is negligibly greater than one. Thus we have e_1 overwhelmingly confirming h_a relative to h_c, and e_4 negligibly confirming h_a relative to h_c. We also have $P(e_5,h_a)/P(e_5,h_c) = 1$, showing that e_4 is not the same as e_5, even if the object observed is the same in both cases, and that e_5 does not confirm h_a relative to h_c at all. This resolves the Hempel raven paradox.

Goodman points out that $P(e_1,h_a)/P(e_1,h_b) = 1$ while $P(e_1,h_a)/P(e_1,h_c) >> 1$. This stresses the importance of clarity as to which alternative hypotheses are being considered. It must be kept in mind that Criteria I and II do not make this decision for you. You must decide what are the seriously-proposed hypotheses to which a given hypothesis is to be compared before the formalism can be applied.

We notice that B^2 indicates that the color of a raven does not depend on its size nor on its being in the Hartford Zoo. Thus we obtain $P(e_1,h_a) = P(e_2,h_a)$. This shows that it is not necessary for the evidence to be the result of random sampling for use in Criteria I and II. In fact Criteria I and II are applicable whenever the hypotheses and the evidence are clearly stated. Also, the evidence statement e describing a process is the total available evidence obtained from that process which is relevant to the comparison of hypotheses h_1 and h_2 iff all currently proposed additional descriptions e_+ of the process are such that $P(e,h_1)/P(e,h_2) = P(e\&e_+,h_1)/P(e\&e_+,h_2)$. Here, e and '$e\&e_+$' are different descriptions of the same piece of evidence and do not refer to different pieces of evidence. Contrary to Popper's claim

([5], p. 418), we have been able to represent the concept of 'total available evidence' in a formalism for the incremental confirmation of one hypothesis relative to another, and no further formalization is required.

In response to another infamous problem, we see that e_A, drawing an ace of spades from a previously unexamined deck, <u>incrementally</u> confirms the hypothesis h_A, that the deck only contains aces of spades, relative to the hypothesis h_R, that the deck is normal. But this does not add h_A to B. In fact, E overwhelmingly confirms the hypothesis h_{MR}, that most decks are normal, relative to all members of $CSPh_{MR}{}^2$, and the addition of e_A to E does not alter this. Since h_{MR} implies the hypothesis h_r, that h_R is highly probable, h_r is also in B. Thus, while e_A incrementally confirms h_A relative to h_R, h_r remains and h_A does not become part of B.

Appendix iii: A Second Proof

Simple calculations give the following results.

$$P(e,h_1{*}h_2) = P(e,h_1)P(h_1,h_1{*}h_2) + P(e,h_2)P(h_2,h_1{*}h_2) - P(e_1,h_1\&h_2)P(h_1\&h_2,h_1{*}h_2).$$

$$P(e,h_1{*}h_2) = P(e,h_1)[P(h_1\&\bar{h}_2,h_1{*}h_2) + P(h_1\&h_2,h_1{*}h_2)] + P(e,h_2)[P(\bar{h}_1\&h_2,h_1{*}h_2) + P(h_1\&h_2,h_1{*}h_2)] - P(e,h_1\&h_2)P(h_1\&h_2,h_1{*}h_2).$$

$$1 = P(h_1\&\bar{h}_2,h_1{*}h_2) + P(\bar{h}_1\&h_2,h_1{*}h_2) + P(h_1\&h_2,h_1{*}h_2).$$

Thus we see that $P(e,h_1{*}h_2)/P(e,h_1) < 1$ iff

$$P(\bar{h}_1\&h_2,h_1{*}h_2)[P(e,h_2)/P(e,h_1) - 1] + P(h_1\&h_2,h_1{*}h_2)/P(e,h_1)[P(e,h_2) - P(e,h_1\&h_2)] < 0.$$

Recalling that $0 \leq P(x,y) \leq 1$,

$$P(h_1\&h_2,h_1{*}h_2) = 0 \quad \text{iff} \quad P(h_1,h_2) = P(h_2,h_1) = 0,$$

and that

$$P(h_1,h_2) \neq 1 \quad \text{implies} \quad P(\bar{h}_1\&h_2,h_1{*}h_2) \neq 0,$$

we thus arrive at the result in the text.

Appendix iv: An Example of a Severe Test

Consider the following situation. The total support of hypothesis h_a, 'All ravens are black,' relative to each member of the set of hypotheses that compete with h_a which have thus far been seriously proposed is very much greater than one. However, no Connecticut ravens have yet been observed and the investigator has a hunch that Connecticut ravens are golden. Obviously, the severe test of randomly sampling the class of Connecticut ravens would be a sincere attempt to refute h_a. A competing hypothesis suggested by this situation is h_g, 'z100% of all Connecticut ravens are non-black and all other ravens are black,' where z is a number between zero and one, but significantly far from zero. If the investigator obtains the evidence e_c, 'A randomly selected Connecticut raven is observed to black,' we have $P(e_c,h_a)/P(e_c,h_g) = 1/(1 - z)$, which is significantly greater than one. Thus we see that according to our criterion, e_c, which is an unsuccessful sincere attempt to refute h_a, incrementally confirms h_a relative to h_g to a large degree, in perfect agreement with our intuitive position. On the other hand, e_d, randomly sampling the class of ravens not from Connecticut and obtaining a black one, gives $P(e_d,h_a)/P(e_d,h_g) = 1/1 = 1$, showing that in this situation e_d is not the result of a sincere attempt to refute h_a relative to h_g. Again this is in perfect agreement with our intuitions.

Notes

[1]I would like to thank Joel Minsky and Philip Mirabelli for the discussion in May 1976 that led to whatever new ideas are contained here, Abner Shimony and Michael Meyer for their help and encouragement, and all my friends, especially W.N., J.P., E.M., B.J.S., R.S., D.C., and T.O., who endured ravens for so long.

[2]P(e,h), sometimes written P(e,h&B) or P(e,h&E&b), is the probability of e given hypothesis h and our belief system B, where e is a sentence giving a particular description of the evidence and B is composed of our background knowledge b and total past evidence E. $P(e) = P(e,h_t)$, where h_t is a tautology. Also note that 'h_1 and h_2', 'h_1 or h_2', and 'not h' are written 'h_1&h_2', 'h_1*h_2', and '$\bar{h}$' respectively, and that $CSPh_1$ denotes the set of seriously proposed hypotheses that compete with h_1.

References

[1] Barnard, G.A. "The Logic of Statistical Inference." _British Journal for the Philosophy of Science_ 23(1972): 123-133.

[2] Birnbaum, A. "Concepts of Statistical Evidence." In _Essays in Honor of Ernest Nagel: Philosophy, Science, and Method_. Edited by S. Morgenbesser, P. Suppes and M. White. New York: St. Martin's Press, 1969. Pages 112-143.

[3] Hacking, I. _Logic of Statistical Inference_. Cambridge: Cambridge University Press, 1965.

[4] Hempel, C.G. "Studies in the Logic of Confirmation." _Mind_ 54(1945): 1-26, 97-120.

[5] Popper, K.R. _The Logic of Scientific Discovery_. New York: Harper and Row, 1968. (Originally published as _Logik der Forschung_. Vienna: J. Springer, 1935.)

[6] -----------. _Conjectures and Refutations_. New York: Basic Books, 1965.

[7] -----------. _Objective Knowledge_. Oxford: Oxford University Press, 1972.

Salmon, Statistics and Backwards Causation

David Papineau[1]

Macquarie University

1. Introduction.

Smokers are more likely to develop lung cancer than non-smokers. This statistical association gives us reason to suppose that smoking can *cause* lung cancer. But we would come to reject this supposition if we discovered that when family histories of lung cancer are taken into account smoking makes no further difference to the probability of lung cancer: that amongst those with a family history of lung cancer (and amongst those without) non-smokers are just as likely to develop lung cancer as smokers. Such a discovery would lead us to conclude that, despite initial appearances, smoking actually makes no causal contribution to lung cancer, but is statistically associated with it only because a genetic predisposition to lung cancer also somehow predisposes people to smoke.

What is the rationale for such inferences? *Why* does a statistical association between two events constitute a *prima facie* reason for supposing one to be the cause of the other? And why is this conclusion discredited when we discover that the statistical association "disappears" when we partition our reference class by "controlling" for some prior factor? Our intuitions here are so strong that these questions seem silly. But remember that causes are standardly supposed to *determine* their effects. How can statistical data show us what determines what? The answer is by no means immediately obvious.

In his recent [3] and [4] Wesley Salmon has developed an approach to explanation and causation which promises to solve this puzzle. Amongst other things Salmon denies that causation need be deterministic. He suggests that it is quite enough for one event to cause another that they simply be statistically associated and this association not be "screened-off" by any prior factors.[2]

In this paper I shall attempt to provide a different kind of solution, one which does not involve rejecting the traditional view of causation as essentially deterministic. In so doing I shall cast some general doubt on Salmon's approach.

PSA 1978, Volume 1, pp. 302-313

To explain the relevance of statistical data to causal conclusions I shall need to refine the orthodox neo-Humean regularity analysis of causation. At the end I shall argue that the refinement in question helps with the problem of "backwards causation"; that is, that it gives us a time-independent analysis of causation which explicates the asymmetry of causal relationships with respect to temporal direction.

2. Barometers and Eclipsing.

The current neo-Humean orthodoxy is that one event is a cause of another if the former is one of a set of factors which is jointly and minimally sufficient for the latter. (For instance, see Mackie [1].) But then there is the problem of the barometer. Whenever the barometer falls, and it hasn't been fiddled with, and there is a certain amount of water vapour in the atmosphere, and various conditions for the formation of water droplets are satisfied, it rains shortly afterwards. The barometer's fall is part of a perfectly good minimally sufficient condition for rain. Yet the barometer's fall does not cause rain. On the contrary, our intuitions are that the barometer's fall and the rain are both independent effects of an earlier factor, namely a fall in atmospheric pressure.

This illustrates a general difficulty. Given the neo-Humean analysis of cause, if A has two successive effects B and C then it will standardly follow that B is a "cause" of C.[3] But we do not standardly want to allow this. To be sure, B can be a cause of C in such cases - A can cause C by causing B. Apart from affecting the barometer, the fall in pressure also causes a condensation of water vapour before the rainfall. And this is an intermediate cause of the later effect, the rainfall. But the difficulty remains. The neo-Humean analysis lacks a way of distinguishing these cases, where an initial cause produces some later effect by causing some intermediate effect, from those where an initial cause produces two effects independently.[4]

Our intuitions are that the barometer's fall, unlike the condensation of water vapour, is in itself irrelevant to the subsequent rainfall. In the circumstances rain would have followed the drop in pressure even if the barometer hadn't fallen. By contrast the rain wouldn't have followed the drop in pressure if there hadn't been any condensation.

Can we find a basis for this distinction? Note that on all occasions where a barometer fall occurs together with the other factors (the barometer's not being manipulated, high humidity, etc.) which make up a guarantee of rain, a drop in atmospheric pressure will also have occurred and, together with some of the same factors (high humidity, etc.), will also have made up a sufficient condition for rain. And what is more this latter set of factors will on occasion occur, and suffice for rain, even when no barometer fall occurs - as when there are no barometers around.

This seems to be characteristic of cases where B and C are independent effects of A. In a sense A eclipses B as a cause of C. Whenever

B "guarantees" C, A is there to suffice for C too; and in similar circumstances A suffices for C even when B is absent. So I would suggest that for B to be a real cause of C, and not an irrelevant side-effect of some prior cause, not only does B need to be part of a minimally sufficient set for C, but in addition there should be no other factors which eclipse B as a cause of C: that is, there should be no other minimally sufficient set not containing B which

(i) is always present when the set containing B is, and

(ii) is sometimes present when the set containing B is absent.

(In this paragraph, and from now on, "A", "B", etc. are to be taken as standing for types of events rather than for individual events.)

Venn diagrammatically, to require that nothing eclipses B as a cause of C is to require that there is no A satisfying the diagram:

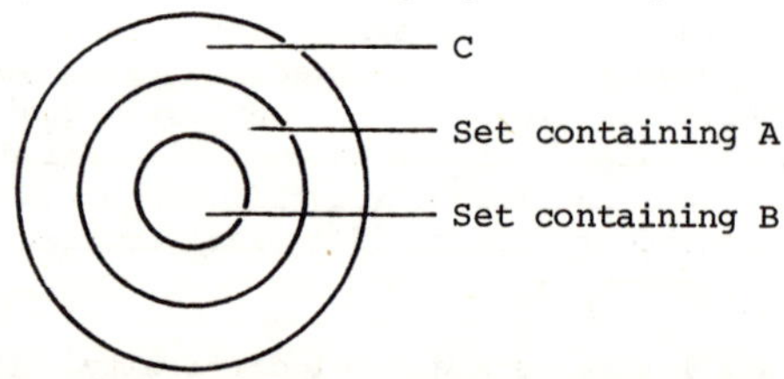

Consider now the case where B is indeed causally intermediate between A and C, Here B does seem to be crucially relevant to C's occurrence. A fall in pressure (A) suffices for rain (C) only insofar as it is accompanied by condensation (B); while condensation can suffice for rain even in the absence of a preceding drop in pressure, as when potential nuclei are artificially or naturally introduced into a supersaturated atmosphere. Indeed in this case it is B that eclipses A as a cause of C. Diagrammatically:

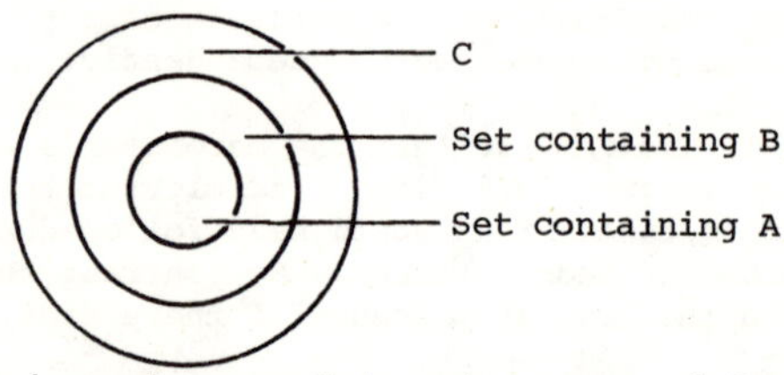

But does this mean that in such cases A is not a cause of C after all? This would be unsatisfactory. The drop in pressure surely causes the rain all right, albeit via causing condensation.

We can sort this out with the following definition. One event is a *direct cause* of another if, as above, it is part of an uneclipsed sufficient condition for it. One event is a *causal ancestor* of another if it is a direct cause or a direct cause of a causal ancestor. Then we can allow that causal ancestors other than direct causes are also causes, albeit indirect ones. So when A causes C through B, B will eclipse A as a direct cause of C - but nevertheless A will be a causal ancestor of

C, through causing B which causes C.

(One can imagine certain awkward cases where the no-eclipsing requirement fails to distinguish straightforwardly between intermediate causes and side-effects. These will be those where A and B always suffice for C together, where the Venn diagrams representing their actual occurrences coincide. For instance, suppose a certain virus can be followed by a skin infection and then by a fever, and suppose further that the virus is never found without the infection nor vice versa. Does the virus cause the fever by causing the infection, or does it cause them both independently? The obvious way to analyse this kind of case is by reference to a theoretical account of the mechanisms behind the co-occurrence of virus, infection and fever. This can be seen as a kind of elaboration of the no-eclipsing requirement. The theoretical explanation allows us to consider (theoretically deviant) contexts in which the virus occurs without the fever, and vice versa, and to use them to show whether the one would still suffice for the fever anyway, even with the other absent. Some further comments on this kind of case are made at the end of the Appendix.)

3. Causes from Statistics.

We are now in a position to understand how statistical associations can give us grounds for causal conclusions. Roughly the idea will be that a statistical association between Z and Y is an initial indication that Y is part of an uneclipsed sufficient condition for Z - but if this association "disappears" when we "control" for some V, then this means that V eclipses Y as a cause of Z.

Suppose[5] Y and Z are statistically associated. That is, Prob (Z/Y) > Prob (Z). The standard objection to concluding that Y is a cause of Z is that the association might be "spurious", a consequence of their both being independently associated with some prior cause. But suppose for a moment that Y is not itself associated with any other causes Z may have. It can then be proved that if Z is caused whenever it occurs Prob (Z/Y) > Prob (Z) if and only if Y is a direct cause of Z. A detailed proof is best left to the Appendix below. But the idea is simple enough. If Y is a direct cause, then when Y occurs we will get Z if either one of Z's other direct causes is present or if the factors which in conjunction with Y suffice for Z are present; while in the absence of Y we will get Z only if one of Z's other direct causes is present. So if Y is a direct cause of Z then Prob (Z/Y) > Prob (Z). Conversely, if Y is not a direct cause of Z then its occurrence will not increase Z's probability. But these inferences clearly require essentially that Y is not statistically associated with Z's other direct causes, for this could produce an association between Y and Z even if Y were not a cause of Z, and it might just lead to the absence of an association between Y and Z even when Y was a direct cause of Z.

What then if Y is associated with other direct causes of Z? To deal with this possibility we need to "control" for these other causes; that is, to consider separately classes of cases which are similar in respect of these other causes. Then, if Y still makes Z more likely *within* any such classes, we can conclude that Y is indeed a direct cause of Z; whereas if it does not we can conclude that despite any original association Y is not after all a direct cause of Z. The detailed substantiation of this principle is once more best left to the Appendix. But again the idea is simple enough. We can go through the same reasoning as before, only now "within" those classes where we "hold constant" all those other causes that are confounding the association between Y and Z. By so doing we effectively ensure that Y is not associated with any "remaining" causes of Z, and so can be definite about what any remaining association between Y and Z shows about their direct causal relationship.

These results explicate our intuitive judgments about data like the hypothetical smoking-cancer statistics with which I introduced this paper. They show why Y's being statistically associated with Z is indeed a *prima facie* reason for thinking Y is a cause of Z. At the same time, they show that this inference presupposes that Y is statistically independent of Z's other causes (or, to be more precise, that Y is statistically independent of all Z's causes other than those causal descendants of Y which are themselves in turn causes of Z - for if Y is an *indirect* cause of Z it is only natural that its association with Z will be through its causal descendants). And so we also see now why we should conclude that Y isn't a cause of Z after all if and when we discover that their association disappears when we "control" for some further cause of Z (excepting, again, those further causes which are intermediate between Y and Z). Of course in practice causal conclusions of the kind in question will always be tentative, even when we have no doubts about the lawlikeness of our statistical associations. To be sure that Y *is* a cause of Z we need to be sure we have taken into account all other confounding influences. Though we could have good grounds for thinking this, it will be difficult to achieve complete certainty. Less obviously, judging that Y is *not* a cause of Z also strictly requires that we have taken into account all confounding influences - for a negative association between Y and some unknown further cause of Z might always just be such as to "cancel out" exactly Y's association with Z even when Y is a cause.

It is worth emphasising that the no-eclipsing addition to the regularity account of causation is essential to our explicating the principles of inference in question. This can be illustrated by considering the barometer example once more. Once we "control" for the prior change in the atmospheric pressure falling barometers don't further increase the chance of subsequent rain. But recall that a falling barometer can be part of a perfectly good minimally sufficient condition for rain. What it fails to be is an *uneclipsed* part of such a set. This shows that it is only because causes are specifically uneclipsed parts of sufficient conditions that we can take the absence of a statistical association to show an absence of causation. (The Appendix shows exactly where the no-eclipsing clause is required in the proof that statistical independence implies absence of causation).

4. Salmon and Indeterministic Causation.

I mentioned in the introduction that Wesley Salmon has recently suggested that for one event to be the cause of another is simply for the former to be relevant to the latter's probability. Antecedent circumstances making an effect certain is for Salmon thus just the limiting case of causation. Of course Salmon has to avoid counting all statistical associations as causal connections. He doesn't want barometer readings causing rain simply on the grounds that they make it more likely. To deal with this he specifies that the explanatory factors we adduce should comprise a "homogeneous reference class" - the probability of the explanadum should not be alterable by including further factors in the explanans. So falling barometers are no good for causing rain, in that adding in prior drops in atmospheric pressure gives a better probability for rain; but drops in atmospheric pressure are not ruled out, since adding barometer readings to them makes no difference to the probability of rain. In such a case Salmon says[6] the atmospheric pressure "screens off" the barometer as a cause of rain.

Clearly there are similarities between Salmon's "screening off" and my "eclipsing". But there remain fundamental differences between his approach and mine. I have been specifically concerned with inferring causes from probabilities when the effect in question is always determined. But for Salmon we will not have a "homogeneous reference class" in such cases until we have included all determining factors. Thus on Salmon's account we are only entitled to infer causes from probabilities when the statistical relationship between cause and effect reflects an actual indeterminacy in the effect's occurrence.[7]

What are we to make of this? We have two superficially similar systems of procedures for selecting legitimate explanations, both hinging on what happens to statistical associations when we "control" for prior factors, but applicable to mutually exclusive sets of situations. I think this should make us suspicious of Salmon's approach. An initial difficulty for his line is that the kind of procedures in question do have an obvious plausibility in cases where our statistical analysis patently does reflect an incomplete knowledge of relevant causal factors, as is testified by much medical and social scientific research. Moreover, I have suggested a justification for using those procedures in such cases. Salmon defends his interpretation of the significance of statistical data largely by reference to illustrative examples. It is very tempting to conclude that his interpretation is plausible only because of its superficial similarity to the one I have defended in this paper, that he is trading on the implicit fact that his procedures would be valid if the statistical associations involved were, contrary to his specifications, only partial reflections of underlying deterministic relationships.

What of events that really are undetermined? I think that the right view of such events is that they simply are not caused, and that as such their occurrence is not explainable. To be sure, we can consider in such cases antecedent conditions which affect the chance of the event's occurrence. Undetermined events can be taken to have objective non-unitary chances of occurrence in the single case, in addition to the frequency-in-a-reference-class probabilities we can attach to determined events in the way I have been presupposing so far. And so we can treat the occurrence of such single-case chances as events in themselves, and ask what caused them. Indeed the discussion of causation in this paper will apply quite straightforwardly to such questions. But why an undetermined event itself occurred, given that it had the chance it did, remains a further question, and one that lacks an answer.

5. Backwards Causation.

The no-eclipsing elaboration of the neo-Humean analysis of causation allows the possibility that effects might precede their causes in time. I shall now attempt to show that it is nevertheless what we need in order to grasp why this possibility is never actualised.

Suppose two determined events, A and B, are statistically associated. This kind of dependency calls for explanation. Single coincidences are only to be expected. But if A and B turn out in general to occur together more often than is indicated by their separate probabilities of occurrence we want some explanation of what is going on.

This demand will be satisfied if we can show there is a "causal net" connecting A and B, where the strands of the net are relationships of direct causation showing either that A is a causal ancestor of B or that B is a causal ancestor of A or that they have a common causal ancestor, and where the "exogeneous" direct causes of the final and the intermediate effects in the net are all singly and jointly statistically independent. For given such a net, the kind of reasoning employed in Section 3. can be used easily enough to account precisely for the difference between Prob (B/A) and Prob (B) solely in terms of the explicit deterministic connections involved and the independent probabilities of the exogenous factors.

It seems to me reasonable to suppose that all statistical associations between determined events can be explained away by reference to such causal nets.[8] For, as was shown in Section 3.,if two determined events, A and B, are statistically associated, then (including the special case where A is itself a direct cause of B) A must be statistically associated with B's direct causes. And, by the same argument, it must be associated with their direct causes. Moreover A will occur if and only if some of its direct causes occur. And they will occur if and only if some of their direct causes do. So A's and B's causal ancestors will be statistically associated backwards *ad infinitum*. Of course if A is a causal ancestor of B, or B an ancestor of A, or if they share a common ancestor, this will be perfectly understandable, for after some point they will actually share causal ancestors. But for A and B to be statistically associated without being causally connected in some such way would propagate an implausibly cosmic succession of general coincidences.[9]

Note now that the association between two events will "disappear" if we "cut" their link by "controlling" for any event intermediate in the net linking them; that is, if we partition our reference class into cases where that event occurs and cases where it does not. This is derivable from the earlier result that if one event is not a direct cause of another, then any association between them will disappear if we control for those direct causes of the second with which the first is associated. Thus in practice we find that the statistical association between falling barometers and subsequent rain disappears when we control for temporally prior drops in atmospheric pressure. Similarly the association between administering penicillin and the subsequent disappearance of the symptoms of an infection will disappear when we control for whether there is a temporally intervening elimination of the relevant germs.

It is however significant that, whenever we find that an association between some A and B disappears on controlling for some C, it is invariably the case that either C comes between A and B temporally or C precedes both A and B temporally - controlling for later events never removes associations between pairs of earlier events. Thus, for instance, the association between smoking and drinking does not disappear when we control for subsequent death by heart failure - even amongst those who do so die (and amongst those who do not) the smokers are more likely to have been drinkers than the non-smokers. Nor will the association disappear when we control for any other subsequent occurrence.

The significance of this is that it gives us reason to suppose that later events are never causes of earlier events; that is, that insofar as later events are parts of sufficient conditions for earlier events they must be eclipsed parts. For suppose that later events did cause earlier events. Then we would expect it to be the case that some of the causal nets accounting for statistical associations between pairs of factors had intermediate links which came after such pairs of factors in time, and consequently that some associations between pairs of earlier events disappeared when certain later events were controlled for. But we don't find this. So it seems that later events don't ever cause earlier events.

It is worth noting that, although this argument does not give any specific illustrations of the way later sufficient conditions are eclipsed as causes of earlier ones, it is not just an attempt to show that there is some "macro-statistical" temporal asymmetry in our universe. On the contrary, the argument takes a certain such asymmetry as a straightforwardly contingent and unexplained premise (earlier associations don't ever disappear when later events are controlled for) and invokes another "macroscopic" premise (any statistical association between events must be explicable in terms of a "causal net" connecting those events) in order to conclude that there must be temporal asymmetries in the specific deterministic causal relationships connecting particular events. The link between the premises and the conclusion is of course provided by the no-eclipsing account of causation which explains the connection between "macro-statistical" associations and causal relationships.

6. Appendix.

The first thing to be shown is that if Z is always caused and Y is statistically independent of any other direct causes Z may have, then Prob (Z/Y) > Prob (Z) if and only if Y is a direct cause of Z.

Let us call a conjunction of direct causes together making up a minimally sufficient condition for some effect a <u>full</u> direct cause. We can express the disjunction of Z's full direct causes in the form YW or X, where X is the disjunction of those full direct causes that do not contain Y as a part, and W the disjunction of those sets of conditions which together with Y give full direct causes of Z. Our question is now whether W has any members - <u>are</u> there any sets of conditions which together with Y makes up full direct causes of Z?

Since, by hypothesis, Z occurs if and only if one of its full direct causes does,

$$\begin{aligned}\text{Prob}(Z/Y) &= \text{Prob}(YW \text{ or } X/Y)\\ &= \text{Prob}(YW \text{ or } YX)/\text{Prob}(Y)\\ &= [\text{Prob}(YW) + \text{Prob}(YX) - \text{Prob}(YWX)]/\text{Prob}(Y).\end{aligned}$$

And so, since Y is independent of X,

$$\text{Prob}(Z/Y) = [\text{Prob}(YW) - \text{Prob}(YWX)]/\text{Prob}(Y) + \text{Prob}(X).$$

Now,

$$\begin{aligned}\text{Prob}(Z) &= \text{Prob}(YW \text{ or } X)\\ &= \text{Prob}(YW) - \text{Prob}(YWX) + \text{Prob}(X).\end{aligned}$$

So, given that Y does not have a probability of 1, which would effectively mean it always happened and so was ineligible as a cause to start with, Prob(Z/Y) > Prob(Z) if and only if Prob(YW) > Prob(YWX).

If Prob(YW) > Prob(YWX), then Y must be a direct cause. For, if it were not, YW and <u>a fortiori</u> YWX would be impossible, and so both would have zero probabilities. So if Prob(Z/Y) > Prob(Z) then Y must be a direct cause of Z.

If, conversely, Prob(YW) ≤ Prob(YWX) - that is, if Prob(YW) = Prob(YWX) - then either Y is not a direct cause and both sides of the equation are zero, or X is certain once YW obtains. But if the latter were the case then X would eclipse YW as a cause anyway. For X and Y are statistically independent, and given that YW's being certain is ruled out to start with this means that X will sometimes occur without YW. So in either case, if Prob(Z/Y) is not greater than Prob(Z), then Y cannot be a direct cause of Z. (It is here that the no-eclipsing analysis of causation is crucial).

That was the first thing to be shown. I now want to consider the situation when Y <u>is</u> associated statistically with X. Let $V_1,\ldots,V_n$ be those parts of the disjuncts of X which are either singly or jointly associated with Y. Then T, the disjunction of sets of factors "left" after we "remove" $V_1,\ldots,V_n$ from X, will be statistically independent of

Y. What we now need to know is whether the presence of Y makes Z more likely <u>within</u> any category of instances which are all similar in respect of $V_1, \ldots, V_n$. That is, is Prob(Z/YV) > Prob(Z/V), for any V which is a combination of the presence and absence of the various $V_1, \ldots, V_n$?

If it is, then we can definitely conclude that Y is a direct cause of Z. For we can go through exactly the same reasoning as before within the class of instances given by our V. Except now we have assured ourselves that Y is statistically independent of the factors in T, the "other causes" of Z "within" our V. So if Y does make Z more likely it must be that, together with certain other factors, which might include elements of V, it makes up a full direct cause of Z.

If Y increases the chance of Z within some combinations of $V_1, \ldots, V_n$ but not within others, then this shows that the relevant elements of such combinations are themselves part of full direct causes containing Y. If, however, Y does not make Z more likely for <u>any such</u> combination - if Prob(Z/YV) = Prob(Z/V) for all possible V's - then we can in general conclude that Y is <u>not</u> a direct cause of Z. Again we can apply our earlier reasoning "within" each V with the assurance that Y is independent of Z's "other" causes: and so we can conclude that there is nothing which together with any combination of $V_1, \ldots, V_n$ can be added to Y to give a full cause of Z.

There is however one exception to the argument of this last paragraph which needs mentioning. Suppose Y <u>is</u> in fact a cause of Z, but that whenever it occurs it is made <u>certain</u> by a given combination of other causes V. Then there is no question of its increasing the probability of Z within that V. Technically Prob(Z/YV) = Prob(Z/V), since the condition YV will be effectively equivalent to the condition V. In the original simple case where we assumed that Y was independent of Z's other causes this difficulty did not arise because if Y were absolutely certain it would be ineligible as a cause. But now we have to worry about a certainty-relative-to-V, which cannot be so ruled out. This is in effect a version of the difficulty which arises with the kind of virus-infection-fever case mentioned at the end of Section 2. We can't tell directly whether or not Y (skin infection) is eclipsed by V (virus) because they always occur together. Correspondingly, in such a case it will not be possible to use statistical data to decide whether or not Y is a cause.

If causally related factors characteristically always occurred together in this manner, we would not have the kind of statistical data discussed in this paper, and the argument mounted in Section 5 above would be inapplicable to our world. But our world is not like that.

Notes

[1]I am grateful to Hugh Mellor and Graham Nerlich for comments on earlier versions of this paper. The argument about backwards causation was originally suggested by a conversation with Dan Gvirtzman. Further treatment of some of the issues herein is given in Chapter 3 of my forthcoming For Science in the Social Sciences, London: Macmillan, 1978.

[2]It should be noted that in [3] Salmon expresses some reservations about actually *defining* causation in terms of statistical relations alone (p.55, p.81). And in [4] he emphasises spatio-temporal continuity and the ability to "transmit marks" as further characteristic features of causal relationships. But the important point is that he allows that causation can be indeterministic. (See Suppes [7] and Sayre [6] for unequivocal defences of statistical definitions of causality.)

[3]See Mackie [2], pp.83-4, for a general demonstration of this point.

[4]The barometer problem is often mentioned in the literature. But so far as I know Sanford [5] is the only account of causation which suggests a satisfactory way of solving it. Though Sanford does not explicitly discuss the barometer problem, I take his account to be similar in spirit to the no-eclipsing account I develop here.

[5]I am interpreting "probability" as the limiting frequency of a kind of event in a hypothetically infinite reference class. I take such probabilities to be evidenced by the frequencies observed in random sampling from such reference classes.

[6]Salmon originally presented the screening off rule as independent of the homogeneity requirement. But, as he notes in his [3], p.105, the former is derivable from the latter.

[7]For Salmon's (admittedly hedged) preference for actual indeterminism over epistemic uncertainty see his [3], pp.79-80.

[8]Salmon similarly suggests that all statistical associations should be explained by causal relationships ("the principle of the common cause",[3], p.70, [4], p.124), and also that this means there is a temporal asymmetry to causation ([3], pp.65-76). But our different views of causation lead us to give a different basis and significance to these ideas.

[9]For simplicity I assume in this section that we are considering a deterministic segment of reality. But if we were not, the argument could still be made to apply by "replacing" any undetermined events by their chances of occurrence.

References

[1] Mackie, J.L. "Causes and Conditions." American Philosophical Quarterly 2(1965): 245-55 and 261-4.

[2] -----------. The Cement of the Universe. Oxford: Oxford University Press, 1974.

[3] Salmon, W.C. "Statistical Explanation." In Nature and Function of Scientific Theories. (University of Pittsburgh Series in Philosophy of Science, Volume 4.) Edited by Robert G. Colodny. Pittsburgh: University of Pittsburgh Press, 1970. Pages 173-231. (As reprinted in Salmon, W.C. et. al. Statistical Explanation and Statistical Relevance. Pittsburgh: University of Pittsburgh Press, 1971. Pages 29-87.)

[4] -----------. "Theoretical Explanation." In Explanation. Edited by S. Körner. Oxford: Basil Blackwell, 1975. Pages 118-145 with discussion to p. 184.

[5] Sanford, D.H. "The Direction of Causation and the Direction of Conditionship." Journal of Philosophy 73(1976): 193-207.

[6] Sayre, K.M. "Statistical Models of Causal Relations." Philosophy of Science 44(1977): 203-14.

[7] Suppes, P. A Probabilistic Theory of Causality (Acta Philosophica Fennica XXIV.) Amsterdam: North Holland, 1970.